This book discusses the structure and dynamical properties of quantized vortex lines in superfluid helium-4 in the light of research on vortices in modern fluid mechanics. It is intended for researchers and graduate students in both fields, and it is the first book to give a comprehensive treatment of the problem.

Vortex motion is a huge subject which has permeated physics for centuries, and the author begins his discussion with a review of the knowledge of classical fluid dynamics which is relevant to the main topics of the book. This is followed by a presentation of basic material on helium II and quantized vortices, giving some historical background and establishing the language and principal phenomenology of the subject. The following chapters deal with different aspects of the subject including vortex dynamics and mutual friction, the structure of quantized vortices and their interaction with helium-3 atoms and ions, vortex arrays, and vortex waves, which are now a major preoccupation in superfluidity. The book concludes with substantial introductions to two currently active topics, namely superfluid turbulence and the nucleation of quantized vortices. Both quantum tunnelling and thermal activation nucleation processes are discussed, including the Kosterlitz–Thouless transition in thin films.

The comprehensive approach provided by the author will make this an invaluable book for students taking advanced undergraduate or graduate courses, and for all those involved in research on classical and quantum vortices.

CAMBRIDGE STUDIES IN LOW TEMPERATURE PHYSICS

EDITORS

A. M. Goldman
Tate Laboratory of Physics, University of Minnesota
P. V. E. McClintock
Department of Physics, University of Lancaster
M. Springford
Department of Physics, University of Bristol

Cambridge Studies in Low Temperature Physics is an international series which contains books at the graduate text level and above on all aspects of low temperature physics. This includes the study of condensed state helium and hydrogen, condensed matter physics studied at low temperatures, superconductivity and superconducting materials and their applications.

CAMBRIDGE STUDIES IN LOW TEMPERATURE PHYSICS 3

Quantized vortices in helium II

Lars Onsager: 1903–76. Photo: Author.

Quantized vortices in helium II

RUSSELL J. DONNELLY

*Department of Physics, University of Oregon
Eugene, Or. 97403*

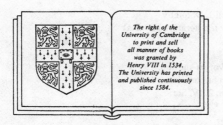

The right of the
University of Cambridge
to print and sell
all manner of books
was granted by
Henry VIII in 1534.
The University has printed
and published continuously
since 1584.

CAMBRIDGE UNIVERSITY PRESS

CAMBRIDGE

NEW YORK   PORT CHESTER   MELBOURNE   SYDNEY

CAMBRIDGE UNIVERSITY PRESS
Cambridge, New York, Melbourne, Madrid, Cape Town, Singapore, São Paulo

Cambridge University Press
The Edinburgh Building, Cambridge CB2 2RU, UK

Published in the United States of America by Cambridge University Press, New York

www.cambridge.org
Information on this title: www.cambridge.org/9780521324007

First published 1991
This digitally printed first paperback version 2005

A catalogue record for this publication is available from the British Library

Library of Congress Cataloguing in Publication data
Donnelly, Russell J.
Quantized vortices in helium II/Russell J. Donnelly.
    p. cm.—(Cambridge studies in low temperature physics)
Includes index.
ISBN 0-521-32400-9 (hardback)
1. Liquid helium. 2. Superfluidity. 3. Vortex-motion.
I. Title. II. Title: Quantized vortices in helium, two.
III. Title: Quantizes vortices in helium, 2. IV. Series.
QC175.4.D58 1990
536'.56—dc20   90-2047   CIP

ISBN-13 978-0-521-32400-7 hardback
ISBN-10 0-521-32400-9 hardback

ISBN-13 978-0-521-01814-2 paperback
ISBN-10 0-521-01814-5 paperback

To the memory of
Lars Onsager and Richard P. Feynman

Contents

x   *Contents*

Richard P. Feynman: 1918-88. Photo: Archives of the California
Institute of Technology.

Preface

Research on the properties of superfluid ^{4}He has fascinated physicists ever since the discovery of superfluidity in 1939 by Allen and Misener and Kapitza. For many years the two-fluid model and its ramifications were a vigorous area of research even though a microscopic theory of the fluid was not forthcoming. The subject has always been attractive to graduate students because of the fascinating combination of skills required for its understanding: quantum mechanics, statistical mechanics and hydrodynamics. With the passage of time and with the advent of new discoveries such as the superfluidity of ^{3}He the study of the two-fluid model and related properties has rightfully lost its place in center stage in low temperature physics, but in its place has emerged a valuable, and I think enduring, discipline: namely the study of specific issues where the unique properties of liquid helium offer a testing ground often unmatched elsewhere in nature. For example, we have the study of critical phenomena near the lambda transition of ^{4}He and its influence on the whole subject of phase transition experiments and theory. We have inelastic neutron scattering and its influence on such matters as the theory of elementary excitations and the question of Bose condensation. We have the study of quantized vortex line turbulence coming along as a new branch of turbulence research. Further examples are Bénard convection experiments in liquid helium and mixtures of ^{3}He in ^{4}He, and the development and testing of rigorous theories of the hydrodynamical properties of helium II, to mention only a subset of the wide range of topics being pursued today.

In the earlier days of liquid helium research there were several attempts to summarize all the known knowledge of the subject, beginning with Keesom's *Helium*. There followed Atkins', Wilks' and Keller's books each of which included extra topics such as mixtures of ^{3}He in ^{4}He and solid helium, and my own brief lecture notes at the University of Chicago written by William Glaberson and the late Peter Parks. The next major attempt to summarize the state of knowledge was that of Benneman and Ketterson, a two volume set with chapters by a group of many authors.

This book is a reflection of the tendency we have mentioned above, namely to consider modern research in ^{4}He physics as pertinent to the development of certain specialized topics. Our subject is vortices. Vortex motion is a huge subject which has permeated physics for centuries. We begin our discussion with a review of the knowledge of classical fluid dynamics relevant to the main topics of the book. Since so much of what we know about quantized vortices comes from classical fluid dynamics, we gain perspective by choosing this as a way to begin. On the other hand, the topics of Chapter 1 are a collection of disparate results which do not easily fit together as a single subject. Readers might well omit some of the specialized topics in Chapter 1 and refer back to them as needed to illustrate specific discussions of vortices in liquid helium. Chapter 2 contains basic material on helium II and quantized vortices, giving some historical background and establishing the language and principal phenomenology of the subject. Chapter 3 considers vortex dynamics and mutual friction, the force coupling the motion of the normal and superfluid components. Chapter 4 provides a discussion of the structure of quantized vortices and the interaction of ^{3}He and ions with vortices. Chapter 5 contains a discussion of distributions of vortices in helium II and various theories of their arrangements in steady flows. Vortex waves have become a major preoccupation in superfluidity which is discussed in Chapter 6. The book concludes with brief introductions to two specialized and currently active subjects, namely superfluid turbulence and the problem of the nucleation of quantized vortices. Thermal activation processes are discussed including the Kosterlitz–Thouless transition in thin films.

The purpose of this book is to acquaint the reader with modern research in vortices in liquid helium in the context of current vortex research in classical fluids. The background assumed is a knowledge of statistical and quantum mechanics at the beginning graduate level as well as some fluid mechanics.

Acknowledgements

Subjects in physics such as superfluidity progress because of the efforts of a very sizeable community of scholars.

Every one of us in indebted to the memory of Lars Onsager and Richard Feynman who almost overnight laid down the basic and enduring concepts of quantized vortices.

William F. Vinen and Henry E. Hall were the pioneering physicists who took the theoretical ideas of Onsager and Feynman designed experiments and interpreted the results in a way which has been of use to the field ever since. As Feynman once remarked to me: 'They made it real.'

I am personally indebted to my teachers and colleagues who have worked with me on vorticity over many years: Cecil T. Lane and Lars Onsager (my advisors), Joe Vinen, Subrahmanyan Chandrasekhar, Paul Roberts, Chris Jones, Ron Hills, Bill Glaberson, Klaus Schwarz, Carlo Barenghi and Tony Maxworthy. My graduate students who have worked on liquid helium with me include Howard Snyder, Brian Springett, David Tanner, Klaus Schwarz, Bill Glaberson, Peter Parks, Don Strayer, Michael Cromar, Bob Walden, Carlo Barenghi, Charles Swanson, Rabi Wang, John Pfotenhauer, Tom Wagner, Ben You, David Samuels, Paul Bunson and Greg Bauer. A number of persons mentioned were kind enough to review parts of the manuscript for me at various stages.

I am grateful to Bill Glaberson and North-Holland Publishing for permission to quote extensively from our joint article on vortices in *Progress in Low Temperature Physics*. This article formed the basis from which this book evolved. Many of the insights in that review and this book came from his ideas, research and graduate students.

I am forever indebted to Paul Roberts and Joe Vinen who have collaborated with my research group throughout my career and whose ideas and critical insight have been invaluable in preparing this account of quantized vortices.

The research in my group has been supported continuously since 1962 by the National Science Foundation Low Temperature Physics Program. NSF support has made it possible to undertake the sustained effort required to work effectively in low temperature physics.

The manuscript, in process for several years, was typed by Marirose Radostitz and Rachel Sewell. Most of the drawings were prepared by Sally Donovan. I have been fortunate in having such expert assistance.

1 Background on classical hydrodynamics

1.1 Equations of motion, potential flow and vorticity

Vortices, vortex motions and vortex interactions have been an important and vigorous branch of fluid mechanics since the early seventies. One reason is the belief that coherent structures exist in many types of turbulent flow and that the representation of turbulence as a superposition of interacting vortices may be useful. While the main subject of this book, quantized vortices, is part of the story of recent progress, and moreover contains a study of a form of turbulence of interacting quantized vortices, there has been remarkably little cross-fertilization of ideas between studies of classical and quantum vortices. Much of the material we wish to study here owes its origin to ideas which come from classical fluid mechanics. We therefore begin our study with a brief overview of some of those ideas. In doing so, we must recognize that the classical hydrodynamical ideas we shall need to refer to come from a wide variety of often rather specialized topics in fluid mechanics.

The equation of conservation of mass for a fluid of density ρ and velocity $\mathbf{v}$ is

$$\partial \rho / \partial t + \operatorname{div} \rho \mathbf{v} = 0 \tag{1.1}$$

In nearly all the discussion in this book, the fluids may be regarded as incompressible, and hence

$$\operatorname{div} \mathbf{v} = 0 \tag{1.2}$$

The equation of motion for an incompressible viscous fluid, the Navier-Stokes equation, is

$$\frac{d\mathbf{V}}{dt} = \frac{\partial \mathbf{V}}{\partial t} + (\mathbf{v} \cdot \operatorname{grad})\mathbf{v} = -\frac{1}{\rho} \operatorname{grad} p + \nu \nabla^2 \mathbf{v} \tag{1.3}$$

where $\mathbf{v}$ is the velocity, p the pressure and $\nu = \eta/\rho$ is the kinematic viscosity, the ratio of the dynamic viscosity η to density ρ. For a derivation of (1.3), and in particular the significance of the substantive derivative dv/dt, one should consult a basic hydrodynamics text such as Landau and Lifshitz (1987) Section 2. The Navier–Stokes equation is usually very difficult to solve, even numerically. In special cases such as where boundaries are absent, there is often much to recommend the study of the flow of an inviscid, incompressible, adiabatic fluid, governed by Euler's equation

$$\frac{d\mathbf{v}}{dt} = \frac{\partial \mathbf{v}}{\partial t} + (\mathbf{v} \cdot \text{grad})\mathbf{v} = -\frac{1}{\rho}\,\text{grad}\,p \qquad (1.4)$$

Equation (1.4) represents an ideal fluid in the sense that we are omitting internal frictional processes and hence viscosity and thermal conductivity. This ideal fluid motion is adiabatic and in many cases the entropy per gram S can simply be taken as a constant everywhere in the fluid. The flow is then described as *isentropic*. If w is the enthalpy per gram of the fluid, the relation $dw = TdS + Vdp$, where T is the absolute temperature and $V\ (=1/\rho)$ the specific volume, reduces to $dw = Vdp = dp/\rho$. We shall discover that liquid helium at absolute zero has many similarities to our ideal fluid. One important difference, however, is that the chemical potential per gram, defined as $d\mu = -SdT + Vdp$ appears in place of the enthalpy.

We can then write Euler's equation in the form

$$\partial \mathbf{v}/\partial t + (\mathbf{v} \cdot \nabla)\mathbf{v} = -\nabla w \qquad (1.5)$$

and because of the vector identity

$$\tfrac{1}{2}\nabla v^2 = \mathbf{v} \times (\nabla \times \mathbf{v}) + (\mathbf{v} \cdot \nabla)\mathbf{v} \qquad (1.6)$$

a further version is

$$\partial \mathbf{v}/\partial t - \mathbf{v} \times (\nabla \times \mathbf{v}) = -\nabla(w + \tfrac{1}{2}v^2) \qquad (1.7)$$

The solution of (1.4) for many flows is part of the background of classical mathematical physics. In some cases there are elegant exact solutions, but to many students it may seem that the ideal fluid is a sterile concept, especially compared to the demands of modern engineering such as aeronautics.

Hydrodynamicists, however, have long understood that classical mathematical fluid mechanics and practical engineering fluid mechanics can be combined by the concept of the boundary layer. That is, at high Reynolds numbers the flow near a boundary is described by (1.3) and that the remainder of the flow can be described by (1.4) with appropriate joining of solutions (we shall mention the boundary layer again shortly).

One can then proceed to include in the flow of the 'inviscid' or 'ideal' fluid such features as circulation and vortices, which dominate the flow and whose generation depends on viscosity as we shall see below. In many cases the effects of viscosity on the evolution and interaction of vortices appear to be relatively slow, making the choice of an inviscid fluid for the discussion of vortices a sensible one.

The curl of the velocity field,

$$\boldsymbol{\omega} = \nabla \times \mathbf{v} \tag{1.8}$$

is known as the vorticity and may be thought of as the circulation per unit area. We shall discuss circulation presently.

A curve drawn from point to point in the fluid, so that its direction is always that of the instantaneous direction of $\boldsymbol{\omega}$ is called a *vortex line*. The differential equation of the line comes from expressing the condition $\boldsymbol{\omega} \times d\mathbf{l} = 0$ in components where $d\mathbf{l}$ is taken along the vortex line. Because of the definition (1.8) we have immediately

$$\text{div } \boldsymbol{\omega} = 0 \tag{1.9}$$

Equation (1.9) shows that vorticity behaves analogously to the velocity field of an incompressible fluid. If we draw a small closed curve C in the fluid and include every vortex line passing through this curve, we have obtained a *vortex tube*. It follows from Gauss' theorem that the integral of the normal component of $\boldsymbol{\omega}$ over any closed surface is zero:

$$\int_S \boldsymbol{\omega} \cdot d\mathbf{S} = 0 \qquad \text{(any closed surface } S) \tag{1.10}$$

and if we apply this result to any two cross-sections $d\mathbf{S}_1$ and $d\mathbf{S}_2$ of the tube, the sides do not contribute and

$$\int_{S_1} \boldsymbol{\omega} \cdot d\mathbf{S}_1 = \int_{S_2} \boldsymbol{\omega} \cdot d\mathbf{S}_2 \tag{1.11}$$

Thus the flux of vorticity across any section of a vortex tube is conserved and represents a characteristic of the tube. Vortex tubes cannot terminate in the fluid, they must be closed or terminate on boundaries. Equations (1.9)–(1.11) are expressions of Helmholtz's first theorem.

The Navier–Stokes equation can be written in terms of the vorticity by taking the curl of (1.3):

$$\partial \boldsymbol{\omega}/\partial t = \nabla \times (\mathbf{v} \times \boldsymbol{\omega}) + \nu \nabla^2 \boldsymbol{\omega}$$

$$= (\boldsymbol{\omega} \cdot \nabla)\mathbf{v} - (\mathbf{v} \cdot \nabla)\boldsymbol{\omega} + \omega \nabla^2 \boldsymbol{\omega} \tag{1.12a}$$

Introducing a characteristic length L and velocity V we can make (1.12a) dimensionless through $\omega \to \omega V/L$, $t \to tL/V$ and $v \to vV$, yielding

$$\frac{\partial \boldsymbol{\omega}}{\partial t} = (\boldsymbol{\omega} \cdot \nabla)\mathbf{v} - (\mathbf{v} \cdot \nabla)\boldsymbol{\omega} + \frac{1}{Re}\nabla^2 \boldsymbol{\omega} \qquad (1.12b)$$

with the Reynolds number $Re = VL/\nu$. Two geometrically similar viscous flows are dynamically similar at the same Reynolds number. For an ideal fluid, (1.12a) simplifies to

$$\frac{\partial \boldsymbol{\omega}}{\partial t} - \nabla \times (\mathbf{v} \times \boldsymbol{\omega}) = 0 \qquad (1.13a)$$

which leads to Helmholtz's second theorem. Consider a surface S enclosed by a contour C. Let $d\mathbf{S}$ be an element of this surface. Multiplying scalarly by $d\mathbf{S}$ and integrating over S, we obtain

$$\int_S (\partial \boldsymbol{\omega}/\partial t) \cdot d\mathbf{S} - \int_S \nabla \times (\mathbf{v} \times \boldsymbol{\omega}) \cdot d\mathbf{S} = 0 \qquad (1.13b)$$

Transforming the second integral by Stokes' theorem, we have

$$\int_S (\partial \boldsymbol{\omega}/\partial t) \cdot d\mathbf{S} + \int_C \boldsymbol{\omega} \cdot (\mathbf{v} \times d\mathbf{l}) = 0 \qquad (1.14)$$

where $d\mathbf{l}$ is an element along the contour C defining the vortex tube at a particular cross-section. A careful argument (see Chandrasekhar (1961) Section 20) now shows that

$$\frac{d}{dt}\int_S \boldsymbol{\omega} \cdot d\mathbf{S} = 0 \qquad (1.15)$$

and that the integral of the normal component of $\boldsymbol{\omega}$ over any surface S bound by a closed curve, remains constant as we follow the surface S with the motion of the fluid elements constituting it:

$$\int_S \boldsymbol{\omega} \cdot d\mathbf{S} = \text{constant} \qquad (1.16)$$

Another statement is that the strength of a vortex tube is an integral of the equations of motion. We see that the vorticity $\boldsymbol{\omega}$ may be changed by 'stretching' the vortex tube, so long as the quantity $\boldsymbol{\omega} \cdot d\mathbf{S}$ remains the same.

Transforming (1.16) by Stokes' theorem we can define the circulation

$$\Gamma = \int_C \mathbf{v} \cdot d\mathbf{l} = \int_S \boldsymbol{\omega} \cdot d\mathbf{S} = \text{constant} \qquad (1.17)$$

This definition shows why the vorticity may be thought of as the circulation per unit area. Equation (1.17) can be put in the form

$$\frac{d\Gamma}{dt} = \frac{d}{dt} \int \mathbf{v} \cdot d\mathbf{l} = 0 \tag{1.18}$$

or

$$\int \mathbf{v} \cdot d\mathbf{l} = \text{constant} \tag{1.19}$$

which is known as Kelvin's theorem, or the law of conservation of circulation.

Streamlines in the flow are made up of lines tangent to $\mathbf{v}$ at all points. Their differential equation comes from the relationship $d\mathbf{l} \times \mathbf{v} = 0$, that is $dx/v_x = dy/v_y = dz/v_z$.

In the particular case of irrotational flow,

$$\boldsymbol{\omega} = 0 \qquad \text{(potential flow)} \tag{1.20}$$

and the circulation around any closed contour is zero:

$$\oint \mathbf{v} \cdot d\mathbf{l} = 0 \tag{1.21}$$

Thus closed streamlines cannot exist in potential flow unless the space involved is multiply-connected. In potential flow in such a region the circulation can be finite if the closed contour around which it is taken cannot be contracted to a point without crossing the boundaries of the region. This exception will prove to be all-important for our studies of helium II.

A particularly simple situation arises for potential flow of an incompressible fluid. For if $\nabla \times \mathbf{v} = 0$ and $\nabla \cdot \mathbf{v} = 0$, with the substitution

$$\mathbf{v} = \text{grad } \varphi \tag{1.22}$$

we obtain Laplace's equation

$$\nabla^2 \varphi = 0 \tag{1.23}$$

for the velocity potential φ.

With Euler's equation in the form (1.5) we can substitute $\mathbf{v} = \nabla \varphi$, and obtain

$$\nabla(\partial \varphi / \partial t + \tfrac{1}{2} v^2 + w) = 0 \tag{1.24}$$

which yields the first integral

$$\partial \varphi / \partial t + \tfrac{1}{2} v^2 + w = f(t) \tag{1.25}$$

where $f(t)$ may be taken to be zero without dynamical consequences because the velocity is the space derivative of φ and we can replace φ

by $\varphi + \int f(t)\,dt$ making the right-hand side of (1.25) zero. We see that in steady incompressible flow the greatest pressure occurs at the points where the velocity is least. For an ideal incompressible fluid in steady potential flow

$$\tfrac{1}{2}v^2 + w = \tfrac{1}{2}v^2 + p/\rho = \text{constant} \tag{1.26}$$

throughout the fluid, a statement referred to as Bernoulli's principle. More generally the constant in (1.26) takes different values along different streamlines, even if the flow is rotational. It can also be shown (Landau and Lifshitz, 1959, Section 6) that the quantity

$$\rho\mathbf{v}(\tfrac{1}{2}v^2 + w) \tag{1.27}$$

is the energy flux density vector for an ideal isothermal fluid. The magnitude of this quantity is the amount of energy passing through a unit area perpendicular to $\mathbf{v}$ in one second.

A well-known result of steady potential flow about an object is that it exerts no lift or drag on that object (d'Alembert's paradox). This is because (1.26) does not contain time explicitly and potential flow around a body, for example, depends only on the velocity and not on acceleration. On the other hand, if the flow is unsteady (i.e., the body is accelerating) φ must be taken into account and the body acts as if it had an effective mass equal to its physical mass plus some fraction of the mass of the liquid displaced by the body. For spheres, the fraction is $\tfrac{1}{2}$, for cylinders (per unit length) it is 1 (Landau and Lifshitz, 1959, Section 11).

A special situation arises if we consider potential flow of a fluid which has circulation Γ about a cylinder combined with a uniform steady flow v_∞ at large distances from the cylinder. By integrating the pressure over the cylinder one can show that while d'Alembert's paradox still holds (i.e., there is no drag on the cylinder), a lift force of magnitude $\rho v_\infty \Gamma$ per unit length is developed (the Kutta–Joukowski theorem). In viscous fluids the circulation can be produced by rotation of the cylinder in a fluid stream. This phenomenon was first observed by the German scientist Magnus in 1852 and is often referred to as the Magnus effect. Elementary reasoning from Bernoulli's principle gives the direction of the lift force, which can be put in vector form as

$$f_M = \rho\mathbf{v}_\infty \times \Gamma \tag{1.28a}$$

This result is independent of the shape of the body. In the liquid helium literature this result is applied to vortices (Section 3.1) and is known as the Magnus force.

The analogous result in electromagnetic theory is the Lorentz force df_L on a current element idl (in Gaussian units)

$$\mathbf{df_L} = (i/c)(\mathbf{dl} \times \mathbf{B}) \tag{1.28b}$$

where i is the current, c the velocity of light and $\mathbf{B}$ the magnetic field. Applied to a superconducting wire (no field penetration) this gives a Lorentz force independent of the detailed shape of the conductor.

Fluid motion may be represented as a combination of translation, rotation and deformation. Vorticity is a measure of rotation. We show in Figure 1.1 the behavior of a square fluid element in rotation and distortion, as in a parallel shear flow between two moving planes. The component of vorticity is obtained by considering an infinitesimal plane element normal to this component: its vorticity is twice the average angular velocity of the lines, and if this average is nonzero, the fluid particle is changing its direction in space, i.e., it is rotating. The factor of 2 here is not a mystery. Tritton (1982) gives a useful discussion of vorticity. The circulation per unit area of a fluid in uniform rotation in a circle is $(\Omega R)(2\pi R)/\pi R^2 = 2\Omega$, expressing the fact that the product of the radius and circumference of a circle is twice its area. It is important

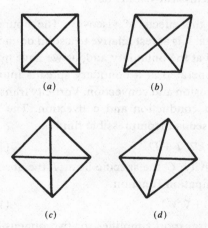

(a)   (b)

(c)   (d)

Figure 1.1 This example shows a square fluid element in rotation and distortion as in the simple shear flow of Figure 1.2(a). The change from (a) to (b) involves rotation of the vertical sides, but not the horizontal. There is thus nonzero average rotation and hence vorticity. The two diagonals have both rotated, but each only half as much (in the limit of small changes) as a vertical side. Diagrams (c) and (d) show the same process, but with the sides and diagonals interchanged (after Tritton (1982)).

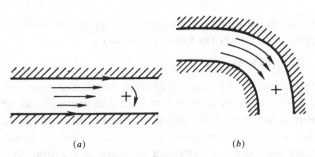

(a)   (b)

Figure 1.2 (a) A parallel shear flow (with the lower plane at rest
and the upper plane in motion) has vorticity whereas (b) potential
flow about a corner does not. In (a) a pair of diagonals will rotate as
shown in Figure 1.1, while in (b) the orientation of diagonals in
space does not change (after McCormack and Crane (1973)).

to note that there is a distinction between particle rotation, as indicated
by vorticity, and motion along a curved path, as shown in Figure 1.2.
One can have irrotational flow with curved particle paths and rotational
flow with straight ones.

1.2   Creation of vorticity in classical viscous flows

The origin of vorticity lies in the effects of viscosity. The boundary
condition for a viscous fluid is that it is at rest relative to a solid boundary.
With motion, shear is produced at the boundary and, as we have noted,
shear has vorticity. Vorticity generated at a boundary spreads into the
bulk of the fluid by means of diffusion and convection. Vorticity transport
is somewhat analogous to heat conduction and convection. The heat
equation (Serrin, 1959) for a viscous incompressible fluid is

$$\rho C[\partial T/\partial t + (\mathbf{v} \cdot \nabla)T] = k\nabla^2 T + D \tag{1.29}$$

where k is the thermal conductivity, C the specific heat, T the absolute
temperature and where the dissipation function

$$D = \eta[\mathbf{\omega}^2 - 2 \operatorname{div}(\mathbf{v} \times \mathbf{\omega}) + \nabla^2 \mathbf{v}^2] \tag{1.30}$$

Equation (1.12) for vorticity transport simplifies for two-dimensional
flows in the (x, y)-plane to

$$\partial \omega/\partial t + (\mathbf{v} \cdot \nabla)\omega = \nu\nabla^2\omega \tag{1.31}$$

where $\omega = \omega_z$ is the vorticity component normal to the (x, y)-plane.
Equation (1.31) is analogous to the heat equation for vanishing dissipa-
tion D. The operation $(\mathbf{v} \cdot \nabla)$ represents convection and ∇^2 represents
diffusion.

Convection of vorticity has the property that vorticity is conserved on a particle path. Thus vorticity can be transferred to neighboring paths only by diffusion, i.e., by the effect of viscosity. At a boundary where the fluid is at rest, vorticity can be transferred only by diffusion.

A particularly simple example of generation of vorticity at a boundary occurs in the flow near an oscillating flat plate (Stokes' second problem). Consider a flat plate lying in the (x, z)-plane with a semi-infinite layer of fluid above it extending in the positive y-direction. If x represents the coordinate parallel to the direction of oscillation and y the coordinate perpendicular to the wall, the fluid must stick to the wall and

$$u(0, t) = U_0 \cos \omega t \qquad y = 0 \tag{1.32}$$

The fluid velocity $u(y, t)$ is the solution to the reduced Navier–Stokes equation

$$\partial u / \partial t = \nu (\partial^2 u / \partial y^2) \tag{1.33}$$

(the heat conduction equation). The solution is

$$u(y, t) = U_0 \exp(-ky) \cos(\omega t - ky) \tag{1.34}$$

where

$$k = (\omega / 2v)^{1/2} \tag{1.35}$$

In this solution, two fluid layers a distance $2\pi / k = 2\pi (2\nu / \omega)^{1/2}$ apart oscillate in phase. Thus a distance k^{-1} is sometimes called the 'penetration depth' of the viscous wave. It is analogous to the temperature distribution of the earth's surface subject to diurnal heating from the sun and to the penetration of electromagnetic radiation into a conductor (skin effect).

The detailed way vorticity enters the fluid is complicated and geometry-dependent. One example is shown in Figure 1.3 where streamlines and equivorticity lines are shown for flow about a sphere at different Reynolds numbers $Re = vD / \nu$. For $Re \to 0$, the familiar Stokes solution is shown in Figure 1.3(a). There is fore–aft symmetry to the flow. But at $Re = 5$ we see this symmetry is broken (Figure 1.3(b)) and, even more so at $Re = 40$ (Figure 1.3(c)). By $Re = 100$ a vortex ring forms behind the sphere, as shown in the temporal series of flows in Figure 1.3(d).

Flow near a flat plate at high Reynolds numbers can be divided into two regions, a thin boundary layer where vorticity exists, and the region outside which can be considered inviscid. The concept is illustrated in Figure 1.4 which shows the velocity and vorticity distributions in a laminar boundary layer.

Boundary layers can detach from the wall at which they are formed. Such a tendency for a sphere is shown in Figure 1.3. There are then,

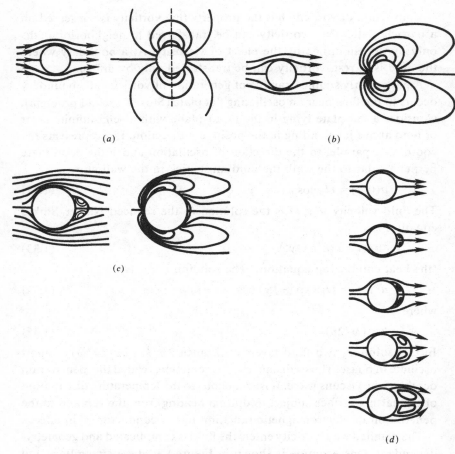

Figure 1.3   Showing the flow past a sphere of a viscous
incompressible fluid at various Reynolds numbers. The tendency to
form a wake behind the sphere is evident. Diagrams (a) and (b)
show streamline and equivorticity lines at *Re* = 0 and 5 (after Jenson
(1959)). Diagram (c) shows streamlines and equivorticity lines at
Re = 40 (after Dennis and Walker (1971)). Diagram (d) shows the
temporal development of streamlines about a sphere at *Re* = 100. The
formation of a vortex ring is clearly illustrated (after Rimon and
Cheng (1969)).

within the fluid, regions where the velocity varies spatially. Note that the
shear flows of Figure 1.4(b) differ only by a constant parallel stream,
and hence have the same vorticity distribution. This demonstrates that
the vorticity field is invariant to changes in inertial frame, unlike velocity.

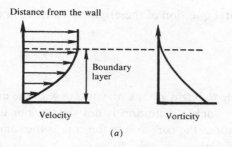

Distance from the wall

Boundary
layer

Velocity

Vorticity

(a)

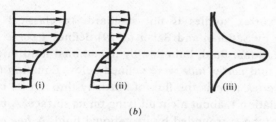

(i)   (ii)   (iii)

(b)

Figure 1.4 (a) Velocity and vorticity distributions in a boundary
layer. (b) Two shear flows (i) and (ii) with equal vorticity
distributions (iii) (after Lugt (1983)).

Figure 1.5 shows that the vorticity associated with a velocity discontinuity spreads out in time and in space. A physical example is the spreading of a submerged jet.

These examples do not begin to convey the rich variety of flows which occur in the laboratory and in nature, but they do give suggestion of an

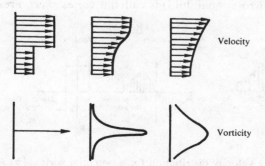

Velocity

Vorticity

Figure 1.5 Decay of a discontinuity line in space and time (after
Lugt (1983)).

answer to the fundamental question of the origin of vorticity in classical fluid mechanics.

1.3   Vortices

One would like to be able to begin a book on vortices with the definition of a vortex. Surprisingly such a definition is not simple, nor unique if the fluid in the motion about the core and in the core is the same (Lugt, 1983). In most cases of interest in this discussion, however, the core material has separate physical properties from the fluid circulating outside, and there is no difficulty in principle in identifying where the vortex is located.

Terminology in vortex studies is not standard. In classical fluid mechanics a review by Saffman and Baker (1979) defines a *vortex* as a finite volume of rotational fluid, bounded by irrotational fluid or solid walls. A *vortex line* and *vortex tube* were defined below Equation (1.8). The *strength of a vortex tube* is the flux of vorticity through its cross-section, or the circulation Γ about a circuit lying on its surface. A *vortex filament* is a vortex tube surrounded by irrotational fluid. A *line vortex* is a vortex filament of finite strength and zero cross-section: it is a singular distribution of vorticity. The last two definitions are close to convention in the discussion of quantized vortices. When the core structure is unimportant, the terminology *vortex line* and *line vortex* are used interchangeably in the helium literature. The surface between rotational and irrotational fluid can be a sharp discontinuity in the vorticity, which is called a vortex jump (see Figure 1.6). The vorticity is bounded and the velocity and pressure are continuous at the jump. A surface across which the longitudinal components of velocity are discontinuous (usually separating regions of irrotational fluid) is called a *vortex sheet*. Pressure

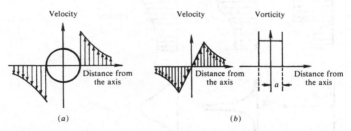

(a)   (b)

Figure 1.6   (*a*) Velocity distribution for a potential vortex. (*b*) A Rankine vortex. A vortex jump marks the edge of the core in the Rankine vortex (after Lugt (1983)).

and the normal component of velocity are continuous across it. The discontinuity in velocity is the strength of the sheet and is a vector parallel to the sheet.

Continuing these definitions, Saffman and Baker (1979) note that since $\boldsymbol{\omega} \cdot \mathbf{n} = 0$ at the surface of a vortex, the total vorticity in a vortex completely surrounded by fluid is zero, i.e., $\int \omega \, dV = 0$, and the flux of vorticity across any simply connected surface intersecting the vortex also vanishes. This makes two-dimensional vortices different from three-dimensional ones, because in a two-dimensional flow the vortex lines are all straight and parallel and the vorticity reduces to a scalar, whose integral over the plane of a vortex need not vanish and defines its total strength. A two-dimensional flow can be regarded either as lying between planes parallel to the flow or as being of infinite extent. Line vortices in two dimensions are referred to as point vortices since they are characterized by a point in the plane. Their strength is the circulation about them. A uniform vortex is a two-dimensional vortex with constant vorticity.

Suppose for the moment that we have a cylindrical rod of radius a in an otherwise unbounded ideal fluid. If the velocity distribution of the fluid around this rod is given by

$$v = \Gamma/2\pi r \tag{1.36}$$

where $\Gamma = 2\pi\Omega a^2$, the fluid motion defines a *potential vortex*, since there is no vorticity in this flow (see Figure 1.6(a)).

If the solid rod is replaced by a cylinder of the same fluid rotating with uniform angular velocity Ω, then we have a Rankine vortex. A Rankine vortex is a uniform circular vortex. The velocity and vorticity distributions for a potential vortex and a Rankine vortex are shown in Figure 1.6. The rotating core has vorticity

$$\boldsymbol{\omega} = \nabla \times \Omega\mathbf{r} = 2\Omega\hat{z} \tag{1.37}$$

where $\hat{z}$ is a unit vector parallel to the axis of rotation of the fluid.

The hollow potential vortex and the Rankine vortex have the advantage that the core radius a is a definite quantity. Classical vortices tend to have relatively thick cores with distributions of vorticity which have been modelled in various ways. The boundary between the cores and the potential flow is then more difficult to characterize.

The kinetic energy per unit length of the potential flow of Figure 1.6(a) is simply

$$K = \int_{a}^{b} \tfrac{1}{2}\rho v^2 \, dr^2 = (\rho\Gamma^2/4\pi) \ln(b/a) \tag{1.38}$$

If a is taken to be the radius of a small hollow vortex core, and b some characteristic distance such as the size of a container, K becomes the energy per unit length of a hollow vortex filament, i.e., a potential vortex. The centrifugal force on each successive ring of fluid surrounding the core is balanced by a pressure gradient

$$\mathrm{d}p/\mathrm{d}r = \rho v^2/r = \rho \Gamma^2/4\pi^2 r^3 \qquad (1.39)$$

Note that if we consider the kinetic energy per unit length of a Rankine vortex, the kinetic energy of the rotating core must be added and (1.38) is amended to

$$K = (\rho \Gamma^2/4\pi)[\ln(b/a) + \tfrac{1}{4}] \qquad (1.40)$$

The other very familiar vortex in nature is the vortex ring shown in Figure 1.7(a). When the radius R is much larger than the core radius a, the kinetic energy of such a ring is (assuming the core is hollow)

$$K = \tfrac{1}{2}\rho\Gamma^2 R[\ln(8R/a) - 2] \qquad (1.41)$$

The vortex ring moves forward with its own self-induced velocity v in a way which we shall discuss below:

$$v = (\Gamma/4\pi R)[\ln(8R/a) - 1] \qquad (1.42)$$

and the momentum, or more properly impulse, of such a ring is

$$P = \rho\Gamma\pi R^2 \qquad (1.43)$$

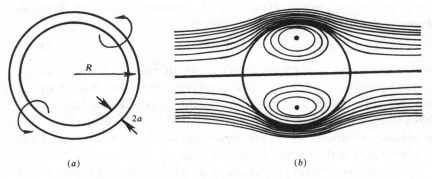

(a)   (b)

Figure 1.7 (a) A vortex ring of radius R and core radius a. The forward motion of a vortex ring is a consequence of self-induced velocity. (b) Hills' spherical vortex. The vortices in (a) and (b) are extremes in a sequence of vortex models referred to in Section 1.6

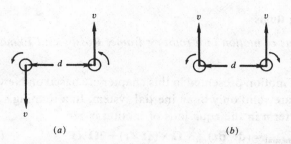

(a)   (b)

Figure 1.8 Behavior of parallel vortex filaments in each other's flow fields: (a) filaments with circulation in the same direction; (b) filaments with opposite circulation.

A derivation of these ring formulae is given by Fetter (1976). The impulse is the time and space integral of a set of forces that will generate the required motion of the fluid. It is not always uniquely defined, since a given flow pattern can sometimes be produced by different force distributions having different impulses. Fetter also discusses the concept of impulse for vortex rings, observing that the rate of change of impulse exactly equals the rate of change of total momentum of the fluid. Confusion over the meaning of impulse occurs when one attempts to interpret it in terms of total linear momentum. It is changes in momentum and impulse that are significant.

Next suppose we have an isolated pair of hollow vortex filaments a distance d apart. If the circulations of the pair are equal and in opposite directions the kinetic energy per unit length of the pair is given by

$$K = (\rho \Gamma^2 / 2\pi) \ln (d/a) \tag{1.44}$$

whereas if the circulations are parallel the energy per unit length is given by

$$K = (\rho \Gamma^2 / 2\pi) \ln (b^2 / ad) \tag{1.45}$$

where b is some outer distance of the system. The situation is shown in Figure 1.8. Neglecting any effective mass associated with the cores, each vortex filament will move with the velocity from the opposite one so that the parallel filaments will rotate about a point halfway between them with an angular velocity $\omega = \Gamma / \pi d^2$ whereas the antiparallel filaments will move through the fluid with velocity $v = \Gamma / 2\pi d$. The 'impulse' of the lines, or the effective momentum per unit length associated with the vortex motion is $\rho \Gamma d$ for the oppositely directed pair.

1.4   Rotating fluids

1.4.1   *Equations of motion in a rotating frame, Rossby and Ekman numbers*

The equations of motion presented in this chapter are based on Newton's second law and are valid only in an inertial system. In a rotating system the acceleration term in the equations of motion is

$$(d\mathbf{v}/dt)_{\text{inertial}} = (d\mathbf{v}/dt)_{\text{rot}} + \mathbf{\Omega} \times (\mathbf{\Omega} \times \mathbf{r}) + 2\mathbf{\Omega} \times \mathbf{v} \qquad (1.46)$$

(with $\mathbf{\Omega}$ = constant). The last two terms on the right are the centripetal and the Coriolis accelerations, respectively. With a change in sign, they may be considered as fictitious forces, i.e., centrifugal and Coriolis forces in a rotating system. The Navier–Stokes equations are then

$$\frac{\partial \mathbf{v}}{\partial t} + (\mathbf{v} \cdot \nabla)\mathbf{v} + 2\mathbf{\Omega} \times \mathbf{v} + \mathbf{\Omega} \times (\mathbf{\Omega} \times \mathbf{r})$$

$$= -\frac{1}{\rho}\nabla p + \nu\nabla\mathbf{v}^2 \qquad (1.47)$$

One immediate consequence of (1.47) is that the equation of motion in a rotating fluid allows periodic solutions representing the propagation of waves. It can be shown (see Chandrasekhar (1961) Section 23) that waves can exist which are transverse, circularly polarized and damped. In the limit of zero viscosity the velocity of propagation of these 'inertial waves' is $U = \omega/k = 2\Omega \cos \theta k$ where θ is the polar angle from the rotation ($\hat{z}$) axis. The dispersion relation for general $\mathbf{k}$ is

$$\omega = 2\Omega\hat{k} \cdot \hat{z} \qquad (1.48)$$

If $\hat{k}$ is perpendicular to $\hat{z}$, these modes do not propagate. Since the centrifugal force has a potential, the vorticity-transport equation in a rotating frame does not contain the contribution from this force:

$$\partial\boldsymbol{\omega}/\partial t + (\mathbf{v} \cdot \nabla)\boldsymbol{\omega} - [(\boldsymbol{\omega}+2\mathbf{\Omega}) \cdot \nabla]\mathbf{v} = \nu\nabla^2\boldsymbol{\omega} \qquad (1.49)$$

or

$$\frac{\partial}{\partial t}(\boldsymbol{\omega}+2\mathbf{\Omega}) + (\mathbf{v} \cdot \nabla)(\boldsymbol{\omega}+2\mathbf{\Omega}) - [(\boldsymbol{\omega}+2\mathbf{\Omega}) \cdot \nabla]\mathbf{v} = \nu\nabla^2(\boldsymbol{\omega}+2\mathbf{\Omega})$$

$$(1.50)$$

which has the same form as Equation (1.12b) in an inertial frame.

Dimensionless quantities can be defined for (1.49) as was done for (1.12). Putting $t \rightarrow t/\Omega$, $\mathbf{\Omega} = \Omega\hat{k}$ where $\hat{k}$ is a unit vector parallel to the axis of rotation gives

$$\partial\boldsymbol{\omega}/\partial t + Ro[(\mathbf{v} \cdot \nabla)\boldsymbol{\omega} - (\boldsymbol{\omega} \cdot \nabla)\mathbf{v}] - 2\hat{k} \cdot \nabla v = Ek\nabla^2\boldsymbol{\omega} \qquad (1.51)$$

where the Rossby number

$$Ro = V/L\Omega \tag{1.52}$$

and the Ekman number

$$Ek = v/L^2\Omega \tag{1.53}$$

appear. The Rossby number is a ratio of convective acceleration to the Coriolis force. It is a measure of the relative importance of nonlinear terms. The Ekman number is the ratio of frictional force to Coriolis force.

The flow in a boundary layer such as is shown in Figure 1.4(a) is modified by rotation. The Coriolis force will twist the boundary layer which is then called an 'Ekman layer'. If the axis of rotation is parallel to the wall, a 'Stewartson layer' develops.

1.4.2 Spin-up

Spin-up of the fluid in a bucket is important in many contexts. Consider a cylindrical bucket filled with fluid impulsively set into rotation. The fluid sticks to the walls and forms a thin boundary layer on the sides and top and bottom. Within a few revolutions, rotation influences these layers and boundary layers of the Ekman type on the top and bottom and Stewartson type on the sides develop as shown in Figure 1.9(a). Thus the fluid in the Ekman layers at the bottom and top of the vessel is deflected in the radial direction toward the side wall and in the direction parallel to the side wall in the Stewartson layer. A meridian section of

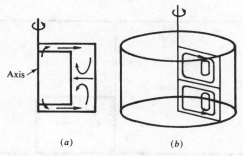

(a)   (b)

Figure 1.9 The 'spin-up process': a cylindrical vessel is suddenly put in constant rotation. (a) Ekman and Stewartson layers develop in which the anular momentum is brought into the nonrotating fluid. (b) In the meridional plane a secondary circulation is visible, which dissipates in the course of time. Finally, the fluid rotates like a solid body (after Lugt (1983)).

the fluid shows secondary flow in the form of two circulation cells as seen in Figure 1.9(*b*). The fluid particle path consists of the primary flow about the axis and the meridional circulation. Further we can have inertial waves from the impulsive start, which decay in time. The angular momentum of the container is transferred to the fluid in the boundary layers and in the course of time the secondary flow vanishes and solid body rotation remains. The transient process taking the fluid from rest to a finite angular velocity, or from one angular velocity to a higher one is called 'spin-up'. The opposite process 'spin-down' differs essentially from spin-up because the spinning mass of fluid on the inside with fluid at rest on the cylindrical wall can develop Taylor vortices (see Section 1.8) which expedite the transfer of angular momentum to bring the fluid to rest.

If the nonlinear terms in (1.51) are neglected or the Rossby number is considered small, the vorticity transport equation in a rotating system becomes

$$(\partial/\partial t - \nu\nabla^2)\boldsymbol{\omega} = (2\boldsymbol{\Omega} \cdot \nabla)\mathbf{v} \tag{1.54}$$

operating on the left with $(\partial/\partial t - \nu\nabla^2)\nabla^2$ eliminates $\mathbf{v}$

$$(\partial/\partial t - \nu\nabla^2)^2\nabla^2\boldsymbol{\omega} + (2\boldsymbol{\Omega} \cdot \nabla)^2\boldsymbol{\omega} = 0 \tag{1.55}$$

A steady inviscid fluid flow is described according to (1.54) by

$$(\boldsymbol{\Omega} \cdot \nabla)\mathbf{v} = 0 \tag{1.56}$$

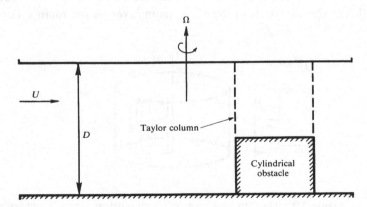

Figure 1.10 A Taylor–Proudman column. If the cylindrical obstacle is attached to the apparatus rotating at angular velocity Ω, flow with velocity U will not pass over the obstacle, but flow around it as if an imaginary cylinder extends from the bottom to the top of the fluid layer as indicated by the dashed lines.

The velocity field is independent of the coordinate parallel to the rotation axis. If Ω is directed along the z-axis, (1.56) yields

$$\Omega\, \partial v / \partial z = 0 \tag{1.57}$$

where v is a function only of x and y. This theorem shows why vortices with small friction are columnar. Changes in velocity in the z-direction are reduced, which means that particle paths are approximately circular.

Taylor (1921) performed the experiment in Figure 1.10. He drew a body along the bottom of a rotating bucket of water and by means of dye tracer showed that the fluid did not flow over the body, but around it as if an imaginary circular cylinder ran from the body at the bottom to the free surface: a Taylor–Proudman column.

1.5   Laboratory generation of thin line vortices

We shall encounter arrays of quantized vortex lines aligned parallel to the axis of rotation in a rotating bucket of helium II (see Figure 2.6). It is only recently that methods have been devised to make such arrays in rotating viscous fluids.

Hopfinger, Browand and Gagne (1982) have constructed the experiment pictured in Figure 1.11(a). A transparent cylindrical tank 40 cm in diameter and 80 cm in length is placed on a turntable. A turbulent velocity field is generated by means of a square-bar grid located some 15 cm above the bottom. The amplitude of oscillation could be adjusted up to 2.5 cm at frequencies up to 20 rad/s, rotation could be up to 2π rad/s. The most striking effect of rotation was the production of a random array of thin vortices roughly aligned with the axis of rotation and with concentrated vorticity up to 40 times the tank vorticity (Ω). The core radius $a = 0.1$–0.15 cm. The circulation in one case studied in detail was $\Gamma = 35$ cm^2/s and the strength was such as to produce a dimple on the free surface of the fluid at the top. The number of vortices generated increased with the rate of rotation and ranged from about 5 to 20. Another important consequence was the observation of travelling wave disturbances. Insofar as the spacing between vortices is fairly large, the authors compared these disturbances to solitary wave (soliton) propagation and identified the waveform with that proposed by Hasimoto (see Section 1.7.2) by means of detailed measurements.

In 1985, Maxworthy, Hopfinger and Redekopp were able to modify the apparatus of Figure 1.11(a) to produce a single vortex filament. The change, shown in Figure 1.11(b), is primarily replacement of the grid

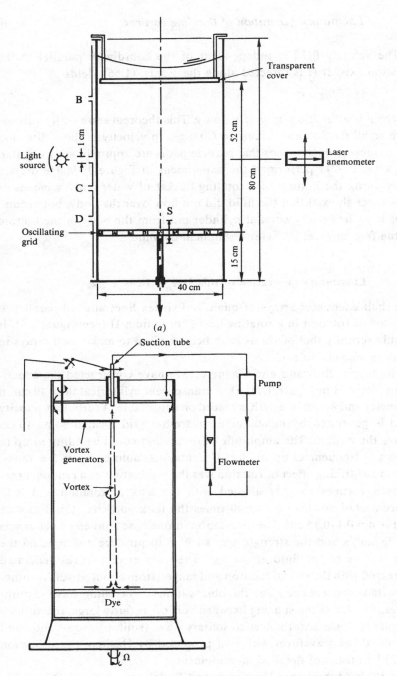

Figure 1.11 (a) Apparatus constructed by Hopfinger, Browand and Gagne (1982) for generating an array of thin line vortices.
(b) Modification to the apparatus above by Maxworthy, Hopfinger and Redekopp (1985) to produce a single vortex filament.

mechanism by a plastic plate and provision of a central suction tube 1.27 cm in diameter to draw fluid upward at the location of one central concentrated vortex. The vortex can be perturbed by cutting the vortex with a moveable thin metal rod which stops the vertical flow for an instant and bands the core. A second strategy was to oscillate the suction tube back and forth. A third method was to stop the exit flow for ~0.1 s, allowing the initial perturbation to develop into a sequence of solitary waves. On restarting the flow the vortex immediately reformed by propagating a wave of contraction along the vortex core.

Vortex rings can be produced in the laboratory by a variety of simple means. We show in Figure 1.12 an apparatus constructed by Maxworthy (1972) in a 12-gallon water tank. It consisted of a half-inch diameter sharp-edged hole drilled through a sheet of brass and connected by a manifold to a hypodermic syringe as shown in Figure 1.12. A short rubber diaphragm percussion head could be used in place of the syringe. Vortex rings were produced by ejecting a puff of fluid through the hole and observed using dye and hydrogen bubbles.

Pairs of line vortices of opposite circulation such as we show in Figure 1.8(*b*) are provided by the trailing vortices of large aircraft. If the air is homogeneous, the vortex pair will descend by self-induction to the ground at constant spacing, the pair will then divide and move across the ground at half the height of the initial separation. However a pair of line vortices is unstable to three-dimensional disturbances, as we shall see in Section 1.7.3.

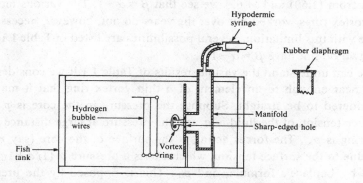

Figure 1.12 Apparatus constructed by Maxworthy to produce vortex rings. A plunger, manifold and hydrogen-bubble wire set were installed in a water tank. The detail shows an alternative device to impart an impulse to the ejected fluid: a rubber diaphragm is glued to the top of the manifold.

1.6   Dynamics of classical vortex rings and the localized induction approximation

It is possible to gain some insight into vortex dynamics and the structure of the cores of vortices of constant circulation by considering the classical expressions for the energy and velocity of vortex rings in an inviscid fluid. We shall need to use such classical expressions in order to interpret the results of experiments in helium II to be described in Chapter 4. We shall show that such rings can be described by a total energy formally equivalent to a Hamiltonian H and that the velocity and impulse of the vortex rings are connected by Hamilton's equation (Roberts and Donnelly, 1970)

$$v = \partial H / \partial P \tag{1.58}$$

The simplest situation occurs when the core radius a is negligible in size compared to the radius of the vortex ring. Then the case for any core model can be written

$$H = \tfrac{1}{2}\rho\Gamma^2 R[\ln(8R/a) - \alpha] \tag{1.59}$$

$$v = (\Gamma/4\pi R)[\ln(8R/a) - \beta] \tag{1.60}$$

$$P = \rho\Gamma\pi R^2 \tag{1.61}$$

Application of (1.58) to (1.59) using (1.61) gives

$$v = \frac{\partial E}{\partial P} = \left(\frac{\Gamma}{4\pi R}\right)\left[\ln\left(\frac{8R}{a}\right) + 1 - \alpha\right] \tag{1.62}$$

Thus from (1.60) and (1.62) we see that $\beta = \alpha - 1$. The various models for vortex rings worked out over the years do not, however, necessarily agree with this limitation. Several possibilities are listed in Table 1.1: the first two do not have $\beta = \alpha - 1$.

We can understand the various results of Table 1.1 if we consider the flow near enough to an element of a thin vortex ring that it may be considered to be straight. Suppose the pressure in the core is p_c, the surface tension of the fluid is σ and the pressure at large distance from the ring is p_∞. The forces acting at the surface of the core (say $r = a$) are due to the surface tension, which exerts a pressure $\sigma\,(1/a + 1/R) \approx \sigma/a$ by Laplace's formula, and p_∞. This is opposed by the pressure gradient of the circulating fluid, given by (1.39). The balance of forces is then

$$p_\infty + \sigma/a - p_c = \int_a^\infty (\rho\Gamma^2/4\pi^2)r^{-3}\,\mathrm{d}r = \rho\Gamma^2/8\pi^2a^2 \tag{1.63}$$

Table 1.1. *Values of α and β for classical vortex rings with different core models*

Model	α	β
Solid rotating core, constant volume	$\frac{7}{4}$	$\frac{1}{4}$
Hollow core, constant volume	2	$\frac{1}{2}$
Hollow core, constant pressure	$\frac{3}{2}$	$\frac{1}{2}$
Hollow core with surface tension	1	0

Suppose the vortex ring is considered to be hollow, the surface tension and core pressures both zero, and the fluid is assumed to be incompressible. This well-known model has $\alpha = 2$, and the energy is purely the kinetic energy of motion K. Then if the volume V of fluid plus core is kept constant, $V = V_f + V_c = $ constant. To the order we are working $V_c = 2\pi^2 a^2 R$ and

$$dV = d(2\pi^2 a^2 R) = 0 \tag{1.64}$$

since V_f is constant. Upon applying an impulse dP to the ring, the radius will increase by dR according to (1.61) but a will decrease according to (1.64). Since we kept V fixed, we find p_∞ has increased by (1.63), but for an incompressible fluid this does not change the energy of motion. The total energy of the system H is equal to the kinetic energy K to within an additive constant and the velocity comes from differentiating H at constant volume, which can be done by writing

$$H = \tfrac{1}{2}\rho\Gamma^2 R\{\ln[8\pi(2R^3)^{1/2}/V_c^{1/2}] - 2\} \tag{1.65}$$

and

$$v = (\partial H/\partial P)_{V_c} = (\partial H/\partial R)_{V_c}(dR/dP)$$
$$= (\Gamma/4\pi R)[\ln(8R/a) - \tfrac{1}{2}] \tag{1.66}$$

which confirms the result $\beta = \tfrac{1}{2}$ for a hollow core at constant volume (Table 1.1).

If the vortex core is considered to contain uniformly rotating fluid, as in a Rankine vortex, the same particles must always move with the vortex filament, so that again the volume of the core is preserved. The moment of inertia of the core, considered as a line of length $2\pi R$ is $I = \rho\pi^2 Ra^4$, and the angular velocity of the fluid in the core is (to leading order) $\omega = \Gamma/2\pi a^2$. Thus the angular momentum of the core is $\tfrac{1}{2}\rho\Gamma\pi Ra^2$, which must be conserved, and this conservation is equally expressed by (1.64). The kinetic energy of the core is $\tfrac{1}{2}I\omega^2 = \tfrac{1}{8}\rho\Gamma^2 R$ which is the difference between the kinetic energies of the solid and hollow core energies of Table 1.1. The value $\beta = \frac{1}{4}$ is confirmed by calculating $v = (\partial H/\partial P)_{V_c}$.

Now consider the case of constant pressure and a hollow core of fixed radius a. We take $p_c = \sigma = 0$ and hence keep p_c fixed upon applying an impulse dP. As the core is lengthened, the external surface of the liquid is displaced against the pressure p_∞, doing work which may be retrieved when the core is shortened. This means the system has, in a formal sense, a potential energy U given by

$$U = p_\infty V_c = \tfrac{1}{4}\rho\Gamma^2 R \tag{1.67}$$

Using (1.59) for the kinetic energy K with $\alpha = 2$,

$$H = K + U = \tfrac{1}{2}\rho\Gamma^2 R[\ln(8R/a) - \tfrac{3}{2}] \tag{1.68}$$

and the velocity of the ring

$$v = (\partial H/\partial P)_a = (\Gamma/4\pi R)[\ln(8R/a) - \tfrac{1}{2}] \tag{1.69}$$

giving α and β for the third entry in Table 1.1.

Finally, consider the case of a core with surface tension. Here $p_c = p_\infty$ and (1.63) gives

$$\sigma = \rho\Gamma^2/8\pi^2 a \tag{1.70}$$

and the core radius is entirely determined by surface tension. Again we have a constant core radius model. This core has a surface energy σA_c where $A_c = 4\pi^2 aR$ is the surface area of the core (we neglect the surface entropy contribution to the energy, $T(d\sigma/dT)A_c$). Considering the surface energy formally as a potential energy

$$U = \tfrac{1}{2}\rho\Gamma^2 R \tag{1.71}$$

the total energy and velocity of a vortex whose core is governed by surface tension is

$$H = K + U = \tfrac{1}{2}\rho\Gamma^2 R[\ln(8R/a) - 1]$$
$$v = (\partial H/\partial P)_\sigma = (\kappa/4\pi R)\ln(8R/a) \tag{1.72}$$

giving the final entry in Table 1.1.

Rectilinear vortices are also affected by core considerations. For a hollow vortex in a container of radius b the kinetic energy per unit length is given by (1.38). If one wishes to describe a hollow core governed by surface tension, the energy per unit length is modified to

$$H = (\rho\Gamma^2/4\pi)[\ln(b/a) + 1] \tag{1.73}$$

Roberts and Donnelly's (1970) demonstration that large circular rings ($a \ll R$) obey Hamilton's equation (1.58) leads one to enquire whether the equation of motion of rings of arbitrary a/R can also be written in canonical form. Rings of moderate a/R do not have circular cross-sections, and it is necessary to redefine a and R; e.g., the cross-sectional area can be taken to be πa^2, and R can be chosen to be the mean of the closest and farthest points of the vortex core from the axis of symmetry (Fraenkel, 1970). Roberts (1972) demonstrates that (1.58) is obeyed exactly for circular vortex rings, for all a/R, no matter what their core structure may be. Moreover, Roberts (1972) shows that the governing equations of motion for a set of vortex rings separated by distances large compared to their dimensions can also be written in canonical form. Dyson (1893) and Fraenkel (1972) investigated one particular core structure, the so-called 'standard model' where vorticity in the core increases with radius out to the edge of the core, i.e., $\omega/r = $ constant. A sequence of vortex rings was shown to exist, which range continuously from a large circular vortex at one extreme to Hill's spherical vortex (Figures 1.7(a) and (b)) at the other. Fraenkel (1972) notes, 'This simplest of all admissible vorticity distributions has been a favorite for over a century: It characterizes the rings of small cross-section constructed by Helmholtz, Kelvin and Hicks (see Lamb, 1945, Section 163) when that work is interpreted correctly, it is the vorticity distribution of Hills' spherical vortex, which is the only steady 'ring' represented by an exact solution in closed form; it satisfies this vorticity equation of a viscous fluid . . . ' Further, Fraenkel (1972) showed that for all sufficiently small a/R, similar families of vortex rings exist for a wide variety of core structures.

When a/R starts to become large, the expressions quoted in this section for energy and velocity are not applicable and one should consult the papers by Fraenkel (1972) and Norbury (1973) for details. Plots of energy and impulse for the classical rings of Norbury (1973) are shown in Figure 1.13.

A number of topics in this book will require an understanding of the 'effective mass' or 'effective density' of a vortex core. To see that this may be important, note that if a hollow core vortex is in an unsteady flow, it acts as if it were a cylinder with an effective mass equal to the mass of the liquid displaced. If for any reason the core has, in addition, some physical mass, then the effective density must contain both contributions.

A significant generalization of the vortex ring calculation was obtained by Arms and Hama (1965). Its importance lies in making possible approximate calculations of the motion of arbitrary configurations of very thin vortex lines. The fluid velocity at some point in space, induced

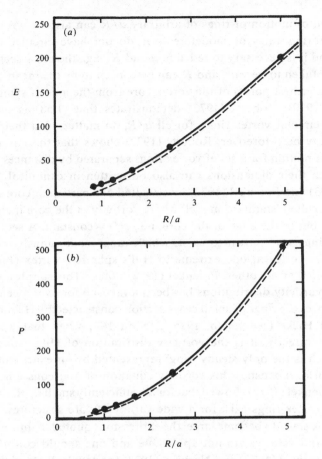

Figure 1.13 (*a*) Energy E and (*b*) impulse P for the classical rings of Norbury (1973) are shown for vortex rings as a function of R/a. The dashed lines represent the equations $E = 2\pi^2 R \left[\ln\left(8R/a\right) - \frac{7}{4}\right]$ and $P = 2\pi^2 R^2 / a^2$. The units of energy and momentum are $\rho\kappa^2 a / 4\pi^2$ and $\rho\kappa a^2 / 2\pi$ to make these plots directly comparable to those for quantum rings in Figure 4.1. The circled points are the calculated results (after Muirhead, Vinen and Donnelly (1984)).

by a vortex line, is given by an exact analogy in the theory of electromagnetism called the Biot–Savart law. In this analogy the fluid velocity corresponds to the magnetic field **H** and vorticity **ω** to the current density **j**. The equation **ω** = curl **v** is analogous to Maxwell's equation curl **H** = $(4\pi/c)$**j** (for limitations on this analogy, see Donnelly and Fetter (1966)). The integration is over all of the vortex singularities in the fluid; boundary effects are included by extending the integral to the images of the singularities.

If an arbitrary segment of vortex line is parametrized as in Figure 1.14, then the Biot–Savart law can be written

$$\mathbf{v} = (\Gamma/4\pi) \int (\mathbf{s}_0 - \mathbf{r}) \times d\mathbf{s}_0 / |\mathbf{s} - \mathbf{r}|^3 \qquad (1.74)$$

where $\mathbf{r}$ is any point in the fluid and the integral is over the relevant line segments. If $\mathbf{r} = \mathbf{s}_0$ is a point *on* the line, the integral in (1.74) diverges. If we expand $\mathbf{s}$ in a Taylor series about $\mathbf{s}_0$, $\mathbf{s} \approx \mathbf{s}_0 + \mathbf{s}'\xi + \frac{1}{2}\mathbf{s}''\xi^2$ and (1.74) becomes

$$\mathbf{v} \approx (\Gamma/4\pi) \int (d\xi/2\xi) \mathbf{s}' \times \mathbf{s}'' \qquad (1.75)$$

where the integral is over the whole vortex array except for a distance of the order of the core radius a on either side of $\mathbf{s}_0$. Ignoring 'nonlocal' portions of the vortex (more than some distance L away from $\mathbf{s}_0$ measured

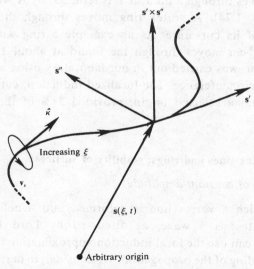

Figure 1.14 The space curve represents a vortex line with its position described as $\mathbf{s}(\xi, t)$. The local tangent $\mathbf{s}' = \hat{\kappa}$, where $\hat{\kappa}$ is a unit vector along the vortex line in the direction of Γ. Here $\mathbf{s}' = d\mathbf{s}/d\xi$ and $\mathbf{s}'' = d^2\mathbf{s}/d\xi^2$ is the local curvature vector (whose magnitude is $1/R$), and the binormal $\mathbf{s}' \times \mathbf{s}''$ is approximately in the direction of the local induced velocity $\mathbf{v}_i$, and also has the magnitude $1/R$. The instantaneous velocity of the line is given by $\mathbf{v}_L = d\mathbf{s}(\xi, t)/dt = \dot{\mathbf{s}}$.

along the arc, and approximating the cross-product by its value at s_0, we get the local self-induced velocity

$$\mathbf{v}_i \approx (\Gamma/4\pi) \ln (L/a)\mathbf{s}' \times \mathbf{s}'' \tag{1.76a}$$

where $\mathbf{s}' \times \mathbf{s}''$ has the magnitude $1/R$, R being the radius of curvature at s_0. In scalar terms

$$v_i \approx (\Gamma/4\pi R) \ln (L/a) \tag{1.76b}$$

The choice of L to best approximate (1.74) depends on the details of the vortex configuration.

If we wish to include the 'far field' of the vortex we can make a rough approximation by integrating (1.74) over the line omitting a region along the vortex of length L on either side of s_0. Then

$$\mathbf{v}_{(\text{nonlocal})} = (\Gamma/4\pi) \int' (\mathbf{s}_0 - \mathbf{r}) \times d\mathbf{s}_0/|\mathbf{s} - \mathbf{r}|^3 \tag{1.77}$$

where the prime on the integral indicates that the local line element is omitted.

An immediate application of (1.76) is an intuitive understanding of how a vortex ring moves through a fluid: if L is replaced by R we have an approximation to (1.74). A vortex ring moves through the fluid principally because of its curvature. As an example a ring with $a = 10^{-8}$ cm and $R = 10^{-3}$ cm moves through the liquid at about 1 cm/s. A numerical simulation was carried out in our laboratory using a mesh of 100 points on the circumference. The localized induction, cut off at nearest neighbors from a selected origin, provided 73% of the total induced velocity.

1.7   Waves on vortex lines and rings: stability of vortices

1.7.1   *Helical waves of constant amplitude*

In a situation in which a vortex line is deformed into a helix, the deformation propagates as a wave, as discussed by Lord Kelvin (Thomson, 1880). We can use the local induction approximation to gain an intuitive understanding of the propagation. We consider, in particular, a helical deformation of wave vector k and amplitude d, where $d \ll k^{-1}$ (see Figure 1.15).

Ignoring the nonlocal contribution and core mass, the line moves with Arms–Hama velocity (1.76) where L is reasonably taken as being of order k^{-1}. When $d \ll k^{-1}$, this velocity is perpendicular to the undisturbed line and to the displacement vector from the undisturbed line to the

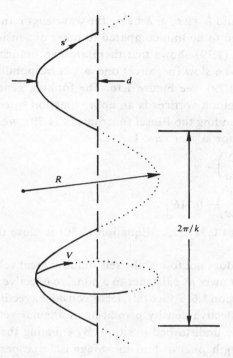

Figure 1.15 A vortex line deformed into a helix. The amplitude of the deformation is d and its wavenumber is k; R is the radius of curvature of the line at some point and s' is a unit vector along the vortex line indicating the direction of the circulation Γ (after Glaberson and Donnelly (1986)).

point considered. Each vortex line element therefore executes motion about the undisturbed line in a circle of radius d and with a frequency $\omega = v_i/d$. The radius R is given approximately by $R \sim 1/dk^2$ so that

$$\omega(k) \sim (\Gamma k^2/4\pi)\ln(1/ka) \tag{1.78}$$

It is evident from the analysis that vortex waves are intrinsically polarized, always executing motion about the undisturbed line in a sense opposite to the sense of the velocity field. The correct expression, assuming a thin hollow vortex core, was given by Lord Kelvin (Thomson, 1880)

$$\omega^{\pm} = \frac{\Gamma}{2\pi a^2}\left(1 \pm \left\{1 + ka\left[\frac{K_0(ka)}{K_1(ka)}\right]\right\}^{1/2}\right) \tag{1.79}$$

where K_n is a modified Bessel function of order n. The same formula was derived by Pocklington (1895) for waves on a hollow ring with

$n = kR$ the mode number and $k = 2\pi/\lambda$, λ being the wavelength. In both cases the core is considered to be thin compared to other dimensions.

The dispersion formula (1.79) shows that there are two branches, a fast (positive) wave ω^+ and a slow (negative) one ω^-, corresponding to $+$ and $-$ respectively in (1.79), see Figure 1.16. The formula generally used in the literature of helium vortices is an approximation to ω^- for long wavelengths. On expanding the Bessel functions in (1.79), we have the following expressions for ω^- for $ka \ll 1$

$$\omega^- \approx -\frac{\Gamma k^2}{4\pi}\left[\ln\left(\frac{2}{ka}\right) - \gamma\right]$$

$$\approx -\frac{\Gamma k^2}{4\pi}\left[\ln\left(\frac{1}{ka}\right) - 0.116\right] \tag{1.80}$$

where γ is Euler's constant $0.5772\ldots$. Equation (1.80) is close to the intuitive result (1.78).

The positive branch ω^+ does not follow the self-induced local velocity discussed above, because it owes its existence to a nonzero effective mass in the core discussed in Section 1.6. To see this, let us consider a rectilinear vortex line with core of effective density ρ rotating at angular velocity ω at a distance d from the undisturbed position. By equating the centrifugal force per unit length $\rho\pi\omega^2 a^2 d$ to the Magnus force per unit length $\rho\Gamma\omega d$, one obtains $\omega = \Gamma/\pi a^2$ which is the limit of ω^+ from Equation (1.79) for $ka \ll 1$.

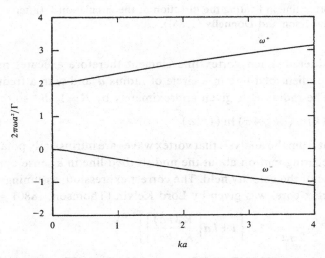

Figure 1.16 The dispersion curves (1.79) for helical vortex waves in a classical fluid (after Glaberson and Donnelly (1986)).

When helical waves of the type shown in Figure 1.15 reflect from a boundary, two (slow) waves of opposite polarization combine to form a plane standing wave called a Kelvin wave which rotates with angular velocity ω given by Equation (1.80) in the long wave limit. The stability of Kelvin waves will be discussed in Section 6.1.3.

There are also axisymmetric (varicose) waves for which the core diameter varies as the wave propagates. These relationships depend on detailed assumptions about the core and the flow at infinity. One such dispersion curve, shown in Figure 1.17, has been suggested to us privately by Mathieu Mory:

$$\omega = \pm \frac{\Gamma k}{2\pi a} [K_1(ka)_1 I(ka)^{1/2}] \tag{1.81}$$

where I and K_1 are modified Bessel functions of the first and second kind.

1.7.2 Solitary kink waves

The possible existence of solitary waves (kink wave solitons) propagating along isolated vortex filaments in an irrotational fluid was first discussed by Hasimoto (1972). Using the space curve parameterization of Figure 1.14 and the localized induction approximation discussed in Section 1.6 (Equation (1.76)) with the logarithm term considered constant, Hasimoto

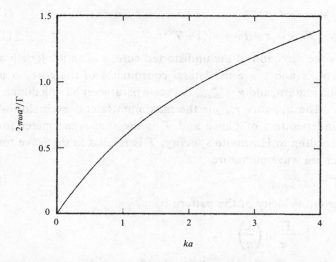

Figure 1.17 Dispersion curve for one model of varicose perturbations on a line vortex.

demonstrated that the filament distortion evolves according to the cubic-nonlinear Schrödinger equation

$$\frac{1}{i}\frac{\partial \psi}{\partial t} - \frac{\partial^2 \psi}{\partial s^2} = \tfrac{1}{2}|\psi|^2 \psi \tag{1.82}$$

where time has been rescaled to $t(\Gamma/4\pi)\ln(L/a)$. The complex wave function ψ is defined in terms of the filament curvature ζ and torsion τ by the relation

$$\psi(s,t) = \zeta \exp\left\{ i \int_0^s \tau \, ds' \right\} \tag{1.83}$$

Maxworthy, Hopfinger and Redekopp (1985) note that (1.81) linearized about an undisturbed state having $\zeta = \tau = 0$ is consistent with the dispersion relation (1.80) and that solitary waves can propagate only if the group velocity is increasing (see Figure 6.3), i.e., the curvature of the dispersion curve is negative.

The theoretical wave shape predicted by Hasimoto is

$$x = r_{max} \operatorname{sech} \eta \cos \phi \tag{1.84a}$$

$$y = r_{max} \operatorname{sech} \eta \sin \phi \tag{1.84b}$$

$$z = s - r_{max} \tanh \eta \tag{1.84c}$$

$$r_{max} = \frac{2}{\nu(1+T^2)} \tag{1.84d}$$

$$\eta = \nu(s - C_g t) \tag{1.84e}$$

$$\phi = T\eta + \omega_p t = T\eta + \nu^2(1+T^2)t \tag{1.84f}$$

where z is the direction of the undistorted core, s is an arc length along the filament, x and y are the lateral coordinates of the core, ϕ is the phase of the pattern, and $\nu = \tfrac{1}{2}\zeta_{max}$ is a scale parameter having dimensions (length)$^{-1}$. The quantity r_{max} is the maximum lateral excursion of the filament independent of phase and T is the shape parameter for the wave. According to Hasimoto's theory, T is related to the wave torsion and maximum wave curvature

$$T = 2\tau/\zeta_{max} \tag{1.85}$$

and the group velocity of the pattern is

$$C_g = 2\tau \frac{\Gamma}{4\pi} \ln\left(\frac{L}{a}\right) \tag{1.86}$$

where we have restored natural time by the transformation mentioned above (1.82). For small r_{max}/λ, $\tau \cong 2\pi/\lambda$ where λ is the wavelength of the central kink so that

$$C_g \cong \frac{\Gamma}{\lambda} \ln \left(\frac{L}{a}\right) = \frac{\Gamma}{2\pi} k \ln \left(\frac{L}{a}\right) \tag{1.87}$$

The wave peaks move with a phase velocity which can be calculated to the present degree of approximation by noting that the rotation rate of the coils of vortex filament is given by

$$\omega_p \approx \frac{-\pi\Gamma}{\lambda^2} \ln \left(\frac{L}{a}\right) \tag{1.88}$$

(when $(\zeta/2\tau)^2 \ll 1$) and is the same as for helical waves (the negative sign indicates rotation opposite to the vortex flow). The maximum and minimum then appear to move at a speed $C_p = \omega_p \lambda / 2\pi$, which becomes

$$C_p \approx \frac{\Gamma}{2\lambda} \ln \left(\frac{L}{a}\right) = \frac{\Gamma}{4\pi} k \ln \left(\frac{L}{a}\right) \tag{1.89}$$

which is equal to $\frac{1}{2}C_g$ and is the same as for long helical waves (1.86).

Hopfinger, Browand and Gagne (1982) and Maxworthy, Hopfinger and Redekopp (1985) were able to use the information given in the development above to establish that they were indeed observing solitary kink waves. We illustrate solitary waves on their vortex cores in Figures 1.18(a) and (b) which are drawn from their discussions.

1.7.3. *Instability of parallel vortices*

A special instability which occurs for two parallel vortex filaments of opposite circulation (Figure 1.8(b)) has been discussed by Crow (1970). In the undisturbed state, the vortices will move with velocity $v = \Gamma/2\pi b$. Helical waves, especially of long wavelength on these vortices, can produce a self-amplifying instability. To see this, note that in Figure 1.14 the elements of line closer than equilibrium to each other have a Magnus force tending to pull them closer, elements further away than equilibrium have less induced velocity than normal, producing an outward Magnus force. Crow's analysis shows that antisymmetric perturbations of short wavelength are also unstable but those shown in Figure 1.19 are the most unstable.

Vortex rings in nature have generally been considered to be both stable and persistent flows. Indeed the work of Lord Kelvin, J. J. Thomson (1883) and others was directed towards establishing the stability of vortex rings and calculating their frequencies of oscillation in order to investigate atomic structure. In practice, however, vortex lines (such as aircraft

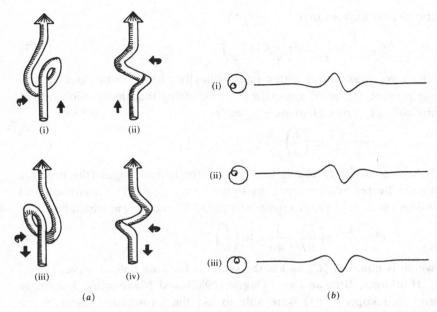

Figure 1.18 (*a*) Schematic representation of possible wave shapes:
(i) left kink; (ii) left antikink; (iii) right kink; (iv) right antikink.
Arrows indicate the direction of propagation and rotation of the
wave patterns. (*b*) Constant torsion soliton shape given by Hasimoto
(1972): (i) $t = 0$; (ii) $t = \frac{1}{4}$ period; (iii) $t = \frac{1}{2}$ period. The view along
the vortex from above is shown for each period (after Hopfinger,
Browand and Gagne (1982) and Maxworthy, Hopfinger and
Redekopp (1985)).

trailing vortices) and vortex rings are often unstable. An accessible review
of the subject has been given by Widnall (1975). The instability of a thin
vortex ring of constant vorticity ($\tilde{\omega}^{-1}$ curl $\mathbf{v} = $ constant as measured from
the center of the core) in an ideal fluid has been considered in detail by
Widnall and Tsai (1977). They found that such a vortex is unstable to
short azimuthal bending waves ($ka = 2.5$ appears to be dominant, where
k is the wavenumber) but the lowest mode which is unstable is the
so-called second radial mode where the core moves in a direction opposite
to the outer flow. A photograph of such an instability is shown in Widnall
(1975), Widnall and Tsai (1977).

It seems reasonable to assume that the instability described by
Widnall's group is associated with vorticity in the core and therefore will
not be expected to occur either for hollow classical vortices, or for
quantum vortices.

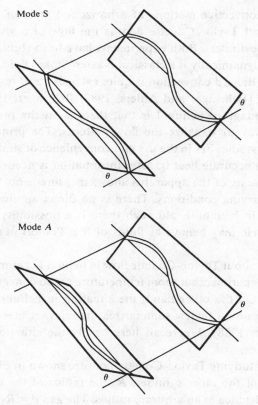

Mode S

Mode *A*

Figure 1.19 Shapes of unstable modes in the Crow instability (after Crow (1970)).

Vortex rings in an inviscid fluid have a total energy proportional to their circumference (see Equation (1.59)). Thus if some frictional force acts upon them, conservation of energy shows that their radii will decrease, and their velocities will, as a result, increase (see Equation (1.60)). We shall see in Section 3.5 that this is what happens to vortex rings in helium II. Vortex rings in real fluids, however, grow to larger diameter and slow down as time progresses. The reader is referred to a paper by Maxworthy (1972) discussing why this happens.

1.8 Bénard convection and Taylor–Couette flow

We note here briefly the problems of Bénard convection and Taylor–Couette flow. In their simplest manifestations, Bénard convection is the field of study of the convective motions of a horizontal layer of fluid heated from below, and Taylor-Couette flow is the flow of a viscous fluid between rotating cylinders. Both experiments have been studied in helium I, which hydrodynamically is a classical Navier-Stokes fluid. The literature in cryogenic Bénard convection is quite extensive and reviews are starting to appear (Behringer and Ahlers, 1982; Behringer, 1985). The principal disadvantage of helium I is that there is, at the present time, no convenient way to visualize the flow patterns. The principal scientific value of such studies lies in the use and convenience of studying certain problems when accurate heat transfer information is needed, or when the low heat capacity of the apparatus allows measurements to be made under rapidly varying conditions. There is no direct application of Bénard convection in helium II, although there is a possibility that mixtures of ^{3}He and ^{4}He may behave as fluids of low Prandtl number (Fetter, 1982).

There is little known about Taylor-Couette flow in helium I (Donnelly, 1959) simply because experiments at room temperature are so convenient and easily visualized. On the other hand, the situation in helium II is potentially challenging and a review summarizes the relevant literature (Donnelly and LaMar, 1988). We recall here briefly the situation in classical fluid mechanics.

The coordinates for studying Taylor-Couette flow are shown in Figure 1.20. R_1 is the radius of the inner cylinder, R_2 the radius of the outer cylinder, and r is the distance to an arbitrary radius. The gap $d = R_2 - R_1$ and the angular velocities of the cylinders are Ω_1 and Ω_2. The ratio of the length of the apparatus L to the gap width d is called the aspect ratio

$$\Gamma = L/d \tag{1.90}$$

For a classical fluid in laminar flow the velocity distribution for infinite cylinders rotating at angular velocities Ω_1 and Ω_2 is

$$v = Ar + B/r \tag{1.91}$$

a combination of solid body rotation and potential flow, where

$$A = \frac{R_2^2 \Omega_2 - R_1^2 \Omega_1}{R_2^2 - R_1^2} = -\Omega_1 \frac{\eta^2 - \mu}{1 - \eta^2} \tag{1.92}$$

and

$$B = -\frac{R_1^2 R_2^2 (\Omega_2 - \Omega_1)}{R_2^2 - R_1^2} = \Omega_1 \frac{R_1^2 (1 - \mu)}{1 - \eta^2} \tag{1.93}$$

where

$$\mu = \Omega_2 / \Omega_1 \quad \text{and} \quad \eta = R_1 / R_2 \tag{1.94}$$

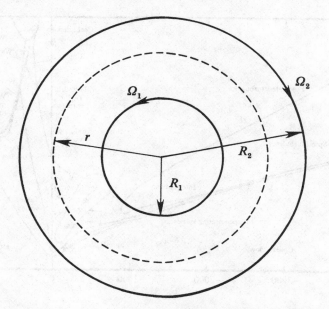

Figure 1.20 Geometry for discussing stability and flow between rotating cylinders.

The vorticity ω for this flow is given by curl **v**:

$$\omega = 2A = \frac{2(R_2^2\Omega_2 - R_1^2\Omega_1)}{R_2^2 - R_1^2} \tag{1.95}$$

The torque G transmitted to a length h of one cylinder as a result of rotation of the other is given by

$$G = \frac{4\pi\eta R_1^2 R_2^2 h(\Omega_1 - \Omega_2)}{R_2^2 - R_1^2} \tag{1.96}$$

where h is the length of the suspended cylinder and η is the viscosity of the fluid.

The problem of the stability of flow between concentric cylinders was first worked out by Taylor and described in a remarkable paper published in 1923 containing both theory and confirming experiments.

A stability diagram for classical Taylor–Couette flow is shown in Figure 1.21. When the gap is narrow, as it is in this case, the Reynolds' numbers for the inner and outer cylinders can be defined as

$$Re_i = \Omega_1 R_1 d / \nu \tag{1.97}$$

$$Re_o = \Omega_2 R_2 d / \nu \tag{1.98}$$

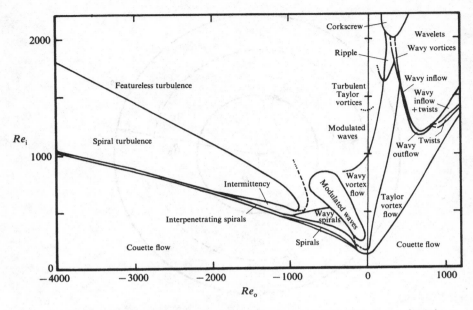

Figure 1.21 Stability diagram by Andereck, Liu and Swinney (1986) showing the astonishing variety of vortex motion which occurs in the flow between concentric cylinders. Here $\eta = 0.85$.

When $Re_o = 0$, we see that a finite Re_i is required to produce an instability: the famous Taylor vortex flow consisting of rolls spaced regularly along the gap with wavelength $\sim 2d$. At a slightly higher Re_i, the wavy mode is observed, consisting of azimuthal variations of the basic Taylor vortex flow. At still higher Re_i, a second wavy mode appears, modulating the first at an incommensurate frequency. Eventually a form of turbulence is observed, which retains the basic axial periodicity of the original Taylor vortex flow.

Rotation of the outer cylinder in the Reynolds number range of Figure 1.21 initially stabilizes the flow. But the picture of what instabilities occur and their patterns is complicated indeed.

1.9   Brownian motion and escape over barriers

The theory of Brownian motion is surely one of the most elegant chapters in modern physics. Brownian particles were first recognized as small grains or particles of colloidal size immersed in a fluid. These exhibit perpetual irregular motions owing to molecular agitation from the fluid. The detailed treatment of Brownian motion in physics is best learned from Chandrasekhar's well-known review article of 1943. We begin the present discussion by recapitulating Kramers' (1940) sophisticated derivation of the escape of Brownian particles over a potential barrier. We shall use this result in discussing the interactions of ions and vortices in Chapter 4, as well as the nucleation of quantized vortices in Chapter 8. Significant parts of Section 8.7 on vortices in thin films depend on this discussion and its application to nucleation in Sections 8.3 and 8.4.

Consider the density distribution of particles $w(\mathbf{r}, t)$ in the presence of an external field of force characterized by an acceleration $K(\mathbf{r})$ and frictional force β characterized by a relaxation time β^{-1}, i.e., the drag force on the particle in steady motion at velocity v is $m\beta v$. The quantity $q = \beta k_B T/m$ where k_B is Boltzmann's constant and m is the mass of the Brownian particle. The diffusion coefficient for free particles is

$$D = q/\beta^2 = k_B T/m\beta \qquad (1.99)$$

Under the conditions that $K(\mathbf{r})$ is slowly varying and that velocities are small compared to $(k_B T/m)^{1/2}$ we obtain Smoluchowski's equation

$$\frac{\partial w}{\partial t} = \operatorname{div}\left(\frac{q}{\beta^2}\operatorname{grad} w - \frac{\mathbf{K}}{\beta} w\right) \qquad (1.100)$$

If a steady state prevails there is a diffusion current $\mathbf{j}$ obeying the law.

$$\mathbf{j} = -\frac{q}{\beta^2}\operatorname{grad} w + \frac{\mathbf{K}}{\beta} w = \text{constant} \qquad (1.101)$$

If $\mathbf{K}$ can be derived from a potential V so that

$$\mathbf{K} = -\operatorname{grad} V \qquad (1.102)$$

then (1.101) can be rewritten in the form

$$\mathbf{j} = -\frac{q}{\beta^2}\exp\left(\frac{-\beta V}{q}\right)\operatorname{grad}\left[w\exp\left(\frac{\beta V}{q}\right)\right] \qquad (1.103)$$

Integrating (1.103) between any two points A and B we obtain

$$\mathbf{j}\cdot\int_A^B \beta \exp\left(\frac{\beta V}{q}\right)d\mathbf{s} = \frac{k_B T}{m} w\exp\left(\frac{\beta V}{q}\right)\Big|_A^B \qquad (1.104)$$

an important equation first derived by Kramers (1940).

Suppose we restrict ourselves to one dimension and consider a potential field $V(x)$ of the type shown in Figure 1.22. Suppose the particles are initially caught in the potential well near A and we wish to compute the rate of passage over the potential barrier at C. Since by (1.99) $\beta V/q = mV/k_B T$ we can see that if $mV_c \gg k_B T$ the rate of diffusion will be small and the problem can be considered quasi-stationary: in particular the

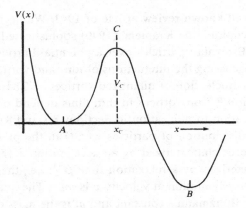

Figure 1.22   Potential barrier used in the discussion of Brownian motion over barriers in one dimension.

distribution of particles at A can be considered to be given by the Maxwell–Boltzmann distribution. Then (1.104) may be written

$$j = \frac{k_B T}{m\beta} \frac{w \exp\left[mV(x)/k_B T\right]\big|_A^B}{\int_A^B \exp\left[mV(x)/k_B T\right] dx} \tag{1.105}$$

Now the number of particles ν_A in the vicinity of A is accurately given by the Maxwell–Boltzmann distribution

$$d\nu_A = w_A \exp\left[-mV(x)/k_B T\right] dx \tag{1.106}$$

Assuming $V(x)$ near A can be expanded by a Taylor series about A, retaining only one term

$$V(x) \cong \tfrac{1}{2}\omega_A^2 x^2 \qquad (\omega_A = \text{constant}, \ x \sim 0) \tag{1.107}$$

so that (1.106) yields

$$\nu_A = w_A \int_{-\infty}^{\infty} \exp\left(-m\omega_A^2 x^2/2k_B T\right) dx \tag{1.108}$$

where the limits of integration reflect the fact that most particles are near A ($x = 0$), hence

$$\nu_A = \frac{w_A}{\omega_A} \left(\frac{2\pi k_B T}{m}\right)^{1/2} \tag{1.109}$$

Returning to (1.105) if we suppose that at least initially the density of particles near B is small, and use (1.107) near A, then to sufficient accuracy

$$j \approx \frac{k_B T}{m\beta} w_A \left\{ \int_A^B \exp\left[\frac{-mV(x)}{k_B T}\right] dx \right\}^{-1} \tag{1.110}$$

Equations (1.109) and (1.110) may now be used to compute the rate at which a particle, initially in the well at A, will escape over the barrier at C:

$$P = \frac{j}{\nu_A} = \frac{\omega_A}{\beta} \left(\frac{k_B T}{2\pi m}\right)^{1/2} \left\{ \int_A^B \exp\left[\frac{mV(x)}{k_B T}\right] dx \right\}^{-1} \quad (1.111)$$

The value of the integral in (1.111) depends largely on the small region near C, since we have had to assume $mV(x) \gg k_B T$. Expanding $V(x)$ as a Taylor series near C,

$$V(x) \approx V_C - \tfrac{1}{2}\omega_C^2(x - x_C)^2 \quad (\omega_C = \text{constant}, \; x \sim x_C) \quad (1.112)$$

Then it is sufficiently accurate to put

$$\int_A^B \exp\left[mV(x)/k_B T\right] dx$$

$$\approx \exp\left(mV_C/k_B T\right) \int_{-\infty}^{\infty} \exp\left[-m\omega_C^2(x - x_C)^2/2k_B T\right] dx$$

$$= \exp\left(mV_C/k_B T\right)(2\pi k_B T/m\omega_C^2)^{1/2} \quad (1.113)$$

Combining Equations (1.111) and (1.113) we have Kramers' famous result

$$P = \left(\frac{\omega_A \omega_C}{2\pi\beta}\right) \exp\left(\frac{-mV_C}{k_B T}\right) \quad (1.114)$$

giving the probability per unit time that a particle originally in the well at A will escape over the barrier at C to B.

The probability estimate (1.114) is adequate if the problem truly approximates a one-dimensional calculation. A little consideration will show, however, that the geometry at C is actually a saddle point, and if the effective width of the saddle point is much different than the width of the well, we must use a solution in two dimensions. Defining the perpendicular direction as y and putting near A

$$V \approx \tfrac{1}{2}\omega_A^2 x^2 + \tfrac{1}{2}s_A^2 y^2 \quad (s_A, \omega_A \text{ constants}, \; x, y \approx 0) \quad (1.115)$$

and near C, defining $\xi = (x - x_C)$, $\eta = y$

$$V \approx V_C - \tfrac{1}{2}\omega_C^2\xi^2 + \tfrac{1}{2}s_C^2\eta^2 \quad (\omega_C, s_C \text{ constants}, \; \xi, \eta \sim 0) \quad (1.116)$$

Donnelly and Roberts (1969) obtained

$$P = \frac{\omega_A \omega_C}{2\pi\beta} \left(\frac{s_A}{s_C}\right) \exp\left(\frac{-mV_c}{k_B T}\right) \quad (1.117)$$

which differs from Kramers' result (1.114) only by the factor (s_A/s_C) expressing the effect of the width of the lip.

2   Background on liquid helium II

The purpose of this chapter is to introduce the reader to the ideas of quantized circulation and quantized vortices in helium II. Some background in the ideas of superfluidity and the two-fluid model would be helpful, and we introduce a number of ideas without presentation of much detail. By following, roughly, the early historical development of the subject of quantized vortices, the principal ideas and notation should become clear. We discuss first the two-fluid model, followed by an account of early ideas and early experiments on vortices.

The critical temperature for liquid helium is 5.20 K and the normal boiling point is 4.215 K. Under saturated vapor pressure, liquid helium is an ordinary classical viscous fluid called helium I down to the lambda temperature $T_\lambda = 2.172$ K. The lambda temperature is named because of the shape of the large specific heat anomaly at T_λ. Below T_λ superfluid properties appear and from T_λ down to $T = 0$ K the liquid is called helium II. It requires about 25 bar of pressure to solidify liquid helium.

There are many general references on liquid helium. Some of the more accessible volumes are those by Atkins (1959), Benneman and Ketterson (1976, 1978), Donnelly, Glaberson and Parks (1967), Keesom (1942), Keller (1969), Khalatnikov (1965), Lifshitz and Pitaevskii (1980), London (1954), McClintock, Meredith and Wigmore (1984), Putterman (1974), Tilley and Tilley (1986), Wilks and Betts (1987) and Wilks (1967).

2.1   The two-fluid model

Helium II is regarded as a mixture of normal fluid and superfluid with total density ρ:

$$\rho = \rho_s + \rho_n \tag{2.1}$$

The densities of the superfluid, ρ_s and the normal fluid ρ_n can be measured in a variety of ways and are known quite accurately as a function of temperature and pressure. Their values at the saturated vapor pressure are shown in Figure 2.1. The temperature variation is such that at $T = T_\lambda$, $\rho_n = \rho$, $\rho_s = 0$ and at $T = 0$ $\rho_s = \rho$ and $\rho_n = 0$. The total density ρ is nearly constant in the helium II temperature range. The flow of the fluid is characterized by normal and superfluid velocity fields, $\mathbf{v}_n$ and $\mathbf{v}_s$. At sufficiently low velocities the superfluid is expected to flow as a fluid obeying Euler's equation (1.4), and the normal fluid as a viscous fluid with viscosity η obeying the Navier–Stokes equation (1.3). The peculiar thermal properties of helium II, characterized by the entropy per gram

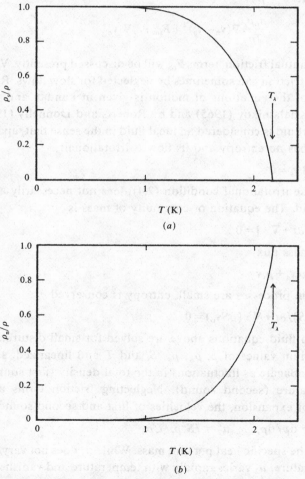

Figure 2.1 Temperature dependence of (a) ρ_s/ρ and (b) ρ_n/ρ at the saturated vapor pressure.

S and temperature T, add a thermomechanical term $\rho_s S \nabla T$ to the equations of motion. The simplest form of the two-fluid equations (neglecting both bulk viscosity and thermal conductivity) is

$$\rho_s \frac{dv_s}{dt} = -\frac{\rho_s}{\rho} \nabla p + \rho_s S \nabla T$$

$$+ \frac{\rho_n \rho_s}{2\rho} \nabla (v_n - v_s)^2 - \mathbf{F}_{ns} \qquad (2.2)$$

$$\rho_n \frac{dv_n}{dt} = -\frac{\rho_n}{\rho} \nabla p - \rho_s S \nabla T$$

$$- \frac{\rho_n \rho_s}{2\rho} \nabla (v_n - v_s)^2 + \mathbf{F}_{ns} + \eta \nabla^2 v_n \qquad (2.3)$$

where the 'mutual friction' terms $\mathbf{F}_{ns}$ will be discussed presently. Viscosity and mutual friction can sometimes be neglected for slow flows. Rigorous discussion of the equations of motion is given in Landau and Lifshitz (1987), by Khalatnikov (1965) and by Roberts and Donnelly (1974).

The superfluid is considered an ideal fluid in the sense that, apart from vortices, it has no entropy and its flow is irrotational:

$$\text{curl } v_s = 0 \qquad (2.4)$$

Note that the irrotational condition (2.4) does not necessarily apply to an ideal fluid. The equation of continuity of mass is

$$\partial \rho / \partial t + \nabla \cdot \mathbf{j} = 0 \qquad (2.5)$$

where the mass flux

$$\mathbf{j} = \rho_s v_s + \rho_v v_n \qquad (2.6)$$

If dissipation processes are small, entropy is conserved

$$\partial (\rho S) / \partial t + \nabla \cdot (\rho S v_n) = 0 \qquad (2.7)$$

If the two fluid equations above are solved for small disturbances to the equilibrium values of ρ, ρ_n, ρ_s, S, and T and linearized, sound is found to propagate as fluctuations in the total density (first sound) and the temperature (second sound). Neglecting friction terms and the coefficient of expansion, the velocities of first and second sound are

$$u_1^2 \approx dp/d\rho, \qquad u_2^2 \approx TS^2 \rho_s / C \rho_n \qquad (2.8)$$

where C is the specific heat per unit mass. While u_1 does not vary rapidly with temperature, u_2 varies rapidly with temperature and vanishes at T_λ. Both u_1 and u_2 are accurately known as functions of temperature and pressure (Brooks and Donnelly, 1977).

The discovery of second sound, which was predicted theoretically before the mutual friction terms were introduced, was such a dramatic success that there was considerable confidence that the Equations (2.2) and (2.3) without F_{ns} were complete. Such equations, however, admit a solution for a rotating container of helium II where the superfluid remains at rest and the normal fluid rotates. This would imply that the shape of the free surface of rotating helium II should be given by $z = (\rho_n/\rho)(\Omega^2 r^2/2g)$ where Ω is the angular velocity of rotation and g is the acceleration due to gravity. Since ρ_n/ρ is a rapidly varying function of temperature, it was a simple matter to check the result. This was done by Osborne in 1950, who found the classical result, $z = \Omega^2 r^2/2g$.

The explanation for this striking result is that the rotating superfluid is threaded by an array of quantized vortex lines in a manner which we shall discuss below (see Figure 2.6), and these vortex lines interact with the normal fluid to give rise to the mutual friction terms F_{ns}.

2.2 Elementary excitations in helium II

The superfluid is considered to be the 'background fluid' in the sense that at $T = 0$ K helium II is completely superfluid and, at low temperatures at least, the normal fluid is considered to consist of thermal excitations, phonons and rotons, of the superfluid. If thermal neutrons from a nuclear reactor with energies typically $E_i = 40$ millielectron volts (meV) are scattered inelastically from density fluctuations in helium II, there is an energy transfer $\varepsilon = \hbar\omega$ to the fluid, resulting in the neutron of mass m_n emerging with lower energy E_f and momentum $\hbar k_f$:

$$\left.\begin{array}{l} E_i - E_f = \hbar\omega \\ \hbar k_i - \hbar k_f = \hbar k \end{array}\right\} \quad (2.9)$$

Since neutron energies $E_i = \hbar^2 k^2/2m_n$ have wavelengths near the interatomic spacing, they provide valuable probes of the structure of the fluid. It is found that the energy and momentum transfer to the fluid occurs at sharply defined energies, especially at temperatures below 1.2 K. A composite fit of such data is reproduced in Figure 2.2. Here the wavenumber k is expressed in reciprocal angstroms, $Q = 10^8 k$. The population of these excitations is large near $Q \to 0$ and $Q \approx 2$ Å^{-1} where the dispersion curve has a pronounced minimum. These two regions of phase space are called the 'phonon' and 'roton' regions and it is roughly true that phonons and rotons comprise most of the elementary excitations of the fluid represented by Figure 2.2. Phonons are density fluctuations, quantized sound waves with energies $\varepsilon = cp = c\hbar k$ where c is the velocity of sound

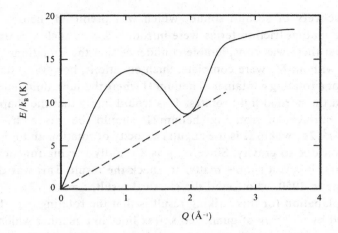

Figure 2.2 Neutron scattering data presented as the energy of elementary excitations in degrees Kelvin ($\hbar\omega/k_B$) as a function of wavenumber Q in reciprocal Ångstroms ($Q = 10^8 k$). The roton energy at the minimum near $Q = 1.9\ \text{Å}^{-1}$ is called Δ and Δ/k_B is about 8.7 K. The dashed line corresponds to the Landau critical velocity $d\varepsilon/dp = \varepsilon/p$ (after Donnelly, Donnelly and Hills (1981)).

($\sim$238 m/s at SVP) and rotons are also density fluctuations with energies near 8.7 K at low temperatures. Rotons do not possess angular momentum as their name might imply. They have been considered to be a vortex ring shrunk to atomic scale, or a fast moving helium atom with backflow, but whatever their origin, they possess a dipolar flow pattern and have the peculiar property that their group velocity $d\varepsilon/dp$ can be positive or negative with respect to the momentum p.

The Landau model of helium II, then, considers the superfluid as an inviscid, irrotational 'background fluid' with thermal excitations, the phonons and rotons, moving upon that background. The elementary excitations constitute the normal fluid and represent *all* the thermal energy of helium II. The normal and superfluid components would not interact except for the presence of vortex lines. This is the more micro-scopic picture of the two-fluid equations (2.2) and (2.3) with $\mathbf{F}_{ns}$ brought about through the interaction of the elementary excitations with vortices.

The simple excitation picture just described does not work at temperatures very near the lambda transition, T_λ. Fluctuations in the order parameter (see Section 2.8) occur near the lambda transition, and the distance over which these fluctuations are correlated is called the *correlation length* or *coherence length* ζ. This is the order of the interatomic spacing at low temperatures and near T_λ, diverges. Its importance to our subject is that the vortex core radius a is considered to be of order ζ and hence diverges also as $T \to T_\lambda$. A plot of a is shown in Figure 2.3.

When the nature of elementary excitations was first put forward, Landau examined by means of energy and momentum conservation the conditions under which a single roton could be created by superflow through a tube. Presumably any faster flow would be slowed by creation of more and more rotons. His observation put an upper limit on superflow connected with the spectrum of elementary excitations

$$d\varepsilon/dp \geq \varepsilon/p \tag{2.10}$$

where $p = \hbar k$. The marginal case is shown by the dashed line in Figure 2.2 and corresponds to a velocity $v_L \sim 60$ m/s. The importance of (2.10)

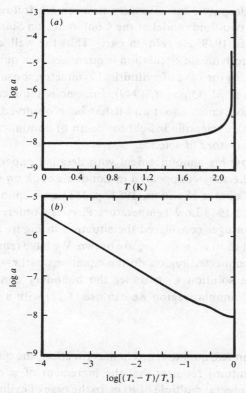

Figure 2.3 Variation of the vortex core parameter a in helium II with (a) temperature and (b) reduced temperature $(T_\lambda - T)/T_\lambda$. This variation reflects the behavior of the coherence length ζ.

is that whatever spectrum of excitations is proposed to model liquid helium, it must have a finite slope as $p \to 0$. While the condition (2.10) is certainly a limit which must exist (and indeed has been experimentally verified by McClintock's group (Ellis and McClintock, 1985)), many critical flow velocities are very much smaller (of order millimeters per second), and it is this discrepancy that owes its origin to the action of quantized vortices.

2.3   First ideas about quantized circulation and vortices

2.3.1   *Onsager's quantization of circulation*

The idea of quantized circulation in superfluid helium was first put forth to students and colleagues at Yale University by Lars Onsager beginning about 1946. Onsager enjoyed the drama of an important scientific announcement and made public his discovery in a remark following a paper by Gorter on the two-fluid model at the Conference on Statistical Mechanics in Florence in 1949. He said, in part, 'Thus the well-known invariant called the hydrodynamic circulation is quantized; the quantum of circulation is h/m ... In the case of cylindrical symmetry, the angular momentum per particle is $\hbar$' (Onsager, 1949). Enormous ramifications of this single remark have come about and it has been observed more than once that the ratio of scientific insight to length of announcement must be a record in the history of science.

Onsager did not follow his announcement with detailed papers. His next useful remark on the subject appeared in London's book on superfluids in 1954. London quotes (London, 1954, p. 151) an unpublished remark by Onsager at the 1948 Low Temperature Physics Conference at Shelter Island. There Onsager considered the situation in Figure 2.4. A series of annular rings of radii $r_1, r_2, \ldots, r_n$ are drawn. We have remarked that only for a simply-connected region do the equations curl $\mathbf{v} = 0$ and div $\mathbf{v} = 0$ imply the sole solution $\mathbf{v} = 0$ under the boundary condition $v_\perp = 0$. For a cylindrical annular region we can use (1.22) with a multi-valued potential

$$v = \text{grad } \varphi \qquad\qquad (2.11)$$

As we shall see in Section 2.8, quantum mechanics (in effect, the quantization of angular momentum) requires that the increment of φ over a closed path must be an integral multiple of h/m. In the case of cylindrical symmetry, the potential is given by

$$\varphi = k\hbar\theta/m \qquad\qquad (2.12)$$

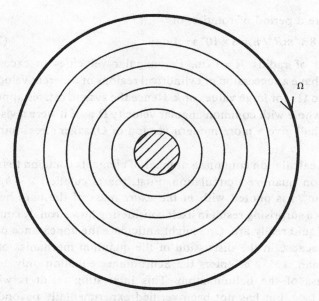

Figure 2.4 Onsager's early model for the rotation of a bucket of superfluid (London, 1954). The central region, shown cross-hatched, does not rotate. The regions of superflow in each successive ring of increasing radius have quanta of circulation $k = 1, 2, 3 \ldots$.

where k is an integer and θ is the azimuthal angle. (We omit the subscript on the mass m of the ^{4}He atom.) The velocity has a θ-component given by

$$v_k = \frac{1}{r} \frac{\partial \varphi}{\partial \theta} = \frac{k\hbar}{mr} \qquad (k = 0, \pm 1, \pm 2, \ldots) \qquad (2.13)$$

Now consider a uniformly rotating bucket of radius R, filled with liquid helium at $T = 0$ K. Imagine that the liquid can be divided up into a series of concentric cylindrical regions of radii $r_1, r_2, \ldots, r_n$. In the region (r_k, r_{k-1}) we assume irrotational circulation v_k but allow discontinuities in the velocity at the interfaces. In the simply-connected central region $(r < r_1)$ the velocity v_0 must vanish. The minimum energy for such a concentric structure is then calculated by London (1954) using a Lagrange multiplier method for N atoms in a Bose–Einstein degenerate state (see Section 2.8). It is concluded that for an angular velocity smaller than critical

$$\Omega_c = \hbar/2mR^2 \qquad (2.14)$$

the entire container will be filled with the stationary central region. This would require a period of rotation of

$$P > 8\pi^2 mR^2/\hbar = 8 \times 10^4 \, \text{s} \sim 1 \, \text{day} \tag{2.15}$$

for a bucket of radius $R = 1$ cm. For angular velocities in excess of $\hbar/2mR^2$ we have a succession of cylindrical regions of different velocities converging to Ωr for large values of k. Hence the system rotates approximately as a whole with common angular velocity Ω as if it were a viscous liquid. We shall give a more modern version of Onsager's reasoning in Section 2.8.

Onsager's calculation contains a number of elements common to many discussions on quantized circulation. First, the circulation $\kappa = h/m \approx 9.97 \times 10^{-4}$ cm^2/s is quoted with m the *exact* mass of the bare helium atom. That is a surprising result in itself considering how strongly coupled atoms in a liquid really are. One might anticipate the appearance of the bare mass because in the discussion of the quantum mechanics of the fluid (Feynman, 1955) m enters the Schrödinger equation only as the physical mass of the helium atom. This interesting result is widely accepted as true, but has not been verified experimentally beyond the level of order 1% obtained in determining the quantum of circulation κ.

Second, for moderate values of Ω the array of Figure 2.4 is nearly indistinguishable from a classical solid body rotation, which has a continuous distribution of vorticity. This tendency for the superfluid to adopt an approximately classical flow distribution proved to be a barrier in establishing the correctness of these ideas experimentally.

2.3.2   Feynman vortices

As it happened Feynman (1955) was working on the same problem and came to a somewhat different conclusion. He considered that the vortices in the superfluid might take the form of a vortex filament with a core of atomic dimensions, truly a vortex line. On this picture, the multiple connectivity of a vortex arises because the superfluid is somehow excluded from the core and circulates about the core in quantized fashion. Feynman considered what core radius would result if the core were hollow and determined by the surface tension of the fluid. This gave him a core radius of $a \approx 0.5$ Å. (Equation (1.70) with $\sigma = 0.354$ dyn/cm). He estimated the energy per unit length (tension) of such vortices as (Equation (1.38) with $\Gamma = \kappa$)

$$\varepsilon = \int \tfrac{1}{2}\rho_s v_s^2 \, \mathrm{d}r^2 = (\rho_s \kappa^2/4\pi) \ln (b/a) \tag{2.16}$$

where ρ_s is the superfluid density, b is the radius of the bucket, or the mean distance between vortices, and v_s is the superfluid velocity. This is an enormous energy: assuming $b/a = 10^7$ it amounts to 1.85×10^{-7} erg/cm or 13.4 K/Å. The centrifugal force on each ring of fluid surrounding the core is balanced by a pressure gradient given by (1.39) with $\Gamma = \kappa$. A sketch of the velocity and pressure distributions around a vortex line is shown in Figure 2.5.

Feynman (1955) also conjectured how the vortex lines might be arranged. First, note that since the circulation κ enters squared in the energy (2.16), a doubly quantized vortex line would have four times the energy of a singly quantized line, and would likely be unstable to breakup into four separate lines. No accepted experimental evidence for multiple

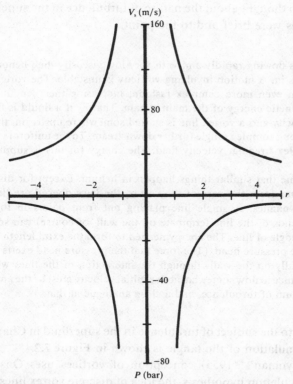

Figure 2.5 Superfluid velocity and pressure distributions about a rectinlinear vortex line located at $r = 0$, having a core radius of 1 Å and circulation $\kappa = h/m$. In an actual quantized vortex, the velocity is unlikely to exceed the Landau critical velocity of 60 m/s (after Glaberson and Donnelly (1966)).

or fractional quantization of vortex lines has ever appeared. Then Feynman noted that in uniform rotation the curl of the velocity is the circulation per unit area, and the curl is 2Ω, hence there should be

$$n_0 = \frac{\text{curl } v_s}{\kappa} = \frac{2\Omega}{\kappa} \sim 2 \times 10^3 \Omega \text{ lines/cm}^2 \qquad (2.17)$$

arranged parallel to the axis of rotation and with approximately uniform density. Equation (2.17) applied to more general flows will be called 'Feynman's rule'. It states that the number density of quantized vortex lines is the ratio of the vorticity in the superfluid to the quantum of circulation. Experimental evidence for Feynman's rule will be discussed in Section 5.3.1.

Feynman also thought about the nature of turbulence in the superfluid. His speculations were brief and to the point:

In ordinary fluids flowing rapidly and with very low viscosity the phenomenon of turbulence sets in. A motion involving vorticity is unstable. The vortex lines twist about in an even more complex fashion, increasing their length at the expense of the kinetic energy of the main stream. That is, if a liquid is flowing at a uniform velocity and a vortex line is started somewhere upstream, this line is twisted into a long complex tangle further downstream. To the uniform velocity is added a complex irregular velocity field. The energy for this is supplied by pressure head.

We may imagine that similar things happen in helium. Except for distances of a few angstroms from the core of the vortex, the laws obeyed are those of classical hydrodynamics. A single line playing out from points in the wall upstream (both ends of the line terminate on the wall, of course) can soon fill the tube with a tangle of line. The energy needed to form the extra length of line is supplied by the pressure head. (The force that the pressure head exerts on the lines acts eventually on the walls through the interaction of the lines with the walls). The resistance to flow somewhat above initial velocity must be the analogue in superfluid helium of turbulence, and a close analogue at that.

We shall return to the subject of turbulence in the superfluid in Chapter 7. A numerical simulation of the tangle is shown in Figure 7.2.

Although Feynman's (1955) conception of vortices uses Onsager's (1949) basic circulation hypothesis, the idea of discrete vortex lines leads to a vastly different phenomenology which forms the principal subject of this book. The logarithmic factor which we see in (2.16) will appear frequently in our discussions and arises because we are dealing with vortex lines and not simply quantized circulation.

2.4 Early experimental evidence for quantized vortices and circulation

2.4.1 *Minimization of free energy*

Hall and Vinen (1956b), who were then research students at Cambridge, suggested that in uniform rotation the arrangement of vortex lines can be found by minimizing the free energy in the rotating system.

In mechanics the internal energy $E(p, q)$ of a body in uniform rotation around a fixed axis with angular velocity Ω and angular momentum $\mathbf{M}$ can be calculated in a coordinate system rotating with the body by the transformation

$$E'(p, q) = E(p, q) - \Omega \cdot \mathbf{M}(p, q) \tag{2.18}$$

As Landau and Lifshitz (1969) observed, the same relations apply to the operators of the corresponding quantities in quantum mechanics. In statistical mechanics the quantities E and M are averaged over the statistical distribution. The condition of thermodynamic equilibrium of a system placed in a bucket rotating at fixed angular velocity Ω is obtained by minimizing the free energy $F - \Omega \cdot \mathbf{M}$ where $F = E - TS$, but since vortex calculations of the type we are considering here are valid at absolute zero, it is equivalent to minimize E.

The simplest case is to find the critical angular velocity Ω_{c_1} at which the first vortex line will appear in a rotating container. The energy per unit length of a vortex line is given by (2.16) and the angular momentum per unit length is

$$M = \int_a^b \rho_s v_s r \, dr^2 \approx \tfrac{1}{2}\rho\kappa b^2 \tag{2.19}$$

assuming $b \gg a$. Thus the free energy of the superfluid at $T = 0$ K will be lower than the free energy of the superfluid at rest with a single vortex line in the center if Ω exceeds Ω_{c_1} given by

$$F = (\rho_s \kappa^2/4\pi) \ln (b/a) - \tfrac{1}{2}\rho_s \kappa b^2 \Omega_{c_1} = 0 \tag{2.20}$$

or, setting, $b = R$, the radius of the bucket

$$\Omega_{c_1} = (\kappa/2\pi R^2) \ln (b/a) \tag{2.21}$$

which differs from Onsager's (1949) result (2.14) by the logarithmic factor. Note that this argument gives us no information about how the vortex appears on increasing Ω, nor how it disappears on reducing Ω below Ω_{c_1}.

The authors then showed that an array of vortex lines has a lower free energy than vortex sheets of the form suggested in Figure 2.4. The basis of their calculation (Hall, 1960) is shown in Figure 2.6 where the rotating bucket contains a uniform array of N vortices with a 'vortex-free strip' on the outside of the array having quantized circulation. The radius of the array of vortex lines is given by

$$r^2 = N\kappa/2\pi\Omega' \tag{2.22}$$

where Ω' need not necessarily be the angular velocity Ω of the bucket. The energy

$$E = \tfrac{1}{4}\rho_s \pi \left(\frac{N\kappa}{2\pi\Omega'}\right)^2 \Omega'^2 + \rho_s \frac{N\kappa^2}{4\pi} \ln\left(\frac{b}{a}\right)$$

$$+ \rho_s \frac{N^2\kappa^2}{4\pi} \ln\left(\frac{R}{r}\right) \tag{2.23}$$

and the angular momentum is

$$M = \tfrac{1}{2}\rho_s \pi \left(\frac{N\kappa}{2\pi\Omega'}\right)^2 \Omega' + \tfrac{1}{4}\rho_s \kappa \left(\frac{N\kappa}{2\pi\Omega'}\right) + \tfrac{1}{2}\rho_s N\kappa \left(R^2 - \frac{N\kappa}{2\pi\Omega'}\right) \tag{2.24}$$

where b is now a distance of the order of the line spacing. The first term in Equations (2.23) and (2.24) is the contribution from solid body rotation within the radius r; the second term is the correction to this for the small radius about the vortex filaments; and the third term is the contribution

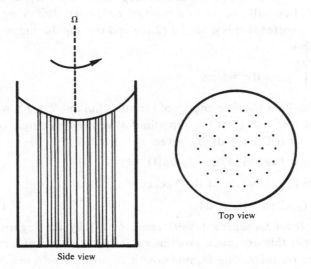

Figure 2.6 Arrangement of a continuous array of vortices and a vortex-free strip envisioned by Hall and Vinen (1956*b*) and Hall (1960).

from the irrotational flow outside the radius r. Differentiating with respect to Ω' and recalling that $b \sim \Omega'^{-1/2}$, we find that

$$\frac{\partial F'}{\partial \Omega'} = \rho_s \frac{N(N-1)}{8\pi\Omega'} \kappa^2 \left(1 - \frac{\Omega}{\Omega'}\right) \tag{2.25}$$

so that in equilibrium $\Omega' = \Omega$, and the vortex lines rotate with the container.

Setting $\Omega' = \Omega$ and minimizing F' with respect to N, the result is that in equilibrium

$$N = 2\pi R^2 \Omega / \kappa - n \tag{2.26}$$

where n, neglecting terms of order $N^{-1/2}$ is given by

$$n^2 = \frac{4\pi R^2 \Omega}{\kappa}\left[\ln\left(\frac{b}{a}\right) - \tfrac{1}{2}\right] \tag{2.27}$$

The first term in (2.26) is the number of lines required to fill the vessel, Equation (2.27) shows that $n \sim N^{-1/2}$, so that about one layer of vortex lines is missing at the edge of the vessel. Hall and Vinen (1956b) show by the same reasoning that concentric vortex sheets such as are shown in Figure 2.4 have a higher free energy, in part because a term representing a surface energy of the vortex sheets would have to be included (Hall and Vinen, 1956b).

It is possible to find an analogy between vortices in calculations of the type we have been discussing and the behavior of current-carrying wires of the same configuration in a circular conductor. This arises, for example, because the fluid velocity $\mathbf{v}$ and vorticity $\boldsymbol{\omega}$ of a system correspond to the magnetic field $\mathbf{H}$ and current density $(4\pi/c)\mathbf{j}$ of the same configuration of current carrying wires. The vorticity $\boldsymbol{\omega} = \text{curl } \mathbf{v}$ is analogous to Maxwell's equation curl $\mathbf{H} = (4\pi/c)\mathbf{j}$. This analogy, however, must be carried out carefully, as discussed by Fetter and Donnelly (1966).

2.4.2 Second sound and quantized vortices. Mutual friction

The investigation of quantized vortices by second sound began when Vinen (1957) conducted experiments on the flow of heat in a 2.4×6.45 mm channel, closed at one end with a heater. According to the two-fluid model, superfluid will enter the open end of the channel and flow toward the heater. On reaching the heater, normal fluid will be created which will flow back towards the entrance. Thus a heat flow in such a geometry consists of a vigorous counterflow of the two fluids, which is laminar for only very low relative velocities, usually less than 1 cm/s. Vinen discovered that second sound is transmitted across the heat flow with unchanged velocity u_2, but that there is an attenuation of amplitude which Vinen described by modifying the Gorter–Mellink mutual friction force (see Equations (2.2) and (2.3)) which had been proposed to account for the temperature gradient in heat flow in narrow channels:

$$\mathbf{F}_{ns} = A\rho_n\rho_s(v_s - v_n)^3 \qquad (2.28)$$

Denoting $\mathbf{v}_{ns} = \mathbf{v}_n - \mathbf{v}_s$ as the instantaneous counterflow velocity between the two fluids and $v_{ns} = |\langle v_{ns}\rangle|$ as the spatial and temporal average of the velocities across the channels, Vinen proposed that (2.28) be modified to read

$$\mathbf{F}_{ns} = A\rho_n\rho_s v_{ns}^2\mathbf{v}_{ns} \qquad (2.29)$$

where A is a constant which is a function of temperature, and of order 50 cm/s g. The magnitude of v_n is related to the heat flux q down the channel by

$$q = \rho S T v_n \qquad (2.30)$$

and since there is no net mass flow in a counterflow (2.6) gives

$$\rho_n v_n + \rho_s v_s = 0 \qquad (2.31)$$

and hence

$$v_{ns} = q/\rho_s S T \qquad (2.32)$$

Under these circumstances, the solution of (2.2) and (2.3) shows that second sound experiences an attenuation (beyond the natural damping present in undisturbed helium II)

$$\alpha = \frac{A\rho}{2\rho_s^2 S^2 T^2 u_2}q^2 \qquad (2.33)$$

and the velocity is amended to

$$u_2' \approx u_2(1 - \alpha^2 u_2^2/8\omega^2) \qquad (2.34)$$

where ω is the angular frequency of the second sound. Vinen attributed the attenuation α to the presence of quantized vortices forming a dense tangle in the counterflow, and we shall discuss this form of turbulence in Chapter 7.

Hall and Vinen (1956a) realized (before Feynman's publications) that uniform rotation would produce mutual friction, and they then proceeded to study the propagation of second sound in uniformly rotating helium in the apparatus shown in Figure 2.7. In the axial mode resonator (Figure 2.7(a)) the second sound (being a longitudinal wave) is propagated parallel to the axis of rotation and in the radial mode resonator (Figure 2.7(b)) it is propagated perpendicular to the axis of rotation.

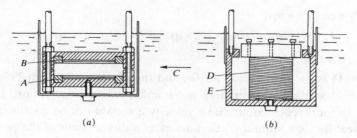

(a)   (b)

Figure 2.7 Second sound resonators used by Hall and Vinen (1956a) in their first investigations of uniformly rotating helium II. Resistance wire was used for the transmitters and receivers. (a) The axial resonator with transmitter and receiver wound on A and B. (b) For the radial mode resonator the transmitter and receiver were wound on the cylinder D. The axial mode resonator can also be used for propagation perpendicular to the axis of rotation by rotating it about the axis C in the figure.

Experiments using rates of rotation from $\Omega = 3$ to 7.5 rad/s do not change the velocity of second sound by more than 0.1%; therefore if the resonator is set on resonance it remains on resonance during rotation and changes in amplitude reflect changes in attenuation. The radial mode resonator showed an equilibrium loss in amplitude found to be accurately proportional to Ω, independent of second sound amplitude and frequency. The authors described the excess attenuation by a linear attenuation coefficient

$$\alpha = B\Omega/2u_2 \tag{2.35}$$

where B is a temperature-dependent dimensionless constant. The axial resonator, however, showed little or no effect on the second sound amplitude. When the axial mode resonator was turned on its side and rotated about the axis (in Figure 2.7(a)) attenuation comparable to the radial mode resonator was obtained. It appeared from the study that the attenuation is strong in any direction perpendicular to the axis of rotation and much smaller for propagation axially. The observed anisotropy was therefore considered to be evidence for a uniform array of vortex lines as suggested in Figure 2.6 and inconsistent with the annular vortex structure pictured in Figure 2.4.

The mutual friction terms in (2.2) and (2.3) can be written in different ways according to the problem at hand. For the case of superfluid turbulence, which we shall discuss in Chapter 7, $\mathbf{F}_{ns}$ has the form (2.29). For the case of helium II rotating uniformly at angular velocity Ω, Hall and Vinen proposed

$$\mathbf{F}_{ns} = -(B\rho_n\rho_s/\rho)\hat{\mathbf{\Omega}} \times (\mathbf{\Omega} \times \mathbf{q}) - B'(\rho_n\rho_s/\rho)(\mathbf{\Omega} \times \mathbf{q}) \qquad (2.36)$$

$$\mathbf{q} = (\mathbf{v}_s - \mathbf{v}_n) \qquad (2.37)$$

where $\hat{\mathbf{\Omega}}$ is the unit vector $\mathbf{\Omega}/|\mathbf{\Omega}|$, and the dimensionless coefficients B and B' describe the dissipative and nondissipative contributions to $\mathbf{F}_{ns}$. (The instantaneous counterflow velocity $\mathbf{q}$ should not be confused with the heat flux q.) Mutual friction arises as a direct result of the presence of quantized vortices, and we shall discuss some of the underlying physics in more depth in Chapter 3. The effect of the mutual friction terms is to make the wave equation for second sound

$$\ddot{q} + (2 - B')(\mathbf{\Omega} \times \dot{\mathbf{q}}) - B\hat{\mathbf{\Omega}} \times (\mathbf{\Omega} \times \dot{\mathbf{q}}) = u_2^2 \nabla^2(q) \qquad (2.38)$$

This equation forms the basis for the measurement of the parameters B and B'. The term containing the parameter B gives rise to a contribution α to the attenuation of second sound given by (2.35) and the term containing $(2 - B')$ gives rise to a measureable coupling between otherwise degenerate modes in a suitably designed resonator.

The attenuation of second sound is a powerful tool for studying quantized vortices. In quantum turbulence experiments, for example, a sensitivity of 20 cm of vortex line per cubic centimeter of helium has been achieved. Since, as we shall see, the core radius is about 1 Å, this resolution corresponds to detecting quantized vortex core material at a volume concentration ratio of 6×10^{-15}.

2.4.3   Experiments on quantized circulation

The first experiment designed to look for quantized circulation was carried out by Vinen (1961). Vinen used free energy arguments to study the conditions for the entry of quantized circulation and free vortices into a rotating annulus of radii R_1 and R_2. Specializing to the limit $R_1 \ll R_2$, he found that as the angular velocity of the container is increased from zero, the circulation Γ increases in a series of equal quantum steps $0, \kappa, 2\kappa, \ldots$ with no free vortices in the liquid until some critical value of Ω is achieved. At this critical value of Ω vortices aligned parallel to the axis appear in the annulus and for further increases in Ω, both Γ and the number of vortices increase. Vinen's result is of fundamental importance in the study of vortices and circulation in helium II. It shows that when the flow takes place in an annulus, a multiply-connected region, both quantized circulation and quantized vortex lines are possible as suggested in Figure 2.8. We shall renew this discussion in Section 2.5.

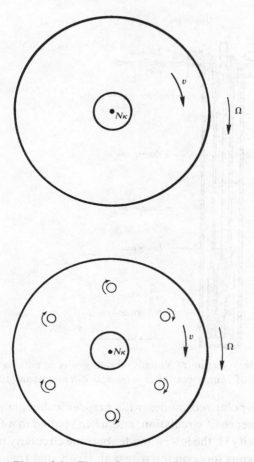

Figure 2.8 Flow of the superfluid in a rotating annulus. (*a*) At very low speeds of rotation, the superfluid remains at rest. After a certain critical velocity quantized circulation $N\kappa$ appears with N increasing, in steps, $N = 1, 2, \ldots$. The situation is equivalent to a fictitious vortex of strength $N\kappa$ located at the center of the container. (*b*) At higher rates of rotation both quantized circulation and free vortices are possible. The equilibrium state of the normal fluid is, in all cases, solid body rotation.

Vinen (1961) constructed an apparatus to detect the quantized circulation by making the inner cylinder a wire of radius $R_1 = 12.5\ \mu$m and the outer cylinder a tube of radius $R_2 = 0.2$ cm. A modern version of the apparatus, shown in Figure 2.9, is placed in a magnetic field so that an oscillating electric current will excite transverse oscillations of the wire. For an ideal geometry, the normal vibrations of the wire can be thought

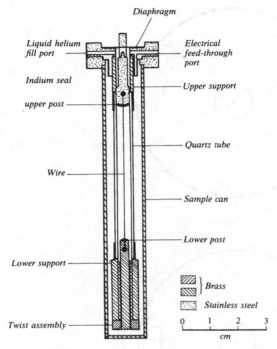

Figure 2.9 Modern version of Vinen's vibrating wire apparatus used at the University of Minnesota (after Kral and Zimmermann (1975)).

of as degenerate plane-polarized modes with perpendicular planes of polarization. In the presence of circulation, and when studied in a frame rotating at angular velocity Ω, the lowest modes become circularly polarized because of the Magnus force on the wire with an angular frequency difference given by $\omega = \rho_s \Gamma / \mu_{eff}$, where μ_{eff} is the sum of the physical mass of the wire per unit length and the hydrodynamic effective mass per unit length, $\pi a^2 \rho_s$ (Section 1.1). Owing to imperfections in the wire or its mounting the normal modes in the absence of circulation are nondegenerate plane-polarized. The addition of circulation then produces elliptically polarized modes.

In the earliest apparatus, Vinen began rotation in helium I, cooled slowly through the lambda point to about 1.3 K. Observations at first failed to show quantization, even though the observed values of Γ were near κ. Vinen interpreted this problem as partial attachment of a free vortex to the wire and subjected the wire to large amplitude vibrations before measuring Γ. The tendency then to record quantized circulation is shown in Figure 2.10(a).

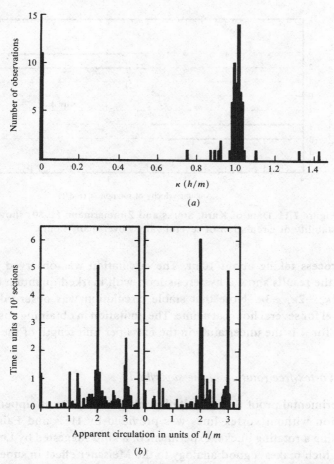

Figure 2.10 Statistics on stability of circulation values obtained (*a*) by Vinen (1961) and (*b*) by Whitmore and Zimmermann (1968).

This experiment has been repeated by Zimmermann and his students at the University of Minnesota. Whitmore and Zimmermann (1968) used thicker wires than in the original experiment and observed stable circulations of $\Gamma = \kappa$, 2κ, 3κ with the apparatus at rest during the observations (see Figure 2.10(*b*). Karn, Starks and Zimmermann (1980) used wires with diameters from 15 to 35 μm, temperatures $0.3 < T < 0.7$ K, and rotation rates $0 < \Omega < 4$ rad/s and found evidence for $\Gamma = \kappa$, 2κ, 3κ, and 4κ. Figure 2.11 reproduces some of their results. The experimental protocol was to accelerate the apparatus slowly from 0 to 4 rad/s, decelerate to 0 and accelerate to -4 rad/s followed by deceleration to rest, the

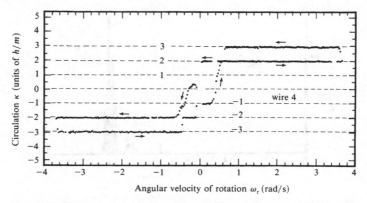

Figure 2.11 Data of Karn, Starks and Zimmermann (1980) showing stability of circulation of several values over many hours.

whole process taking about 10 hr. The circulation was observed every 10 s and the results show a hysteresis loop with marked quantum levels $\Gamma = 3\kappa$, 2κ, -2κ, -3κ. Note that stable circulation was observed in a single level for several hours at a time. The limitation in obtaining absolute accuracy for κ is the uncertainty in the mass per unit length of the wire.

2.4.4   Vortex-free rotation of the superfluid

The experimental proof that a limitation like (2.21) gives an upper limit for rotation without vortex lines was provided by Hess and Fairbank (1967) using a rotating bucket experiment of a kind suggested by London (1954), which makes a good analogy to the Meissner effect in superconductors, i.e., the expulsion of magnetic flux when a type I superconductor is cooled through its transition temperature in a magnetic field. They suspended a 0.089 cm diameter tube containing 7 mg of helium and attached a ferrite slug in a Beams type magnetic bearing. This rotor is thereby suspended in free space inside a 1 K vacuum chamber just below the magnetic pole piece, and may rotate with extremely small dissipative torque. By measuring the rate of rotation of the bucket and raising the temperature through T_λ by means of heat supplied by a beam of light, they were able to determine the fraction of helium which was not rotating below T_λ. Hess (1967) carried out theoretical calculations of the speeds at which various equilibrium configurations of vortices would appear in a rotating bucket. The Hess-Fairbank experiment did not, however, demonstrate the appearance of separate vortices in the bucket. This step was accomplished at Michigan by Packard and Sanders (1972) using negative ions trapped in the vortex cores which we shall discuss presently.

2.5 Ions and vortices: evidence for quantized vortex rings and lines

A further powerful tool for investigations of quantized vortex lines involves the interaction of ions and vortices. Careri, McCormick and Scaramuzzi (1962) investigated the motion of negative and positive ions moving perpendicular to the axis of rotation of a container of helium II. They discovered that while positive ions were unaffected, negative ions were strongly absorbed, at least at low temperatures. The theory of this interaction will be discussed in Section 4.5. Ions are attracted to the cores of vortices by the gradient of pressure about the vortex shown in Figure 2.5; the potential well in which they are trapped depends on the size of the ions, being larger for negative ions (radius ~ 16 Å) than positive ions (radius ~ 6.9 Å). Ions escape from the vortex potential well by thermal activation: above 0.8 K positive ions do not remain trapped and above 1.7 K negative ions do not remain trapped. These limits make investigation of vortices by ions near T_λ impractical.

2.5.1 Experiments on quantized vortex rings

Quantized vortex rings were first studied by Rayfield and Reif (1964) using an ion time-of-flight apparatus shown in Figure 2.12. The ions somehow (see Section 8.5) produce vortex rings and then get trapped in the cores of the rings (see Section 4.5). Their experiment, performed at

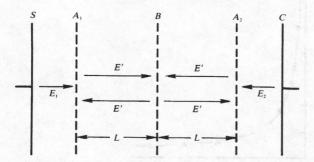

Figure 2.12 Ion time-of-flight spectrometer used by Rayfield and Reif (1964) to determine the dynamics of ion-quantized vortex ring complexes. The ionization was produced by an alpha radioactive source on plate S. The field E' is reversed periodically by applying square wave potential to grid B.

temperatures near 0.3 K, is based on the observation that very little dissipation in the energy of the charge carriers (vortex rings plus charge) occurs in the spectrometer. A potential V is applied between the radioactive ion source S and grid A_1. The charge carrier then arrives at A_1 with an energy $E = eV$ and some velocity v. It is prevented from reaching the collecting electrode C by a retarding potential $(\sim - V)$ applied between A_2 and C. A small square wave potential of frequency ν is applied to grid B, and small electric fields E' alternately directed toward and away from B are produced in the region A_1A_2 (free of dc fields). If ν is such that the time-of-flight L/v of the charge carrier through the distance L from A_1 to B is just equal to the time $(2\nu)^{-1}$ between field reversals, then the carrier remains in synchronism with this field and thus gains from it a small net amount of energy sufficient to overcome the retarding potential between A_2 and C to reach the collector C. At the frequency $\nu = \frac{1}{2}v/L$, therefore, the collected current will exhibit a resonance maximum. It was found that at low enough temperatures and at high enough E the carriers indeed exhibited negligible dissipation in A_1A_2. Plotting E versus v obtained by the above method, we see that v does decrease with E and that the plot agrees extremely well with the theoretical values obtained by eliminating R between (1.68) and (1.69) and by letting $a \sim 1$ Å. A comparison of the theoretical and experimental values are shown in Figure 2.13. The experiment shows that the charge carriers behave exactly like vortex rings with strength $\kappa = h/m$.

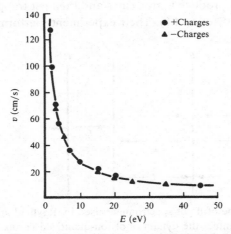

Figure 2.13   Rayfield and Reif's (1964) data on the velocity of ion-ring complexes compared to the classical results (1.68) and (1.69) with $\Gamma = \kappa$.

2.5.2 Experiments on quantized vortex lines

Perhaps the first demonstration of the 'grainy' structure of rotating helium II was made by Tanner, Springett and Donnelly (1965) in a simple rotating rectangular frame (see Donnelly, Glaberson and Parks (1967) p. 55) where a radioactive source and collector were placed on the vertical members of the frame, and an electric field arranged to draw negative ions across the fluid. A collector on the top of the apparatus and a

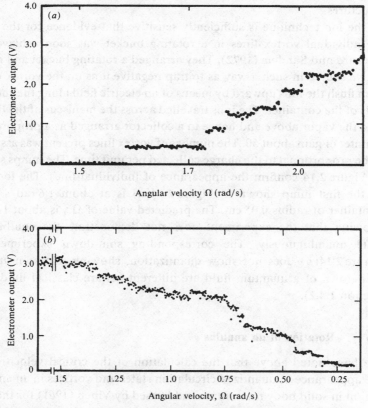

Figure 2.14 (*a*) Ion experiment of Packard and Sanders (1972) showing the formation of singly quantized vortices in rotating helium II which is slowly accelerated. Vortices are counted by passing negative ions transverse to the vortex array which trap ions in their cores. The electrons are subsequently swept out by an axial electric field and the charge collected assumed to reflect the number of vortices present. (*b*) The results on deceleration are qualitatively different and do not show clear steps.

'repeller' electrode at the bottom were used to provide a vertical component of electric field to push trapped negative ions along quantized vortices produced by rotating the entire apparatus. It was found that when the apparatus was set in rotation the horizontal beam of negative ions was attenuated with almost all the lost current appearing at the top collector. This simple experiment shows that ions can charge vortex lines, that these charges can be pushed vertically along the vortex lines, and that the lines therefore must be coherent over the length of the apparatus. This is one reason physicists refer to vortex lines as macroscopic quantum structures.

The ion technique is sufficiently sensitive that evidence for the entry of individual vortex lines in a rotating bucket was soon achieved by Packard and Sanders (1972). They arranged a rotating bucket and radioactive source in such a way as to trap negative ions on the vortex cores, then flush the ions upward by means of an electric field along the rotation axis of the container. The ions travelled across the meniscus of the liquid, into the vapor above and hence to a collector arranged as a proportional counter of gain about 50. The number of vortex lines present was assumed to be proportional to the charge collected per unit time. The jumps shown in Figure 2.14 confirm the appearance of individual lines. The location of the first jump shown in Figure 2.14(a) is at about 1.6 rad/s for a container of radius 0.05 cm. The predicted value of Ω_{c_1} is about 1 rad/s showing that the experiment does not demonstrate the equilibrium state unambiguously. The corresponding spin-down experiment in Figure 2.14(b) does not show quantization, showing that spin-up and spin-down of a quantum fluid are different (as are classical fluids, see Section 1.4.2).

2.6   Rotation of an annulus

We have noted above that the calculation of the critical velocities for the appearance of quantized circulation states and vortices in an annular region in solid body rotation was addressed by Vinen (1961) for the very small radius ratio appropriate to the vibrating wire experiment. Similar calculations for a narrow gap were carried out by Donnelly and Fetter (1966), and later more completely by Stauffer and Fetter (1968). The technique was again free energy minimization. They showed that at low angular velocities the equilibrium flow is a purely irrotational circulation with tangential velocity as close as possible to that of the cylindrical walls. The quantized circulation states form a sequence of equally spaced levels up to a critical angular velocity

$$\Omega_0 = (\kappa/\pi d^2)\ln(2d/\pi a) \tag{2.39}$$

at which point singly quantized vortices appear in the bulk of the fluid. Here $d = (R_2 - R_1) \ll (R_1 + R_2)/2$. For Ω just beyond Ω_0 the vortices are equally spaced on a circle midway between the walls, and their number increases rapidly with Ω. Stauffer and Fetter (1968) were also able to estimate the angular velocity for the appearance of the second row of vortices.

A further remark of Donnelly and Fetter (1966) may be useful here. They noted that quantized vortices appear in the annulus at an angular velocity Ω_0 when vortices can first compensate for the difference in irrotational velocity between the inner and outer walls. This result may

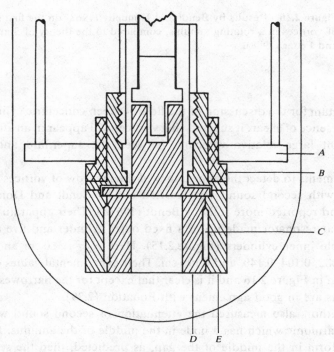

Figure 2.15 Apparatus of Bendt and Donnelly (1967) for determining the first entry of vortices into a rotating annulus. The frame A is stationary, B is a plastic bearing, C is the resonant cavity formed from one of five interchangeable inner cylinders D, and a fixed outer cylinder E. Second sound was generated and received by Aquadag coatings on the walls of C.

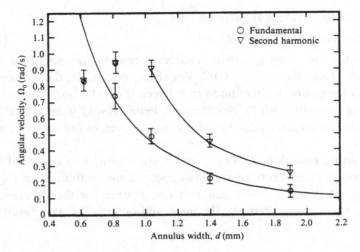

Figure 2.16 Results by Bendt and Donnelly (1967) on the first entry of vortices in a rotating annulus, compared to the theory of Stauffer and Fetter (1968).

be important for the discussion of the flow between concentric cylinders in the presence of shear: it suggests that vortices will appear in an attempt to prevent large relative velocities between the superfluid and the walls.

Experiments to detect the appearance of the first row of vortices in an annulus with second sound were carried out by Bendt and Donnelly (1967) and reported more fully by Bendt (1967*b*). Their apparatus was an annular resonator made up of a fixed outer cylinder and five inter-changeable inner cylinders (Figure 2.15). Here $R_2 = 1.560$ cm and $d = 0.062, 0.082, 0.104, 0.140$, and 0.190 cm. Their experimental values of Ω_0 are shown in Figure 2.16 and it is clear that except for the narrowest gap the results are in good agreement with Equation (2.39).

The authors also measured the attenuation of second sound with a second harmonic which has a node in the middle of the annulus. If the vortices form in the middle of the gap, as predicted, then the second harmonic should be relatively insensitive to their presence. This indeed was the case; attenuation for the second harmonic first appeared at about $1.9\Omega_0$, which the authors attributed to the appearance of a second row of vortices. The appearance of a double row was predicted to be at about $1.85\Omega_0$ by Stauffer and Fetter (1968).

2.7 Vortex waves

We have noted in Chapter 1 that the kinetic energy per unit length of the potential flow about a vortex core of radius a is (1.38)

$$\varepsilon = (\rho\Gamma^2/4\pi)\ln(b/a) \tag{1.38}$$

Assuming quantized circulation for the superfluid and guessing the ρ should be replaced by ρ_s, the corresponding formula for a vortex in helium II should be (see Section 2.3.2)

$$\varepsilon = (\rho_s\kappa^2/4\pi)\ln(b/a) \tag{2.40}$$

Since an object having energy per unit length has tension, this result alone suggests that vortices in helium II should be able to sustain wave motion. Since there was already considerable skepticism about the existence of quantized vortices in the 1950s, it was clear that observation of vortex waves would be crucial in accepting the idea that rotating helium II could have a 'grainy structure'.

The first evidence that the vortex system in helium II could sustain oscillatory modes was obtained by Hall (1958). Hall clamped a pile of disks in a thin-walled aluminum can; the whole disk assembly could be filled with liquid helium and suspended in the vapor. Thus the experimental helium II was isolated from the nonrotating helium, and edge effects were avoided. The torsion system, from which the disk system was suspended, could be rotated uniformly at angular velocity ω_0 by a synchronous motor, and oscillations of the disk at frequency Ω relative to the head were observed photoelectrically.

Experiments were done at 1.27 K, low enough for the excitations of the normal fluid to be negligible. Observations were made on the period of oscillation of the disks, and expressed as that (fictitious) fluid density ρ', which when filling the disks and oscillating with them, would produce the observed change in period of oscillation.

The results for fast rotation and slow oscillation ($\omega_0 \gg \Omega$) showed that for given values of ω_0 and Ω, ρ'/ρ_s increases monotonically as the spacing $2l$ between disks *increases*. One might have expected that dragging of liquid by the disks would be *less* effective as the spacing increases.

In the case of slow rotation and fast oscillation ($\omega_0 \ll \Omega$) smooth mica disks produced no results, but with rough disks, values of ρ' which were greater and less than zero were observed, as if anomalous dispersion were being observed. The publication of these results was significant evidence for the existence of quantized vortices, since the data were

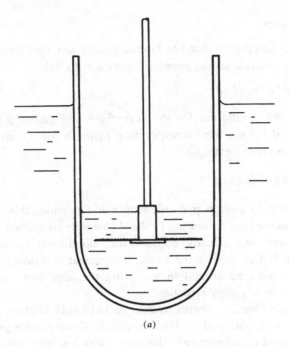

(a)

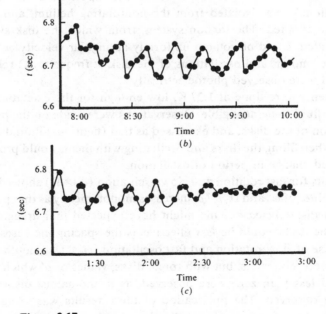

Figure 2.17

sufficiently complex to rule out the idea that simple inertial waves (see Section 1.4.1) were being excited in the system.

A particularly elegant experiment was designed by Hall (1960) in which a single disk, roughened on the upper surface only, is suspended in a beaker of helium II rotating at the same angular velocity as the point of suspension of the disk. Vortex waves excited by oscillation of the upper surface of the disk will be reflected at the free surface of the liquid and resonances would occur for suitable values of the distance from disk to liquid surface (Figure 2.17(a)).

The apparatus shown was immersed in a bath of helium II in such a way that the beaker slowly filled with liquid by superfluid film flow. The period of oscillation of the disk was measured as a function of time as the beaker filled; this scheme gives a continuously variable thickness of fluid layer in one single experimental run. The periodic dependence of period of oscillation on liquid level is easily seen at both 1.3 K and 1.6 K as shown in Figures 2.17(b) and (c). The fact that successive resonances at 1.3 K diminish very little in strength show that the attenuation of vortex waves in the liquid is small. Analysis of the results suggested that the rather broad resonances may be due to imperfect reflection of waves at the disk surface. Work on the stability of Kelvin waves described in Section 6.1.3 may lead to a different interpretation of the resonance shapes.

Andronikashvili and Tsakadze (1960) carried out similar experiments with a more massive disk, which enabled them to study the damping; they found a series of peaks in damping as a function of liquid level, corresponding to resonant absorption of energy by vortex waves. Andronikashvili's group also carried out careful comparisons with classical fluids (water, helium I) to verify that the results are not classical in nature. A review of the Russian work has been prepared by Andronikashvili and Mamaladze (1967).

Caption for Figure 2.17.

Figure 2.17 (a) The single disk apparatus. The beaker surrounding the disk is rigidly connected to the torsion head from which the disk is suspended, and rotates with it. Changes in period are measured as the beaker fills with helium by film flow. (b) Dependence of the single disk period on time, due to changing liquid level at 1.3 K; $\omega_0 = 0.140$ rad/s. (c) *At 1.6 K.* The period axis is placed at the time of day when the disk surface was just covered by liquid (after Hall (1958, 1960)).

In all the work described above, the disk oscillation periods were long so that the corresponding vortex wavelengths were large compared to the distance between vortex lines. We shall observe in Chapter 6 that this implies that interactions among vortices could be important in determining the dynamical behavior in some instances.

2.8   Quantum mechanics and superfluidity. Phase slip

2.8.1   The Bose gas

There is at present no truly fundamental microscopic picture of the superfluidity of helium II. Nevertheless there are ideas about superfluidity based on quantum mechanics which are important for the understanding of the structure and dynamics of quantized vortices.

^{4}He atoms are bosons and hence it is tempting to draw an analogy between the behavior of helium II and an ideal Bose gas. Below a critical temperature T_c, an ideal Bose gas begins to condense particles into the lowest available energy level. T_c is equal to 3.1 K for a gas of the density of liquid ^{4}He. When $T = 0$ K all the particles are in the ground state. At a finite temperature $T < T_B$ the fraction 'condensed' into the ground state is given by (Landau and Lifshitz, 1969)

$$N_0(T)/N = 1 - (T/T_c)^{3/2} \tag{2.41}$$

Since helium II is a liquid the attractive forces between atoms must be important and for close enough approach the atomic interactions must be repulsive.

Below T_c, one phase, the condensed phase, consists of the N_0 particles in the zero-momentum state, while the other phase consists of the $N - N_0 = N(T/T_c)^{2/3}$ uncondensed particles distributed over the excited states. It is tempting to associate the superfluid with the particles in the ground state and the normal fluid with the excited particles. Since the entropy is proportional to the logarithm of the number of available states, and since there is only one state available for the particles in the superfluid state as postulated above, the entropy of the superfluid is zero in agreement with experiments on liquid helium (see, e.g., Donelly, Glaberson and Parks (1967) Section 5). For the ideal Bose gas, however, the energy momentum relation of the thermally excited particles is simply $\varepsilon = p^2/2m$ and Landau criterion in Equation (2.10) $d\varepsilon/dp \geq \varepsilon/p$ will not hold. Thus the ideal Bose gas does not possess one critical ingredient of superfluidity and interparticle interactions must therefore be included.

Bogoliubov (1947) obtained an approximate energy spectrum for a weakly interacting Bose system that predicts the linear phonon dispersion seen at low momenta in Figure 2.2. An account of Bogoliubov's theory is contained in Section 25 of Lifshitz and Pitaevskii (1980) and in Fetter and Walecka (1971) and in Section 21 of Donnelly, Glaberson and Parks (1967). The principal result is that the dispersion relation is

$$\varepsilon_p = \left[\left(\frac{p^2}{2m} \right)^2 + \frac{p^2}{m} NV_p \right]^{1/2} \tag{2.42}$$

For small momenta the interaction potential V_p becomes a δ-function for repulsive interaction of strength V_0. In this case

$$\varepsilon_p = \left(\frac{NV_0}{m} \right)^{1/2} p \tag{2.43}$$

and for momenta greater than about $(2mNV_0)^{1/2}$ we have

$$\varepsilon_p \approx p^2/2m \tag{2.44}$$

which is just the free particle dispersion relation. Thus the interactions allow superfluidity in the sense of (2.10) to occur on this model, and although rotons (the excitations near $Q = 2 \, \text{Å}^{-1}$ in Figure 2.2) are missing, the Bogoliubov result is of fundamental importance in the understanding of our subject. The fact that the lowest lying states correspond to longitudinally polarized phonons, suggests that states of rotation which have vorticity cannot be created, i.e., curl $v_s = 0$. This argument was first put forward by Feynman (1955 pp. 30, 37).

2.8.2   The Bose condensate

As we have noted, the ground state of an ideal Bose gas constitutes a 'condensate' to which all particles belong $N_{p=0} = N$, $N_{p \neq 0} = 0$. In the weakly interacting gas, the condensate has small depletion. For liquid helium $N_{p=0}/N$ is believed to be of order 13% at absolute zero (see, e.g., Sears, Svensson, Martel and Woods, 1982). The effect of interactions is to deplete the condensate, but not to eliminate it. It is widely accepted that this condensate is the essential feature of superfluidity. The situation envisaged is illustrated in Figure 2.18

The existence of the condensate implies the existence of a condensate wave function, and this wave function can be generalized to describe states when the condensate is in motion, provided gradients in velocity do not vary substantially over distances of the order of the coherence length ζ. Indeed, it can be argued that the velocity, which is the gradient

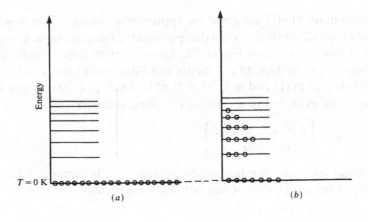

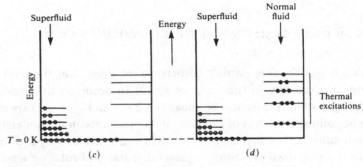

Figure 2.18 (*a*) Population of energy levels in an ideal Bose-Einstein gas at $T = 0$, all particles are in the lowest energy level (*b*) $0 < T < T_c$, some particles occupy excited states, but substantial numbers of particles remain in the condensate, the lowest energy level. (*c*) For helium II the effect of interactions is to promote some particles to states with $p \neq 0$, the condensate being the fraction of particles in the lowest energy state. The superfluid represents all the particles in the condensate and the depleted part. (*d*) At finite temperatures, thermal excitations form, the fraction of superfluid, the condensate becomes temperature-dependent and vanishes at the lambda transition (after Tilley and Tilley, (1986)).

of the phase of the condensate, will determine the flow of the *entire* superfluid, and that these ideas will remain valid for the superfluid all the way to the lambda transition. The condensate wave function is the coherent wave formed as a result of a large number of particles in the same state, much as electromagnetic waves form in a laser. The existence of the condensate wave function, which is of the form of one coherent single-particle wave function extending over the whole volume of the system, makes superflow coherent over an infinite range (neglecting boundaries).

The superfluid cannot change its velocity in a toroidal flow, for example, because the momenta of all the particles in the condensate cannot change simultaneously. It is this property that gives rise to the expression 'macroscopic quantization' often used in discussions of superfluidity.

2.8.3   The Madelung transformation

For the hydrodynamicist, the connection between velocity and phase can be seen quite elegantly by considering first a *single* particle of mass m in a field of fixed potential $w(\mathbf{r})$ (Putterman, 1974; Roberts, 1974). Madelung (1927) appears to have been the first to appreciate that the Schrödinger equation can be expressed in a form more familiar in fluid mechanics, by expressing the wave function $\psi(\mathbf{r}\ t)$ in terms of its amplitude $f(\mathbf{r}\ t)$ and phase $\varphi(r, t)$ where f and φ are real. Thus the familiar wave equation

$$i\hbar\, \partial\psi/\partial t = -(\hbar^2/2m)\nabla^2\psi - w\psi \tag{2.45}$$

with the substitution

$$\psi = f \exp(i\varphi) \tag{2.46}$$

can be divided into real and imaginary parts

$$\hbar\frac{\partial\varphi}{\partial t} - \frac{\hbar^2}{2m}\frac{\nabla^2 f}{f} + \frac{\hbar^2}{2m}(\nabla\varphi)^2 + w = 0 \tag{2.47}$$

and

$$2\frac{\partial f}{\partial t} + 2\frac{\hbar}{m}\nabla f \cdot \nabla\varphi + \frac{\hbar f}{m}\nabla^2\varphi = 0 \tag{2.48}$$

The mass probability density ρ is

$$\rho = m\psi\psi^* = mf^2 \tag{2.49}$$

and the mass current $\mathbf{j}$ is given by

$$\mathbf{j} = (\hbar/2i)(\psi^*\nabla\psi - \psi\nabla\psi^*) = \hbar f^2 \nabla\varphi \tag{2.50}$$

From the last two equations we can define a velocity $\mathbf{u}$ by noting that the mass flux (2.50) must have the form $\mathbf{j} = \rho\mathbf{u}$. Thus

$$\mathbf{u} = (\hbar/m)\nabla\varphi \tag{2.51}$$

which describes a potential flow. Multiplying (2.48) by f and using (2.47) we find the equations of continuity and momentum conservation for the flow (2.51),

$$\frac{\partial \rho}{\partial t} + \nabla \cdot (\rho \mathbf{u}) = 0 \tag{2.52}$$

$$\frac{\hbar}{m} \frac{\partial \varphi}{\partial t} + \tfrac{1}{2} \mathbf{u}^2 + \frac{w}{m} + B = 0 \tag{2.53}$$

where the term

$$B = -\frac{\hbar^2}{2m^2} \frac{\nabla^2 f}{f} \tag{2.54}$$

depends on the bending of the wave function and is unimportant if the scale of variation of f is large. It vanishes if $\hbar \to 0$.

Since our 'fluid' is the wave function of a single particle, one would expect differences from classical potential flow. The quantity of 'fluid' is conserved by (2.52) and fixed by

$$\int |\psi|^2 \, d\mathbf{r} = 1 \quad \text{or} \quad \int \rho \, d\mathbf{r} = m \tag{2.55}$$

If the 'fluid' is confined by some means to a multiply-connected region R, ψ must remain single-valued. From (2.46), around any contour c in R not reducible to a point by a continuous deformation, φ can change by a multiple of 2π only. This condition

$$\int_c \nabla \varphi \cdot d\mathbf{l} = 2n\pi \qquad (n = 0, \pm 1, \pm 2, \dots) \tag{2.56}$$

ensures that the circulation about c is restricted to values given by

$$\Gamma = \int_c \mathbf{u} \cdot d\mathbf{l} = nh/m \qquad (n = 0, \pm 1, \pm 2, \dots) \tag{2.57}$$

Hydrostatic equilibrium comes from setting $\varphi = -\omega t$ and $\mathbf{u} = 0$, then B balances $E - w$ everywhere. For example, (2.53) can then be written as

$$-\hbar^2/2m\nabla^2 f = (E - w)f \tag{2.58}$$

where $E = \hbar\omega$. We can apply (2.58) to a familiar problem in quantum mechanics, a particle in a one-dimensional rectangular potential well. Let

$$w = \begin{cases} w_1 < 0 & -b \le x \le b \\ w_2 = 0 & \text{elsewhere} \end{cases} \tag{2.59}$$

Defining

$$\hbar^2 k^2/2m = |w_1| - |E| > 0 \qquad (2.60)$$

$$\hbar^2 \kappa^2/2m = |E| > 0 \qquad (2.61)$$

we find by elementary methods that (2.58) is described by a ground state of even parity:

$$f = f_1 = A \cos kx \qquad -b \le x \le b \qquad (2.62)$$

$$f = f_2 = A \cos kb \exp\left[-\kappa(x - b)\right] \qquad -b > x > b \qquad (2.63)$$

Matching f_1 to f_2 at $x = \pm b$ quantizes E and normalization by (2.55) fixes A. When $\Delta w = w_2 - w_1$ is large compared to $\hbar^2/2mb^2$, E becomes a low-lying energy level, i.e., $|E| \approx w_1$. In that case we have a characteristic length $1/\kappa \approx \hbar/(2m|E|)^{1/2} \approx \hbar/(2m\Delta w)^{1/2}$ which we denote

$$a_m = \hbar/(2m\Delta w)^{1/2} \qquad (2.64)$$

Hydrodynamically we understand that the step increase of w at $r = b$ is balanced by an increment in B. Moreover, one can see that if a uniform velocity $\mathbf{u}$ is added to the 'fluid' in the well, the boundary layer will be thinner. This is because $\frac{1}{2}\mathbf{u}^2$ adds to w/m in (2.53) and the well depth Δw is increased by $\frac{1}{2}m\mathbf{u}^2$.

A further interesting flow from our point of view is a vortex. Consider Equations (2.52) and (2.53) for a vortex centered at $r = 0$. Since $\partial\varphi/\partial t = 0$ and since by (2.57) we have $u = \hbar/mr$ for a singly quantized vortex, (2.52) and (2.53) become

$$\frac{-\hbar^2}{2m} \frac{\nabla^2 f}{f} + \frac{1}{2} \frac{\hbar^2}{m^2 r^2} + \frac{w}{m} = 0 \qquad (2.65)$$

This equation, in cylindrical coordinates, can be written

$$r^2 \frac{d^2 f}{dr^2} + r \frac{df}{dr} - f + \frac{2mw}{\hbar^2} r^2 f = 0 \qquad (2.66)$$

Putting $r \to (2mw/\hbar^2)r$, (2.66) becomes

$$r^2 \frac{d^2 f}{dr^2} + r \frac{df}{dr} + (r^2 - 1)f = 0 \qquad (2.67)$$

which has the solution $f = j_1(r)$. Thus a vortex in our single-particle 'fluid' has density ρ vanishing at $r = 0$ and increasing as r^2 near the axis. For normalization purposes we could put a container wall at the first zero of $j_1(r)$, and we see then that the 'core' of the vortex must be comparable with the size of the container.

If instead of a single particle we now consider an assembly of N_0 identical bosons of mass m in a potential field $W(\mathbf{r})$. If these bosons do not interact, the wave function $\psi(\mathbf{r}, t)$ of the system would be given by a symmetrized product of N_0 one-particle wave functions with the normalization condition (2.55) replaced by

$$\int |\psi|^2 \, d\mathbf{r} = N_0 \quad \text{or} \quad \int \rho \, d\mathbf{r} = \rho_\infty v \tag{2.68}$$

where v is the volume of the system and $\rho_\infty = mN_0/v$. The ground state of this system for the potential (2.59), however, would have the particles concentrated with high probability near $x = 0$ just as our single particle does. Real liquid helium has a typical interatomic potential including a strong repulsive potential. In this model a short range repulsive potential $V(x - x')$ is introduced which eliminates the unphysical behavior of the perfect Bose gas and the potential

$$\int V(x - x')|\psi(x')|^2 \, dx' \tag{2.69}$$

is added to w which increases as the density of neighboring bosons increases. The simplest case arises when V is taken to be

$$V(x - x') = V_0 \delta(x - x') \tag{2.70}$$

The Schrödinger equation (2.45) is then replaced by

$$i\hbar \partial\psi/\partial t = -(\hbar^2/2m)\nabla^2\psi + (V_0|\psi|^2 + w)\psi \tag{2.71}$$

(Gross, 1961, 1963; Pitaevskii, 1961. See particularly Fetter and Walecka, 1971, Chapters 13 and 14). The Madelung transformation

$$\psi = F \exp (i\varphi) \tag{2.72}$$

proceeds much as with the Schrödinger equation, yielding

$$(\hbar/m)\partial\varphi/\partial t + \tfrac{1}{2}u^2 + w/m + 2p/\rho + B = 0 \tag{2.73}$$

and an equation analogous to (2.53). Equation (2.73) differs from (2.53) principally by the addition of a barotropic 'gas pressure'

$$p = (V_0/2m^2)\rho^2 \tag{2.74}$$

The repulsion potential V_0 and its associated gas pressure restore effects one would expect to see in a fluid, but which are absent in our single-particle model.

For example, the behavior of the condensate in the hydrostatic approximation in the potential well (2.58) now reads

$$d^2F/dx^2 = (2m/\hbar^2)(E - w)F - (2mV_0/\hbar^2)F^3 \tag{2.75}$$

Again if b is large, E is a low-lying energy level and if $N_0 V_0$ and Δw are both large compared with $\hbar^2/mb^2$, (2.75) gives

$$\rho = \rho_\infty = mF_1^2 = mN_0/v \tag{2.76}$$

where v is the volume of the system. The corresponding one-particle energy being given by

$$E - w_1 = \rho_\infty V_0 / m \tag{2.77}$$

Then the condensate is uniformly spread out in the well. When we approach the walls of the well, the derivatives in (2.75) are no longer negligible and the uniform solution (2.76) fails.

In order to examine this situation we introduce a quantity known as a 'healing length', noting that in the well E is very nearly w_1 and thus $E \approx \Delta w$ as before. This new length is

$$a = \hbar/(2m\Delta w)^{1/2} = \hbar/(2\rho_\infty V_0)^{1/2} \tag{2.78}$$

If $F = F_\infty f$ we can rewrite the condensate equation as

$$-a^2 \, \mathrm{d}^2 f/\mathrm{d}x^2 - f + f^3 = 0 \tag{2.79}$$

Consider a semi-infinite mass of condensate next to a wall at $x = 0$. A first integral of (2.79) is (with $\zeta = x/a$)

$$(\mathrm{d}f/\mathrm{d}\zeta)^2 = \tfrac{1}{2}(1 - f^2)^2 \tag{2.80}$$

chosen so that $f' \to 0$ or $f^2 \to 1$ as $\zeta \to \infty$. Then $\mathrm{d}f/\mathrm{d}\zeta(1 - f^2)/2^{1/2}$ and

$$f = \tanh (x/2^{1/2}a) \tag{2.81}$$

which was first derived by Ginzburg and Pitaevskii (1958).

Now suppose there is imposed a uniform flow $\mathbf{u}$ of the condensate in the well. Then we see by (2.73) that $\tfrac{1}{2}mu^2$ must be added to Δw in the healing length. The amplitude of the equilibrium condensate wave function F_∞ is given by

$$F_\infty^2 = \left| \frac{m}{V_0} (\tfrac{1}{2}mu^2 + \Delta w) \right| \tag{2.82}$$

and the length

$$a = \frac{\hbar}{[2m(\tfrac{1}{2}mu^2 + \Delta w)]^{1/2}} \tag{2.83}$$

Thus the effect of this 'dynamic healing' is to increase F_∞ and decrease a as shown in (2.82) and (2.83). Note that healing is a quantitatively different phenomenon than any realized in the two-fluid model. It plays an important part in the considerations of this book.

We should observe here that the phenomenon of healing, as used in the liquid helium literature, refers to the condensate wave function falling to zero inside the liquid. Healing is the result of interparticle interactions.

A vortex line exists in the Bose condensate. Here, as in our single-particle fluid, $u = \hbar/mr$. Scaling distances with a $\zeta = r/a$ and requiring

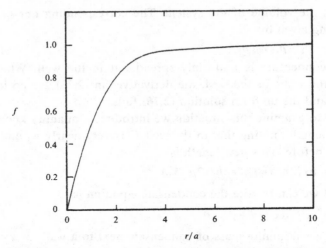

Figure 2.19 Solution for the amplitude of the wave function of a vortex line in a Bose condensate as a function of radial distance from the line. On this model, the vortex line is characterized by a node in the wave function, and a δ-function of quantized vorticity at $r = 0$ (after Kawatra and Pathria (1966)).

$f^2 \to 1$ at large distances, we get

$$\frac{\mathrm{d}^2 f}{\mathrm{d}\zeta^2} + \frac{1}{\zeta}\frac{\mathrm{d}f}{\mathrm{d}\zeta} - \frac{1}{\zeta^2}f + f - f^3 = 0 \tag{2.84}$$

This equation varies as ζ for small distances from the axis and (by a series solution) as $f = 1 - 1/\zeta^2 + \dots$ at large distances. It must be solved numerically and this has been done by Kawatra and Pathria (1966) as shown in Figure 2.19.

The energy of such a vortex is

$$E = \frac{\rho_\infty \kappa^2}{4\pi}\left[\ln\left(\frac{b}{a}\right) + 0.378\right] \tag{2.85}$$

which is to be contrasted with the classical results ((1.31) and (1.33)).

Vortex rings have been investigated by Amit and Gross (1966) and Roberts and Grant (1971). For large rings Roberts and Grant obtain

$$E = \tfrac{1}{2}\rho_\infty \kappa^2 R[\ln(8R/a) - 1.615] \tag{2.86}$$

$$v = (\kappa/4\pi R)[\ln(8R/a) - 0.615] \tag{2.87}$$

$$P = \rho_\infty \kappa \pi R^2 \tag{2.88}$$

Thus rings in the condensate obey Hamilton's equation (1.58). Vortex rings act at great distances as if they have a permanent dipole moment. The same relationship holds for rings in the condensate:

$$M = P/4\pi\rho_\infty \qquad (2.89)$$

The bending of the wave function on a surface given by (2.80) has associated with it a surface energy, or surface tension. Kawatra and Pathria (1966) show that this excess energy is

$$E_1 - E_0 = (2^{1/2}/3)(\hbar^2 N_0/ma), \qquad (2.90)$$

the energy being exactly half kinetic and half potential.

On restoring the time dependence one notes that the gas pressure provides the restoring force necessary for compressional waves. Perturbing about the steady state solution one can show that long wavelength sound propagates at the velocity (see (2.74))

$$c = (dp/d\rho)^{1/2} = (2p/\rho)^{1/2} = \hbar/ma2^{1/2} \qquad (2.91)$$

Note that when variations in the potential are slow in scale compared to a, $(2p/\rho + B)$ is constant. Thus the gradient of (2.73) becomes, with (2.51)

$$\partial \mathbf{u}/\partial t + \nabla(\tfrac{1}{2}\mathbf{u}^2 + \mu) = 0 \qquad (2.92)$$

where we have identified w with the chemical potential μ, which is the equation of motion for classical inviscid irrotational flow (1.7) and also the equation of motion for the superfluid first obtained by Landau in 1941. Equation (2.92) is contained in (2.2) if we note that, neglecting the gradient of second-order terms and $\mathbf{F}_{ns}$ as we do in the absence of vortices, (2.2) reads

$$d\mathbf{v}_s/dt + \nabla\mu = 0 \qquad (2.93)$$

and, using (1.6) for irrotational flow

$$d\mathbf{u}/dt = \partial\mathbf{u}/\partial t + \tfrac{1}{2}\nabla\mathbf{u}^2 \qquad (2.94)$$

We see that (2.92) and (2.93) are the same, with (2.92) of course, restricted to zero temperature. We also must identify the condensate velocity $\mathbf{u}$ with the superfluid velocity $\mathbf{v}_s$ and the density with ρ_s in the sense discussed at the beginning of this section.

The considerations above lead to three of the most important equations used in the study of superfluidity. Equation (2.51) is interpreted to be

$$\mathbf{v}_s = (\hbar/m)\nabla\varphi \qquad (2.95)$$

Equation (2.93) is considered to give Landau's equation for the superfluid

$$\rho_s \frac{d\mathbf{v}_s}{dt} = \frac{-\rho_s}{\rho}\nabla p + \rho_s S\nabla T \qquad (2.96)$$

and Equations (2.49), (2.54) and (2.73) are interpreted to give

$$\frac{\hbar}{m}\frac{\partial\varphi}{\partial t}+\tfrac{1}{2}v_s^2+\frac{\mu}{m}-\frac{\hbar^2}{2m^2}\frac{\nabla^2(\rho_s)^{1/2}}{(\rho_s)^{1/2}}=0 \tag{2.97}$$

where the barotropic pressure for the imperfect Bose condensate (2.74) is considered to be represented by the pressure term for real helium II. If $v_s = 0$ and boundaries are far away, (2.97) becomes

$$(\hbar/m)\,\partial\varphi/\partial t+\mu =0 \tag{2.98}$$

The choice of m as the mass of the ^{4}He atom may not matter for many purposes when working with wave functions. However, it is not arbitrary in (2.95). A Galilean transformation of N particles of ^{4}He must involve the physical mass of the ^{4}He atom. Thus the quantum of circulation $\kappa = h/m$ is defined absolutely, independent of temperature and pressure and external fields such as gravitation.

2.8.4   Ginzburg–Pitaevskii theory

A most valuable approach to the theory of superconductivity was developed when Ginzburg and Landau (1950) applied ideas from Landau's earlier general theory of second order phase transitions to superconductivity. In a number of instances the results of this approach coincide with the more rigorous microscopic theory of superconductivity and have yielded valuable insights into the nature of superconductivity.

Ginsburg and Pitaevskii (1958) appplied Ginzburg–Landau theory to the superfluidity of ^{4}He. Unfortunately the lambda transition is not a perfect second-order phase transition and as a result the theory does not apply rigorously in any range of temperatures. Specifically, it fails to take into account fluctuations in the order parameter. A discussion of Ginzburg–Landau theory applied to superconductivity and superfluidity may be found in the book by Tilley and Tilley (1986). The Landau theory of second order phase transitions is discussed in Landau and Lifshitz (1969).

In the Ginzburg–Pitaevskii (1958) approach, the superfluid is described in terms of a complex order parameter ψ such that $\rho_s = m|\psi|^2$ and $v_s = (h/m)\nabla\psi$. The principal assumption is that the free energy per unit volume can be expanded in the form

$$F_s = F_n - \alpha|\psi|^2+(\beta/2)|\psi|^4+(\hbar^2/2m)|\nabla\psi|^2 \tag{2.99}$$

Minimizing this free energy with respect to variations of ψ yields the

Ginzburg–Landau (1950) equation

$$-(\hbar^2/2m)\nabla^2\psi - \alpha\psi + \beta|\psi|^2\psi = 0 \qquad (2.100)$$

where it is assumed that α is proportional to $(T_\lambda - T)$ and $\beta \sim$ constant. Note that ψ is an order parameter and not the solution of the Schrödinger equation. As was true for superconductors, their model predicts a 'healing length'

$$a = \hbar/(2m\alpha)^{1/2} \qquad (2.101)$$

that diverges as $(T_\lambda - T)^{1/2}$; in particular, from thermodynamic data near T_λ

$$a \sim 4(T_\lambda - T)^{-1/2} \text{ Å} \qquad (2.102)$$

The variation of superfluid density near a wall was calculated and found to be given by the form (2.81). They were the first to obtain a solution which corresponds to a vortex filament in the center of a container of radius b whose energy is quoted in (2.85) above.

2.8.5 Phase slip

The fact that a quantized vortex line forces the wave function to zero (Figure 2.19) has a key consequence in vortex dynamics, as observed by Anderson (1966). An isolated vortex line has lines of equal phase as shown in Figure 2.20(a). If a vortex line moves from one side of a channel to the other, as shown in Figure 2.20(b), the phase of the superfluid wave function between points A and B will change by 2π. This arises because before the vortex moves across the channel the phase can be measured along a path like A C′B. But the phase difference in the two paths differs by the phase going around the vortex once, which is 2π. If n vortices move across per second then the phase difference between A and B slips at the rate

$$\frac{d}{dt}(\varphi_A - \varphi_B) = 2\pi\frac{dn}{dt} \qquad (2.103)$$

If $u = 0$, $2p/\rho + B =$ constant and $\mu = w$ as in (2.92), then (2.73) gives $(\hbar/m)\,d\varphi/dt + \mu = 0$. Applying this result to (2.103) gives

$$\frac{\hbar}{m}\frac{d(\varphi_A - \varphi_B)}{dt} = -(\mu_A - \mu_B) = \frac{h}{m}\frac{dn}{dt} \qquad (2.104)$$

and the chemical potential difference appears perpendicular to the direction of vortex motion (Figure 2.20). For a slowly varying chemical

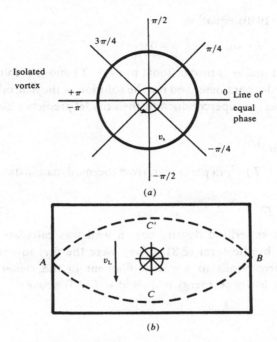

(a)

(b)

Figure 2.20 (*a*) Lines of equal phase round an isolated vortex line. (*b*) Vortex line crossing a channel containing a superfluid. The phase difference between A and B before the vortex crosses (contour ACB) differs by 2π from the phase difference after the vortex has crossed (contour AC′B) (after Tilley and Tilley (1986)).

potential (2.104) can be written

$$\frac{\hbar}{m}\frac{d\Delta\varphi}{dt} = -\Delta\mu = \kappa\frac{dn}{dt}$$

This relationsship can be put into continuous form by dividing by Δx and taking the limit, confining the flow to slow changes:

$$\frac{\hbar}{m}\frac{d\nabla\varphi}{dt} = -\nabla\mu = \kappa\frac{dn}{dx\,dt} = \kappa\frac{dn}{dx^2}\frac{dx}{dt}$$

The left side can be put in terms of superfluid velocity by (2.95). On the right side we recognize $\kappa\,dn/dx^2$ as the vorticity since dn/dx^2 is the area density of lines. Taking $v_L = dx/dt$, we have

$$\frac{\partial\mathbf{v}_s}{\partial t} = \mathbf{v}_L\times\boldsymbol{\omega} + \nabla\mu \tag{2.105}$$

where $\Delta\mu$ is added to include pressure and temperature gradients which also accelerate the superfluid.

As Anderson has observed, the derivation above assumes quantized vortices can cross streamlines of the flow. This cannot occur for vortices in a classical fluid unless a means is found to apply forces to the cores. Quantized vortices are different because they have a true discontinuity in superfluid density at the core. It is this discontinuity – the node in the wave function – which leads to that cornerstone of hydrodynamics: circulation. It is interesting to note that the phase slip equation in the form (2.105) was stated without proof six years earlier by Hall (1960).

Still another way to appreciate the content of phase slip is to note that if the normal fluid is at rest the energy flux density for the superfluid is (Landu and Lifshitz, 1959, Section 130)

$$\rho_s v_s (\tfrac{1}{2} v_s^2 + \mu) \tag{2.106}$$

(which should be contrasted to the classical expression (1.27)) so that the presence of phase slip and hence a chemical potential produces a flow of energy through a surface perpendicular to v_s. We shall encounter phase slip arguments in Sections 5.4, 8.3 and 8.7.

The concept of phase slip, attractive as it is, gives no hint whatsoever as to *how* the vortex motion that produces the phase slip originates. We shall discuss some of the difficulties in Chapter 8.

3 Vortex dynamics and mutual friction

3.1 Vortex dynamics

One of the most remarkable topics in the superfluidity of helium II is our understanding of the motion of quantized vortices undergoing frictional dissipation. We shall see that, on the one hand, our empirical knowledge of vortex dynamics is so complete that we can make progress in understanding a problem as complicated as the turbulent flow of a tangle of vortices (Chapter 7), and on the other hand, that our understanding of the microscopic origin of the drag processes leaves much to be desired.

Although the ideas of mutual friction were originally developed for rectilinear vortices in uniform rotation, it appears that the same ideas can be applied to curved vortices, as least as long as the curvature is, in some sense, not too rapid.

Figure 1.14 shows a segment of a vortex configuration described in parametric form by an arc length ξ (where ξ increases in the direction of the circulation vector $\boldsymbol{\kappa}$), so that $\mathbf{s}(\xi, t)$ is the position of a particular point on the configuration at time t, and $\mathbf{s}'(\xi, t) = \partial \mathbf{s}(\xi, t)/\partial \xi = \hat{\boldsymbol{\kappa}}$ is the unit tangent at that point.

Consider first a segment of vortex line moving in the laboratory frame at absolute zero with velocity $\mathbf{v}_L$ and in the presence of a local superfluid velocity $\mathbf{v}_{sl}$. The velocity $\mathbf{v}_{sl}$ consists of the vector sum of the velocity $\mathbf{v}_s$ of the superfluid at large distances from any vortex line and $\mathbf{v}_i$ the superflow induced by any curvature of the vortex line, in the sense described by Equation (1.76).

We have seen in Section 1.1 that a cylinder of radius a with circulation Γ about it in a flow $\mathbf{v}_\infty$ will experience a lift force $\rho \mathbf{v}_\infty \times \Gamma$. Hall and Vinen (1956b) noted that if a cylinder in helium II at absolute zero is impulsively pulled into motion by a force $\mathbf{f}$ per unit length applied perpendicular to the cylinder, the subsequent motion of the cylinder consists of a steady movement transverse to $\mathbf{f}$ with a velocity $\mathbf{v}$ given by $\mathbf{f} = \rho\mathbf{v} \times \mathbf{\Gamma}$ on which is superimposed an oscillation of amplitude $f/\Gamma\rho\omega$ and frequency $\omega = \Gamma\rho/(M + M')$ where M is the physical mass of the cylinder per unit length, and $M' = \pi a^2 \rho$ is its hydrodynamic mass per unit length (Milne-Thomson, 1968, p. 250). If the circulation about the cylinder $\Gamma = \kappa$, $M \to 0$, and a is shrunk to atomic dimensions, we have a quantized vortex line as envisioned by Feynman (1955).

As an illustration of the situation just described applied to helium II at absolute zero, let a vortex line lie along the z-axis and suppose that at $t = 0$ a flow v_s is impulsively applied parallel to the x-axis (Figure 3.1). The equations of motion of the line are

$$M'\dot{v}_y = -\rho\kappa(v_x - v_s)$$
$$M'\dot{v}_x = \rho\kappa v_y \tag{3.1}$$

Put $v_L = v_x + iv_y$, then the equation of motion becomes

$$M'\dot{v}_L = -i\rho\kappa v_L + i\rho\kappa v_s$$

and

$$v_L - v_s = A \exp(-i\rho\kappa t/M') \tag{3.2}$$

where A is a constant of integration. Thus

$$v_x - v_s = A \cos(\rho\kappa t/M')$$
$$v_y = -A \sin(\rho\kappa t/M') \tag{3.3}$$

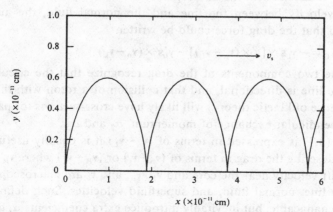

Figure 3.1 Motion of a vortex line oriented along the z-axis in a flow $v_s = 10$ cm/s impulsively applied at $t = 0$. The subsequent motion is a cycloid as shown.

when $t = 0$, $v_x = v_y = 0$, then $A = -v_s$ so the motion is given by

$$v_x = v_s - v_s \cos(\rho \kappa t / M')$$
$$v_y = v_s \sin(\rho \kappa t / M') \tag{3.4}$$

The motion of the line is a cycloid with the frequency $\rho \kappa / M'$

$$\omega = \rho \kappa / M' = \rho \kappa / \pi \rho a^2 = \kappa / \pi a^2 \tag{3.5}$$

The application of flow to the line will cause it to move as shown in Figure 3.1 with oscillation of high frequency $\omega = \kappa / \pi a^2 \approx 3 \times 10^{12}$ rad/s and small amplitude ($v_s / \omega \approx 3 \times 10^{-12}$ cm for $v_s = 10$ cm/s). The authors then supposed that the Magnus force could be applied to quantized vortices even though the structure of the vortices is not completely understood, reasoning that the force is likely dependent upon the circulation away from the core. The assumption, then, is that the Magnus force $\mathbf{f}_M$ on a quantized vortex line can be written $\mathbf{f}_M = \rho \kappa \mathbf{s}' \times (\mathbf{v}_L - \mathbf{v}_{sl})$ recognizing the fact that the line element can have a velocity $\mathbf{v}_L$ different from the local superfluid velocity. The sign convention in superfluidity is that flows and line velocities along the positive x-axis appear as shown.

Now suppose we are at finite temperatures. The only modification to the lift force from absolute zero is to use ρ_s in place of the total density:

$$\mathbf{f}_M = \rho_s \kappa \mathbf{s}' \times (\mathbf{v}_L - \mathbf{v}_{sl}) \tag{3.6}$$

But now the vortex line will be moving through a gas of excitations, phonons and rotons as discussed in Section 2.2. Above 1 K most of the drag is believed due to roton–line collisions. Assuming that the friction owing to collisions with the line depends in a first approximation on the relative velocity between the line and the normal fluid, the authors supposed that the drag force could be written

$$\mathbf{f}_D = -\gamma_0 \mathbf{s}' \times [\mathbf{s}' \times (\mathbf{v}_n - \mathbf{v}_L)] + \gamma_0' \mathbf{s}' \times (\mathbf{v}_n - \mathbf{v}_L) \tag{3.7}$$

where the two components of the drag recognize that the circulation about the line is directional, and that collision of a roton with the line (in the sense of kinetic theory) will likely have cross-sections for parallel and perpendicular exchange of momentum, $\sigma_{\parallel}$ and $\sigma_{\perp}$.

While (3.7) is expressed in terms of $(\mathbf{v}_n - \mathbf{v}_L)$ it is equally useful and valid to describe the drag in terms of $(\mathbf{v}_n - \mathbf{v}_s)$ or $(\mathbf{v}_R - \mathbf{v}_L)$ where $\mathbf{v}_R$ is the roton drift velocity near the core and $\mathbf{v}_L$, $\mathbf{v}_n$, and $\mathbf{v}_s$ are macroscopically averaged line, normal fluid, and superfluid velocities. Such definitions are interchangeable, but inevitably introduce extra coefficients α, α' and D, D_t as follows:

$$\mathbf{f}_D = -\alpha \rho_s \kappa \mathbf{s}' \times [\mathbf{s}' \times (\mathbf{v}_n - \mathbf{v}_{sl})] - \alpha' \rho_s \kappa \mathbf{s}' \times (\mathbf{v}_n - \mathbf{v}_{sl}) \tag{3.8}$$

Table 3.1(a). *Formulae for calculation of various mutual friction parameters from B and B'*

$\alpha = B\rho_n/2\rho$	$\alpha' = B'\rho_n/2\rho$
$\gamma_0 = \rho_s\kappa\alpha/(e^2+\alpha^2)$	$e = 1-\alpha'$
$\gamma_0' = \rho_s\kappa(\alpha^2-\alpha'e)/(e^2+\alpha^2)$	
$\gamma = \gamma_0\rho_s^2\kappa^2/[\gamma_0^2+(\rho_s\kappa-\gamma_0')^2] = \rho_s\kappa\alpha$	
$D = c/(c^2+d^2)$	$c = \alpha/\rho_s\kappa(\alpha^2+\alpha'^2)-1/E$
$D_t = -d/(c^2+d^2)$	$d = \alpha'/\rho_s\kappa(\alpha^2+\alpha'^2)-1/\rho_s\kappa$
$\sigma_\parallel = D/\rho_n V_G$	
$\sigma_\perp = D_t/\rho v_n v_G + \kappa/v_G$	

Table 3.1(b). *Formulae for calculation of mutual friction parameters from* $\sigma_\parallel$, $\sigma_\perp$

$D = \rho_n v_G \sigma_\parallel$	$D' = \rho_n v_G \sigma_\perp$	$D_t = D' - \rho_n\kappa$
$B = \dfrac{2\rho}{\rho_n\rho_s\kappa}\dfrac{a}{a^2+b^2}$		$B' = \dfrac{2\rho}{\rho_n\rho_s\kappa}\dfrac{b}{a^2+b^2}$
$a = \dfrac{D}{D^2+D_t^2}+\dfrac{1}{E}$		$b = \dfrac{-D_t}{D^2+D_t^2}+\dfrac{1}{\rho_s\kappa}$

or
$$\mathbf{f}_D = -D\mathbf{s}'\times[\mathbf{s}'\times(\mathbf{v}_R-\mathbf{v}_L)]+D_t\mathbf{s}'\times(\mathbf{v}_R-\mathbf{v}_L) \quad (3.9)$$
The relationships among coefficients appear in Table 3.1. Equation (3.9) serves to introduce the microscopic friction coefficients D and D_t which are related to the kinetic scattering coefficients $\sigma_\parallel$ and $\sigma_\perp$ by
$$D = \rho_n v_G \sigma_\parallel \quad (3.10)$$
$$D_t = \rho_n v_G \sigma_\perp - \rho_n\kappa \quad (3.11)$$
where v_G is the thermal average group velocity of the rotons. We shall discuss the microscopic theory of mutual friction in Section 3.4.

The excitations that form the normal fluid (primarily rotons above 1 K) are dragged by a vortex line so that their average drift velocity $\mathbf{v}_R$ near the core is not $\mathbf{v}_n$. Hall and Vinen (1956b) define a quantity E by
$$\mathbf{v}_n - \mathbf{v}_R = \mathbf{f}_D/E \quad (3.12)$$
For low amplitude counterflow oscillations with frequency ω and wavelength λ, they find that E is given by
$$E = -4\pi\eta/M \quad (3.13)$$

where

$$M = \ln (l/2\delta) + 1 + i\pi/4 \tag{3.14}$$

and the viscous penetration depth $\delta = (\eta/\rho_n\omega)^{1/2}$, the roton mean free path $l = 3\eta/\rho_n v_G$, the average roton group velocity $v_G = (2k_B T/\pi\mu)^{1/2}$, k_B is Boltzmann's constant, and μ is the roton effective mass. At zero frequency, Equation (3.14) is modified primarily by replacing the frequency-dependent viscous penetration depth δ by a velocity-dependent effective penetration depth $\delta_{eff} = 2\eta/\rho_n|v_n - v_L|$ (see Lamb (1945) Section 343, Vinen (1957) and Yarmchuck and Glaberson (1979)). Equation (3.14) then becomes

$$M = \ln (l/2\delta_{eff}) + \tfrac{1}{2} \tag{3.15}$$

Hall and Vinen (1956b) treated the normal fluid as a continuous fluid to which the localized friction force is applied, but with a correction associated with the finite mean free path l of the rotons. Their calculation was based on an appropriate solution of the Navier–Stokes equation which had been published by Sewell (1910). For calculations of E near T_λ Barenghi, Donnelly, and Vinen (1983) suggest one should replace l with $(l^2 + \xi^2)^{1/2}$ near T_λ. Here ξ is the correlation length for helium II which is of the order of angstroms at low temperatures but diverges near the lambda transition.

Equations (3.7)–(3.15) can be solved to give relationships between the mutual friction coefficients. These are summarized in Table 3.1, showing how the coefficients can be calculated beginning with experimental values of B, B' or calculated values of $\sigma_\parallel$ and $\sigma_\perp$.

If we neglect the inertia of the core, then the sum of the forces on each line element must vanish and we have from (3.6) and (3.8)

$$\mathbf{f}_M + \mathbf{f}_D = \rho_s \kappa \mathbf{s}' \times (\mathbf{v}_L - \mathbf{v}_{sl}) - \alpha \rho_s \kappa \mathbf{s}'$$
$$\times [\mathbf{s}' \times (\mathbf{v}_n - \mathbf{v}_{sl})] - \alpha' \rho_s \kappa \mathbf{s}' \times (\mathbf{v}_n - \mathbf{v}_{sl})$$
$$= \rho_s \kappa \mathbf{s}' \times [(\mathbf{v}_L - \mathbf{v}_{sl}) - \alpha \mathbf{s}' \times (\mathbf{v}_n - \mathbf{v}_{sl}) - \alpha'(\mathbf{v}_n - \mathbf{v}_{sl})]$$
$$= 0 \tag{3.16}$$

Thus the term in square brackets must be in the direction $\mathbf{s}'$ or equal to zero. The latter condition leads to

$$\mathbf{v}_L = \mathbf{v}_{sl} + \alpha \mathbf{s}' \times (\mathbf{v}_n - \mathbf{v}_{sl}) - \alpha' \mathbf{s}' \times [\mathbf{s}' \times (\mathbf{v}_n - \mathbf{v}_{sl})] \tag{3.17}$$

Equation (3.17) is a fundamental result in vortex dynamics which can be used for a variety of purposes including simulations of the development of three dimensional vortex motion in an arbitrary flow.

3.2   Uniformly rotating helium II

The various results and definitions of the previous section, by themselves, give little insight as to their origin and usefulness. The first application of mutual friction was to relate the attenuation of second sound in uniformly rotating helium II to a more microscopic view of frictional forces on an isolated vortex line. Consider rotation of a container which is sufficiently rapid that we have a uniform array of vortices such as is pictured in Figure 2.6. When equilibrium is reached the sum of the force per unit volume on the normal fluid (see (2.2)) plus the drag force per unit volume on the vortices must total zero:

$$\mathbf{F}_{ns} + n_0 \mathbf{f}_D = 0 \tag{3.18}$$

From (3.8) we have

$$\mathbf{f}_D = -\alpha \rho_s \kappa \mathbf{s}' \times [\mathbf{s}' \times (\mathbf{v}_n - \mathbf{v}_s)] - \alpha' \rho_s \kappa \mathbf{s}' \times (\mathbf{v}_n - \mathbf{v}_s) \tag{3.19}$$

where $\mathbf{v}_s$ is used in the place of $\mathbf{v}_{sl}$ because the lines are assumed to be straight and there is no self-induced velocity. Equations (3.18) and (3.19) combine to give

$$\mathbf{F}_{ns} = \omega \rho_s \alpha \mathbf{s}' \times [\mathbf{s}' \times (\mathbf{v}_n - \mathbf{v}_s)] + \omega \rho_s \alpha' \mathbf{s}' \times (\mathbf{v}_n - \mathbf{v}_s) \tag{3.20}$$

where the vorticity $\omega = 2\Omega$. Equation (3.20) is identical to (2.36), the original form proposed by Hall and Vinen (1956*b*). We have described in Section 2.4.2 how second sound resonance experiments give values of B and B'. Once these values are known it is possible to deduce the remaining drag coefficients through the relationships quoted in Table 3.1.

A more general case of mutual friction arises if the vortex lines are not straight, but curved. Under these conditions the self-induced velocity $\mathbf{v}_i$ will not be zero. If the curvature of the vortex lines is slow enough that the local radius of curvature R can be used in the Arms–Hama approximation (1.76), then (with $\Gamma = \kappa$) we can write $v_i = (\kappa/4\pi R) \ln (L/a)$ where L is some characteristic microscopic length: the radius of curvature, or the interline spacing, for example. Referring to Figure 1.14 the self-induced velocity will be in the direction of $\mathbf{s}' \times \mathbf{s}''$ which has the magnitude $1/R$:

$$\mathbf{s}' \times \mathbf{s}'' = (\mathbf{s}' \cdot \nabla)\mathbf{s}' \tag{3.21}$$

and we have approximately

$$\mathbf{v}_i \approx \nu(\mathbf{s}' \cdot \nabla)\mathbf{s}' \tag{3.22}$$

where by (2.16)

$$\nu = (\kappa/4\pi) \ln (L/a) = \varepsilon/\rho_s \kappa \tag{3.23}$$

If the curved lines are dense enough to form a continuum, then (3.8) and (3.18) can be used in the equation of motion with

$$v_{sl} = v_s + v_i = v_s + \nu(s' \cdot \nabla)s' \tag{3.24}$$

and (3.20) is generalized to read

$$F_{ns} = \rho_s \omega \alpha s' \times [s' \times (v_n - v_s)]$$
$$+ \rho_s \omega \alpha' s' \times (v_n - v_s) + \rho_s \omega \alpha \nu \text{ curl } s'$$
$$+ \rho_s \omega \alpha' \nu(s' \cdot \nabla)s' \tag{3.25}$$

where the last two terms arise from v_i. The third term uses the vector relationship $(s' \cdot \nabla)s' = -s' \times \text{curl } s'$.

Under these conditions we must amend the superfluid acceleration condition to recognize that phase slip produces a chemical potential difference even at absolute zero. This relationship is just Equation (2.105). At $T = 0$, $v_L = v_{sl}$ which we now simply call v_s so that the supfluid equation of motion becomes

$$\partial v_s / \partial t = v_s \times \omega + \nu\omega(s' \cdot \nabla)s' + \nabla\mu$$

Absorbing $v_s \times \omega$ into dv_s/dt (2.2) is generalized to read

$$\rho_s \frac{dv_s}{dt} = -\frac{\rho_s}{\rho}\nabla p + \rho_s S\nabla T + \frac{\rho_n\rho_s}{2\rho}\nabla(v_n - v_s)^2$$
$$+ \rho_s \nu\omega(s' \cdot \nabla)s' - F_{ns} \tag{3.26}$$

The normal fluid equation remains unchanged in form. Equations (2.3), (3.25) and (3.26) are called the Hall-Vinen-Bekarevich-Khalatnikov equations (HVBK). The reader is referred to Yarmchuck and Glaberson (1979) Appendix A to see the HVBK equations written in a rotating coordinate system.

3.3   Frequency and velocity dependence of B and B'. Temperature dependence of the coefficients of mutual friction

Values of B and B' are contained in Table 3.2 compiled by Barenghi *et al* (1983) which are adequate for many purposes. However, experiments that have used B in their analyses have been conducted at a wide range of counterflow velocities and frequencies. For example, in quantum turbulence experiments described in Chapter 7 one uses B to convert from measurements of the attenuation of second sound (at some frequency) as a function of heat flux to the vortex line density as a function of heat flux. Experimenters have used resonances varying from 4 Hz (Martin and Tough, 1983) to greater than 23 kHz (Swanson, 1985),

Table 3.2. *Mutual friction coefficients B and B' (Barenghi et al., 1983)*

T (K)	B	B'
1.30	1.52	0.61
1.35	1.46	0.53
1.40	1.40	0.45
1.45	1.35	0.38
1.50	1.29	0.31
1.55	1.24	0.25
1.60	1.19	0.19
1.65	1.14	0.15
1.70	1.10	1.10
1.75	1.06	0.07
1.80	1.02	0.05
1.85	0.99	0.04
1.90	0.98	0.04
1.95	0.98	0.05
2.00	1.01	0.04
2.01	1.02	0.04
2.02	1.04	0.04
2.03	1.05	0.03
2.04	1.07	0.02
2.05	1.10	0.01
2.06	1.13	0.00
2.07	1.16	−0.01
2.08	1.21	−0.03
2.09	1.26	−0.05
2.10	1.33	−0.08
2.11	1.42	−0.12
2.12	1.53	−0.17
2.13	1.69	−0.24
2.14	1.90	−0.36
2.15	2.21	−0.54
2.16	2.67	−0.83
2.161	2.73	−0.94
2.162	2.80	−1.00
2.163	2.88	−1.07
2.164	2.99	−1.15
2.165	3.12	−1.25
2.166	3.28	−1.37
2.167	3.49	−1.51
2.168	3.75	−1.71
2.169	4.13	−1.98
2.170	4.72	−2.40
2.171	5.93	−3.28

Table 3.3. *Mutual friction parameter B versus temperature for various second sound frequencies in the low amplitude limit (after Swanson et al. (1987))*

T (K)	Frequency (Hz)				
	1	10	100	1000	10 000
1	1.507	1.509	1.511	1.513	1.515
1.1	1.549	1.557	1.564	1.572	1.579
1.2	1.520	1.538	1.556	1.574	1.592
1.3	1.431	1.464	1.499	1.534	1.571
1.4	1.268	1.318	1.370	1.427	1.487
1.5	1.098	1.159	1.227	1.302	1.387
1.6	0.958	1.026	1.104	1.194	1.300
1.7	0.855	0.926	1.011	1.112	1.236
1.8	0.788	0.863	0.953	1.063	1.203
1.9	0.760	0.836	0.929	1.045	1.194
2.0	0.788	0.863	0.954	1.067	1.210
2.1	1.106	1.197	1.304	1.432	1.588
2.11	1.192	1.287	1.398	1.531	1.691
2.12	1.299	1.398	1.514	1.651	1.815
2.13	1.436	1.541	1.661	1.802	1.967
2.14	1.623	1.732	1.856	1.998	2.164
2.15	1.902	2.014	2.140	2.282	2.442
2.16	2.415	2.528	2.652	2.787	2.933
2.17	4.600	4.698	4.798	4.899	5.002

a range of three and one half decades. Table 3.3 shows the corresponding range of B: at 1.9 K, B varies by more than 50% from 1 Hz to 10 kHz. The parameter B is also used to convert measurements of temperature gradient to line density. Here B is a function of the steady counterflow velocity. Experiments of interest to cryogenic engineers have been performed with heat fluxes as high as 20 W/cm^2, producing a considerable temperature difference over a length of order 1 cm (Pfotenhauer and Donnelly, 1985). The resulting counterflow velocity varies with position but is everywhere greater than 160 cm/s. In the other extreme, quantum turbulence experiments have been carried out with v_{ns} as low as 0.1 cm/s. Table 3.4 shows that the change in B over the relevant velocity range can be of more than a factor of 2 at 1.9 K.

The theory of the frequency and velocity dependences of mutual friction is contained in Equations (3.12)–(3.15). Using this theory one can develop a method for obtaining the mutual friction parameters at arbitrary frequency or counterflow velocity. One implementation of this

Table 3.4. *Mutual friction parameter B versus temperature for various vortex line-normal fluid relative velocities in the steady state limit (after Swanson et al. (1987))*

T (K)	v_{nL} (cm/s)			
	0.1	1	10	100
1	1.503	1.507	1.511	1.515
1.1	1.539	1.553	1.568	1.583
1.2	1.501	1.536	1.572	1.609
1.3	1.406	1.472	1.543	1.619
1.4	1.244	1.343	1.456	1.585
1.5	1.080	1.205	1.360	1.556
1.6	0.949	1.092	1.285	1.555
1.7	0.856	1.012	1.238	1.590
1.8	0.798	0.967	1.227	1.672
1.9	0.778	0.956	1.239	1.756
2.0	0.812	0.990	1.268	1.763
2.1	1.139	1.350	1.656	2.140
2.11	1.226	1.445	1.760	2.247
2.12	1.334	1.563	1.885	2.369
2.13	1.473	1.711	2.037	2.511
2.14	1.662	1.906	2.231	2.681
2.15	1.942	2.190	2.506	2.914
2.16	2.455	2.700	2.990	3.332
2.17	4.635	4.835	5.039	5.243

idea is contained in the paper by Swanson, Wagner, Donnelly and Barenghi (1987). Some results of this rather technical procedure are contained in Tables 3.3 and 3.4.

The temperature dependence of the various friction coefficients is complicated. They are illustrated in Figures 3.2–3.5, taken from the review by Barenghi *et al.* (1983). The quantities B and B' in Figure 3.2 are represented by cubic splines. These splines together with the formulae of Table 3.1 allow the fits of Figures 3.3–3.5 to be generated.

3.4   Microscopic form of mutual friction and the Iordanskii force

A microscopic form of mutual friction is directly related to roton scattering from an element of line. The drag force on unit length of line due to scattering of excitation is:

$$\mathbf{f}_{ex} = D(\mathbf{v}_R - \mathbf{v}_L) + D'\mathbf{s}' \times (\mathbf{v}_R - \mathbf{v}_L) \qquad (3.27)$$

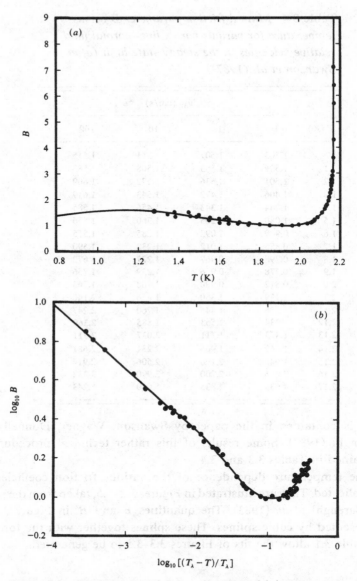

Figure 3.2 The behavior of B as a function of (a) temperature and (b) reduced temperature, and the behavior of B' as a function of (c) temperature and (d) reduced temperature after Barenghi *et al.* (1983). The lines are calculated from the microscopic cross-sections as described in Section 5 of Barenghi *et al.* and represented by a spline in Table VIII of their paper.

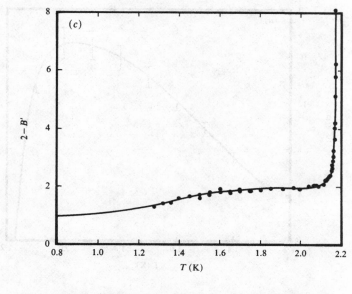

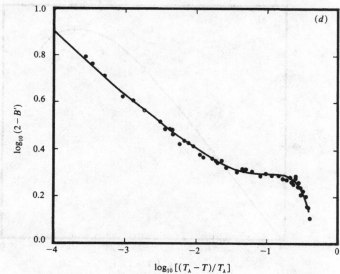

Figure 3.2 (*cont.*)

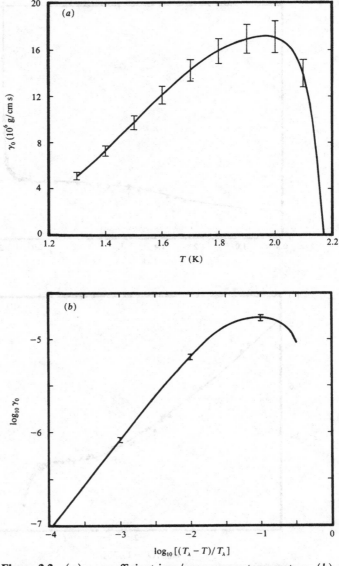

Figure 3.3 (*a*) γ_0 coefficient in g/cm s versus temperature. (*b*) γ_0 coefficient versus reduced temperature. The lines in figures 3.3–3.5 are calculated from the expressions in Table 3.1 and the fits of B and B' shown in Figure 3.2 (after Barenghi *et al.* (1983)).

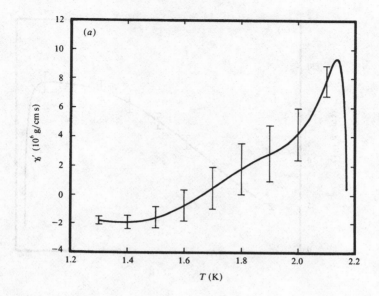

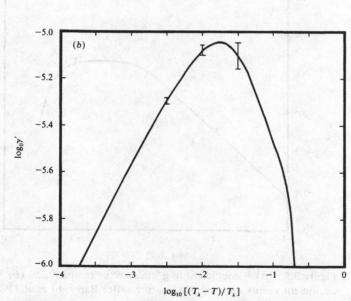

Figure 3.4 (a) γ_0' coefficient in g/cm s versus temperature. (b) γ_0' coefficient versus reduced temperature (after Barenghi *et al.* (1983)).

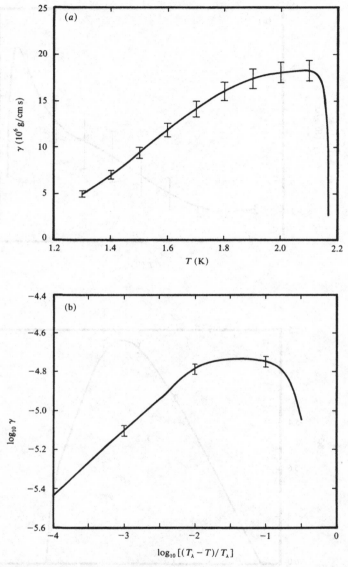

Figure 3.5 (*a*) γ coefficient in g/cm s versus temperature. (*b*) γ coefficient versus reduced temperature (after Barenghi *et al.* (1983)).

where v_R is the normal fluid velocity at the core of the line, and where the miscoscopic parameters D and D' are related to scattering lengths $\sigma_\parallel$ and $\sigma_\perp$ by

$$D = \rho_n v_G \sigma_\parallel, \qquad D' = \rho_n v_G \sigma_\perp \qquad (3.28)$$

v_G being the average group velocity of rotons. Before completing the balance of forces with the Magnus force one must first note the existence of a subtle effect, first discussed by Iordanskii (1964, 1965, 1966), which modifies the microscopic forces by the addition of a new force per unit length of line called the Iordanskii force f_I:

$$f_I = -\rho_n \kappa \times (v_R - v_L) \qquad (3.29)$$

It is important in understanding the Iordanskii force to remember that the normal component in the two-fluid model cannot be identified exactly with the fluid formed from the gas of thermal excitations. The momentum density carried by the normal fluid is $\rho_n v_n$, while that carried by the excitations is $\rho_n(v_n - v_s)$. The *total* momentum density is obtained by either adding that of the superfluid ($\rho_s v_s$) to that of the normal fluid or adding that of the background fluid (ρv_s) to that of the excitation gas.

The existence of the Iordanskii term does not depend on the structure of the core of the vortex, and a discussion (drawn from Appendix C of Barenghi *et al.* (1983)) will demonstrate this fact. We consider a rectilinear vortex, and we place round it, concentrically, an imaginary cylinder of radius R_0, where R_0 is large compared with the core radius a, but small compared with the viscous penetration depth. The normal fluid velocity (equal to the drift velocity of the excitations) can then be taken as constant over the surface of the cylinder and equal to v_R. For convenience and simplicity we work in a frame of reference in which the vortex line is at rest ($v_L = 0$). The superfluid velocity over the surface of the cylinder is composed of two parts: the part v_{s_0} due to the circulation, κ, round the core; and a spatially uniform part, v_{s_1}, due to superfluid flow past the vortex (since $R_0 \gg a$, v_{s_1} is not significantly distorted at the surface of the cylinder by any disturbance to the flow in the neighborhood of the core). The scattering of the excitations inside the cylinder R_0 gives rise to a net force f_{ex} on the unit length of the line, and we write this force from (3.27) in the form for $v_L = 0$

$$f_{ex} = D v_R + D' s' \times v_R \qquad (3.30)$$

Our task is to find an expression relating v_{s_1}, v_R, D, D', which will in turn lead to relations between γ_0, γ_0', D, D'.

We shall make use here of a hypothetical device by means of which we separate the excitations from the rest of the system while they are being scattered. We imagine that there is a second cylinder just outside the cylinder of radius R_0; excitations crossing into this second cylinder are removed from the fluid as they do so, while excitations are restored to the fluid at the surface of the cylinder in such a way as to provide the correct outflow of excitations from the cylinder. Removal of an excitation does not mean the removal of any liquid; it means simply the annihilation of the excitation, its energy and momentum being imparted to some mechanism. Creation of an excitation implies removal of energy and momentum from the mechanism. Elastic scattering of excitations by the vortex implies transfer of an appropriate amount of extra momentum to the mechanism (at a rate $\mathbf{f}_{ex}$), this extra momentum appearing ultimately in the newly created excitations. With this device the fluid flowing at and through the surface of the cylinder of radius R_0 is entirely superfluid of density ρ, the excitations having been removed, and the forces acting on the fluid inside this cylinder are easily calculated. The velocity $\mathbf{v}_{s_0}$ on the surface of R_0 is unaltered by the removal of the excitations, but the velocity $\mathbf{v}_{s_1}$ must be changed to $\mathbf{v}_{s_2} = (\rho_s/\rho)\mathbf{v}_{s_1} + (\rho_n/\rho)\mathbf{v}_R$ in order to conserve mass. The fluid inside R_0 is therefore subject to a Magnus force, given by

$$\mathbf{f}_M = -\rho\boldsymbol{\kappa} \times \mathbf{v}_{s_2} = -\rho_s\kappa\mathbf{s}' \times \mathbf{v}_{s_1} - \rho_n\kappa\mathbf{s}' \times \mathbf{v}_R \tag{3.31}$$

To achieve an overall balance of forces this Magnus force must be equal to the force $\mathbf{f}_{ex}$ due to the scattering of the excitations. It follows that

$$\rho_s\kappa\mathbf{s}' \times \mathbf{v}_{s_1} = D\mathbf{v}_R + (D' - \rho_n\kappa)\mathbf{s}' \times \mathbf{v}_R \tag{3.32}$$

The term $-\rho_n\boldsymbol{\kappa} \times \mathbf{v}_R$ is the Iordanskii force. The microscopic balance of forces is now $\mathbf{f}_M + \mathbf{f}_I + \mathbf{f}_{ex} = 0$: i.e.,

$$\rho_s\kappa\mathbf{s}' \times (\mathbf{v}_s - \mathbf{v}_L) = -D\mathbf{s}' \times [\mathbf{s}' \times (\mathbf{v}_R - \mathbf{v}_L)] + D_t\mathbf{s}' \times (\mathbf{v}_R - \mathbf{v}_L) \tag{3.33}$$

where the transverse friction constant D_t introduced in Section 3.1 is defined as

$$D_t = D' - \rho_n\kappa \tag{3.34}$$

and where the right-hand side of (3.33) defines our $\mathbf{f}_D$ in Equation (3.9). Values of D, D_t, D', $\sigma_\parallel$, and $\sigma_\perp$ are subject to considerable experimental uncertainties, and the interested reader should consult Barenghi *et al.* (1983) for recommended values of these quantities. We illustrate in Figures 3.6 and 3.7 values of $\sigma_\parallel$ and $\sigma_\perp$; there are considerable uncertainties in the magnitudes of these quantities which should be appreciated before making use of them. Details of these plots, their uncertainties, the origin of the discontinuity at $T = 1.748$ K are discussed in the review by Barenghi *et al.* (1983).

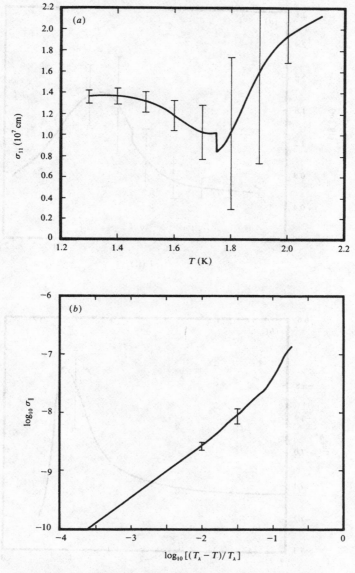

Figure 3.6   (*a*) Parallel scattering length $\sigma_\parallel$ in cm versus temperature. (*b*) Parallel scattering length $\sigma_\parallel$ versus reduced temperature (after Barenghi *et al.* (1983)).

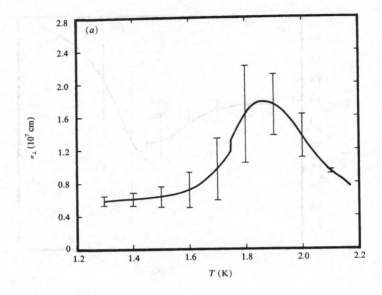

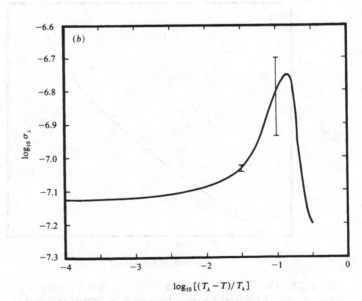

Figure 3.7 (*a*) Transverse scattering length $\sigma_\perp$ in cm versus temperature. (*b*) Transverse scattering length $\sigma_\perp$ versus reduced temperature (after Barenghi *et al.* (1983)).

3.5   The lifetime and range of vortex rings

So far we have used the ideas of vortex dynamics only for rectilinear vortex lines where $v_{sl} = v_s$. Let us now examine an interesting case where the self-induced velocity of the ring is important. This problem is relevant to the vortex ring experiment discussed in Section 2.4.

Assume that the vortex ring is oriented along the z-axis as in Figure 3.8, and assume further that the ring carries a charge e and is in an electric field E. Then in cylindrical coordinates (r, θ, z), the force balance equation now reads

$$2\pi R(\mathbf{f}_M + \mathbf{f}_D) + e\mathbf{E} = 0 \qquad (3.35)$$

with the various components

$$
\left.
\begin{aligned}
\hat{\boldsymbol{\omega}} &= (0, 1, 0) \\
\mathbf{v}_L &= (\dot{R}, 0, v) \\
\mathbf{v}_s &= (0, 0, v_s) \\
\mathbf{v}_n &= (0, 0, v_n) \\
\mathbf{v}_i &= (0, 0, v_i) \\
\boldsymbol{\kappa} &= (0, \kappa, 0) \\
e\mathbf{E} &= (0, 0, eE)
\end{aligned}
\right\} \qquad (3.36)
$$

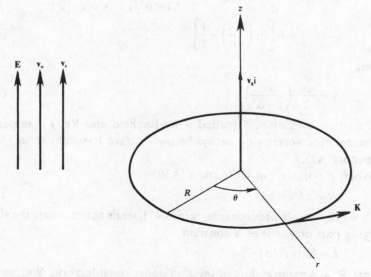

Figure 3.8 Vortex ring oriented along the z-axis. The electric field $\mathbf{E}$ and counterflow velocities $\mathbf{v}_n$ and $\mathbf{v}_s$ are all supposed to be in the z-direction (after Barenghi *et al.* (1983)).

where $\dot{R} = dR/dt$. Resolving the forces in the r- and z-directions we have

$$\left.\begin{aligned} \rho_s \kappa (v - v_s - v_i) - \gamma_0 \dot{R} - \gamma_0'(v - v_n) = 0 \\ \gamma_0' \dot{R} - \rho_s \kappa \dot{R} + eE/2\pi R - \gamma_0(v - v_n) = 0 \end{aligned}\right\} \quad (3.37)$$

which can be rearranged to read

$$\left.\begin{aligned} (\rho_s \kappa - \gamma_0')(v - v_n) - \rho_s \kappa (v_s + v_i - v_n) - \gamma_0' \dot{R} = 0 \\ -\gamma_0(v - v_n) + eE/2\pi R - (\rho_s \kappa - \gamma_0') \dot{R} = 0 \end{aligned}\right\} \quad (3.38)$$

Solving for $\dot{R}$ by eliminating v from (3.38) gives

$$\dot{R} = \alpha(v_n - v_s - v_i) + \frac{1 - \alpha'}{\rho_s \kappa} \frac{eE}{2\pi R} \quad (3.39)$$

where, as we have seen in Equation (1.69)

$$v_i = \frac{\kappa}{4\pi R}\left[\ln\left(\frac{8R}{a}\right) - \frac{1}{2}\right] \quad (3.40)$$

and where we have used relations between (γ_0, γ_0') and (α, α') derivable from Table 3.1. Several interesting results come from (3.39). If an experiment is done in which an electric field is used to prevent the rings from decaying (Rayfield and Reif, 1964), then $\dot{R} = 0$ and the force required is (for the case $v_s = v_n = 0$)

$$F = eE = \frac{2\pi R \rho_s \kappa \alpha}{1 - \alpha'} v_i = \frac{\rho_s \kappa^2}{2}\left(\frac{\alpha}{1 - \alpha'}\right)\left[\ln\left(\frac{8R}{a}\right) - \frac{1}{2}\right]$$

$$= \delta\left[\ln\left(\frac{8R}{a}\right) - \frac{1}{2}\right] \quad (3.41)$$

where

$$\delta = \frac{\rho_s \kappa^2}{2}\left(\frac{\alpha}{1 - \alpha'}\right) \quad (3.42)$$

The friction coefficient δ (called α by Rayfield and Reif) is important in interpreting vortex ring decays below 1 K (see Barenghi *et al.* (1983) Appendix A2).

When $E = 0$, we can write from (3.39)

$$\dot{R} = \alpha(w - v_i) \quad (3.43)$$

with $w = v_n - v_s$. In order to integrate (3.43), let us approximate the slowly varying part of (3.40) as a constant

$$\bar{L} = \ln(8R/a) - \tfrac{1}{2} \quad (3.44)$$

where R is a mean radius of the ring under consideration. We can then define

$$\beta = \kappa \bar{L}/4\pi \quad (3.45)$$

so that (3.40) reads

$$v_i = \beta/R \tag{3.46}$$

then

$$\frac{dR}{dt} = \alpha\left(w - \frac{\beta}{R}\right) = -\frac{\alpha\beta}{R}\left(1 - \frac{wR}{\beta}\right) \tag{3.47}$$

which can be integrated:

$$\int_{R_0}^{a} \frac{R'\,dR'}{1 - R'w/\beta} = -\alpha\beta \int_0^t dt' \tag{3.48}$$

The ring dies (i.e., $R \approx a$) at a certain time $t = \tau$, so that the lifetime τ of a ring of initial radius $R_0 \gg a$ is

$$\tau = \frac{R_0 - a}{\alpha w} + \frac{\beta}{\alpha w^2} \ln\left|\frac{\beta - wR_0}{\beta - wa}\right| \tag{3.49}$$

If the counterflow velocity is small, the logarithm may be expanded to yield

$$\tau = \frac{R_0^2 - a^2}{2\alpha\beta} + \frac{w(R_0^3 - a^3)}{3\alpha\beta^2} + \cdots \tag{3.50}$$

the first term of which is the lifetime in the absence of counterflow, which if a is negligible becomes

$$\tau = \frac{R_0^2}{2\alpha\beta} \tag{3.51}$$

We can also compute the distance a vortex ring will travel in the presence of friction. We have

$$\dot{R} = \frac{dR}{dt} = \frac{dR}{dz}\frac{dz}{dt} = \frac{dR}{dz}v \tag{3.52}$$

where v is the velocity of the vortex ring in the z-direction (once the ion drag force, if any, has been taken into account). In the case $\mathbf{E} = 0$ from (3.38) and (3.45) we have

$$v = (1 - \alpha')(v_i + v_s) + \alpha' v_n \tag{3.53}$$

By substituting (3.53) and (3.45) into (3.52) we obtain

$$dz = -\frac{\beta(1 - \alpha') + (v_s + \alpha'w)R}{\alpha(\beta - wR)}dR \tag{3.54}$$

By integrating (3.54) from $R = R_0$ to $R = a$ we obtain the distance z travelled by the vortex ring of initial radius R_0:

$$z = \frac{v_s + \alpha'w}{\alpha w}(R_0 - a) + \frac{\beta(w + v_s)}{\alpha w^2} \ln\left|\frac{\beta - wR_0}{\beta - wa}\right| \tag{3.55}$$

The results obtained above all use the approximation (3.44). If higher accuracy is required, the integration should be performed numerically. In absence of counterflow ($w = v_n = v_s = 0$) (3.52) becomes

$$\dot{R} = -\alpha v_i = (1 - \alpha') v_i \frac{dR}{dz} \qquad (3.56)$$

and we have exactly (since v_i cancels)

$$z = \frac{\alpha}{1 - \alpha'} (R_0 - a) \qquad (3.57)$$

3.6   Axial mutual friction, the vortex-free strip and metastability

The experiment reported by Hall and Vinen (1956a) on the attenuation of second sound in uniformly rotating helium II showed a substantial anisotropy between the attenuation of second sound propagating radially (Figure 2.7(b)) and axially (Figure 2.7(a)). This anisotropy was taken as strong evidence for the arrangement of vortex lines envisioned in Figure 2.6. Surprisingly, this seemingly straightforward observation has proved to be an enduring mystery: the evidence for zero axial attenuation has never been certain.

In order to address the experiment problem formally, one can generalize the second sound wave Equation (2.38) to include still another friction term B'' which would give rise to attenuation of second sound transmitted parallel to the axis of rotation (see Donnelly, Glaberson and Parks (1967) Section 18):

$$\ddot{q} + (2 - B')\Omega \times \dot{q} - B\hat{\Omega} \times (\Omega \times \dot{q}) + B''\hat{\Omega}(\Omega \cdot \dot{q}) = u_2^2 \nabla^2 q \qquad (3.58)$$

For second sound transmitted in the z-direction, we let

$$q = (q_x, q_y, q_z) \exp(i\sigma t - ikz - \alpha z) \qquad (3.59)$$

where q_x, q_y and q_z are constants, σ is the angular frequency of the second sound, k is the second sound wave number and α is a second sound attenuation coefficient (due to vortex lines only). We also let $\Omega = \Omega(\sin \theta, 0, \cos \theta)$, so that θ is the angle between the transmission (z-direction) and the vortex lines. Then, to second order in Ω/σ,

$$\alpha = (\Omega/2u_2)(B \sin^2 \theta + B'' \cos^2 \theta) \qquad (3.60)$$

In 1966, Snyder and Putney investigated the angular dependence of mutual friction, confirming that B depends on $\sin^2 \theta$, where θ is the angle of tilt of the resonator with respect to the axis of rotation.

Mathieu, Plaçais and Simon (1984) have carefully restudied this problem with a tilting rectangular cavity. They address the important question of the behavior of vortices near the boundaries when $\theta \neq 0$. They advance a model for this behavior based on the fact that vortices must touch a boundary normally (see Figure 3.9). They find that the curvature of the vortices takes place over a characteristic length $d_0 \cong 0.29 \Omega^{-1/2}$ mm (here 0.29 is a dimensional constant). They also examine the mutual friction as a function of angle, reconfirming the Snyder and Putney (1966) result, and giving some evidence for a nonzero value of the mutual friction coefficient parallel to the axis of rotation. Their data are shown in Figure 3.10. The authors avoid interpreting their finite value of B'' in terms of axial mutual friction since they found, as did Snyder and Linekin (1966), that the ratio B''/B can vary over a substantial range. The nature of B'', then, remains an open problem.

We shall show in Section 5.1 that the arrangement of vortices in a uniformly rotating layer of fluid constitutes a triangular lattice. The angular velocity of this array is, as we have suggested in Section 2.4.1 just the angular velocity of the container.

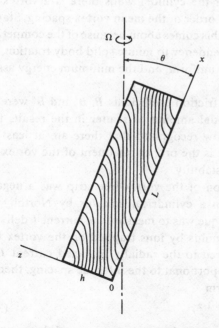

Figure 3.9 Conjectural drawing of vortices in a tilted rectangular second sound resonator (after Mathieu *et al.* (1984)).

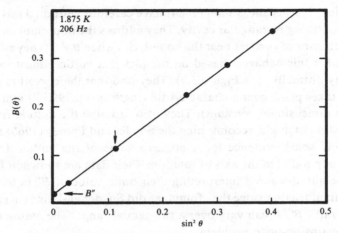

Figure 3.10   Measurements of B as function of $\sin^2\theta$ by Mathieu *et al.* (1984). The intercept marked as B'' is evidence for some axial mutual friction effect not yet completely understood.

There are interesting deviations from a regular triangular vortex array that warrant discussion. Near the cylinder walls there is a vortex-free strip having a thickness of the order of the mean vortex spacing. Stauffer and Fetter (1968) suggest that this comes about because of the competition that takes place between the tendency to mimic solid body rotation, and hence keep the vortex density uniform, and the minimum energy associated with each vortex.

When values of the mutual friction coefficients B, B' and B'' were first reported there was a substantial amount of scatter in the results from different laboratories. It is now recognized that there are at least two reasons for this problem: one is the proper treatment of the vortex-free strip, and the second is metastability.

An early report on detection of the vortex-free strip was a negative ion absorption experiment in a cylindrical annulus by Northby and Donnelly (1970). Their technique was to measure the current I delivered to the top of the rotating annulus by ions trapped on the vortex lines produced by rotation compared to the radially collected current I_0. If there is a vortex-free strip proportional to the interline spacing, then the current ratio will have the form

$$(I/I_0)/\Omega = A(1 - B\Omega^{-1/2}) \tag{3.61}$$

where A and B depend on the trapping cross-section and the geometry. Northby and Donnelly (1970) arranged their collector to study only the vortex-free strip on the outer cylinder. Their data established that the form (3.61) is obeyed, and the results correspond to the absence of two rows of vortices at the outer cylinder instead of the expected single row.

Northby and Donnelly (1970) speculated that their results are consistent with the notion that the liquid may well have on average the equilibrium distribution of vortices, but that the lifetime of the vortices in the outer row is short compared to the time for an ion to migrate to the top collector (a matter of less than a minute).

Second sound is also an important tool for studying the vortex-free strip. Here, of course, the wavelength is long enough that the average properties of the vortex distribution are probed, although some spatial information can be obtained by using different modes (Bendt and Donnelly, 1967; Bendt, 1967). A careful study of the attenuation of second sound in rotating rectangular cavities has been reported by Mathieu, Marechal and Simon (1980). They are able to distinguish three vortex distributions for any given value of the rotation rate Ω: an equilibrium state with N_0 vortices, and two limiting metastable states containing the minimum number N_1 and maximum number N_2 vortices. For $\Omega \leq 10^{-1} \mathrm{s}^{-1}$ they find $\Delta N/N_0 \sim 1$ where $\Delta N = N_2 - N_1$, and for $\Omega \gtrsim 1\,\mathrm{s}^{-1}$ they find $\Delta N/N_0 \leq 10\%$. In the higher velocity range ΔN can be related to a variation in thickness d of the vortex-free strip near the walls: d varies between external values d_1 and d_2 which appear to be independent of the boundary geometry. From free energy considerations, they infer that the equilibrium vortex-free width should be

$$d = (b/2^{1/2})[\ln (b_0^*/a)]^{1/2} \qquad (3.62)$$

where $b^* = e^{-3/4}b$ and $b^2 = \kappa/2\pi\Omega$. They consider Equation (3.62) to be valid for any arbitrary-shaped cylinder with walls parallel to the axes of rotation. They report that the vortex distribution they achieve can be reduced to the equilibrium one with N_0 and d_0 by strong perturbations of the cryostat, or feeding large heat flux into the cavity.

The results reported above make clear why values of B are likely to have been subject to scatter. The experimental measurement of B requires an investigation of the attenuation of second sound over a wide range of Ω and efforts to check against metastability. And of course the value of B depends upon the frequency of the second sound (see Section 3.3).

4 The structure of quantized vortices

4.1 Vortex rings in helium II

We have indicated in Chapter 2 that the core structure of classical vortex rings is not well understood in many cases. Considering how small the vortex core in helium II is, i.e., of order an angstrom, it would seem that one either ought to know how it is constructed, or one ought to find a way to ignore it. Unfortunately neither goal has been achieved. Indeed, today, there is no theory of the structure of quantized vortex lines which can describe the situation at all temperatures from $T = 0$ to $T = T_\lambda$. We shall take note of hints from several directions, including experiment, and we suggest an interim picture to use until more definite information becomes available.

Rayfield and Reif's 1964 experiment was interpreted by means of the solid rotating core model discussed in Section 1.6. The solid rotating core is a constant volume core model and cannot possibly apply to vortices generated by ions which start at angstrom size and expand to micron size. Instead the core radius a is taken as a variable set by the thermodynamic properties of the fluid, i.e., the temperature and pressure of interest. The exact choice of α and β which we have discussed in Section 1.6 does not matter particularly unless one is trying to deduce the core parameter a from experiment. Most classical theories of vortices useful in the study of quantum vortices are appropriate to hollow vortices with no potential energy explicitly associated with the core, i.e., the free energy per unit length of the vortex is specified by the core parameter. Such expressions are Equations (1.59) and (1.60) for vortex rings with $\alpha = 2$, $\beta = 1$ and (1.38) for vortex lines.

Having made these choices Rayfield and Reif's experiment can be reanalyzed assuming κ is known *a priori*, with the result $a = (0.81 \pm 0.08)$ Å (Glaberson and Donnelly, 1986, Section 2.2). Rayfield

Table 4.1. *Temperature dependence of the vortex core parameter (after Glaberson and Donnelly (1986)).*

T (K)	0.28	0.35	0.40	0.45	0.50	0.55	0.60
a (Å)	0.81±0.08	0.77±0.02	0.77±0.02	0.80±0.02	0.79±0.02	0.82±0.02	0.84±0.02

Table 4.2. *Pressure dependence of the relative core parameter at $T = 0.368$ K (after Glaberson and Donnelly (1986)).*

p (atm)	0	3.7	4.4	6.9	7.5	10.5	11.1	13.7	14.6	18.0	20.1	21.0	24.3
a/a_0	1.0	1.06	1.05	1.11	1.10	1.15	1.17	1.16	1.21	1.21	1.23	1.24	1.29

(1968) went on to demonstrate that the vortex core increases with pressure and temperature. The increase in core radius with pressure is surprising in view of the fact that the only comparable physical length, the interatomic spacing, decreases with increasing pressure. Steingart and Glaberson (1972) performed an accurate series of vortex core parameter experiments using an ion time-of-flight technique. They allowed carefully for the effects of drag on the vortex core and ion, which is measureable even at the lowest temperatures. Their data can also be interpreted on the classical model with $\alpha = 2, \beta = 1$ (Glaberson and Donnelly, 1986, Section 2.2). The combined core parameter data of Rayfield and Reif and Steingart and Glaberson are presented in Table 4.1 as a function of temperature and as a function of pressure in Table 4.2.

The astonishingly small size of quantized vortex cores in helium II makes one appreciate that the classical vortex experiments described in Section 1.5 are done in a regime where the vortex cores are a substantial part of the entire flow. In helium II this would correspond either to flows in the extreme quantum limit (for example $ka \to 1$ for helical waves in Figure 6.1), or to the situation very near T_λ where the core size has increased to macroscopic dimensions.

4.2   Vortex rings in a Bose condensate

In Chapter 1 the vortex core is described as being 'hollow' or 'rotating'. These descriptions are clearly inadequate for dealing with the structure of an object whose size is of the order of the interparticle spacing. It is a surprising and rewarding fact in superfluid physics that simple phenomenological concepts taken from the hydrodynamics of an ideal fluid work as well as they do on a microscopic scale. A first principles quantum mechanical description of the vortex core in helium II has yet to be devised.

We have discussed the Bose condensate model, which can have vortex lines and rings in Section 2.8. The energy per unit length (2.85) is

$$E = (\rho_\infty \kappa^2/4\pi)[\ln (b/a) + 0.378] \tag{4.1}$$

and the formulae for thin rings are given by (2.86)–(2.88):

$$E = \tfrac{1}{2}\rho_\infty \kappa R^2[\ln (8R/a) - 1.615] \tag{4.2}$$

$$v = \frac{\kappa}{4\pi R} [\ln (8R/a) - 0.615] \tag{4.3}$$

$$P = \rho_\infty \kappa \pi R^2 \tag{4.4}$$

These formulae are analogous to (1.59)-(1.61) for classical vortex rings and are valid only for small a/R. When a/R starts to grow, corrections such as we have discussed in Section 1.6 and illustrated in Figure 1.13 start to become significant for vortex rings in the Bose condensate.

An investigation of small vortex rings in the condensate was undertaken by Jones and Roberts (1982), with results shown in Figure 4.1. Another view of the results is shown in Figure 4.2, where a comparison is made of thin ring formulae (4.2)-(4.4) and the Jones-Roberts calculations (using $a = 1.25$ Å) with the dispersion curve for elementary excitations.

The Jones-Roberts (1982) investigation, however, revealed a substantial surprise. The authors searched numerically for axisymmetric disturbances that preserve their form as they move through the condensate. A continuous family was obtained whose dispersion curve consists of two branches coming to a cusp which we show in Figure 4.2 for a particular choice of a. The lower branch is (for large enough P) a vortex ring of circulation κ, as $P \to \infty$, its radius $\tilde{\omega} = (P/\pi\kappa)^{1/2}$ becomes infinite and its forward velocity tends to zero. The upper branch has no vorticity and is a rarefaction sound pulse that becomes increasingly one-dimensional as $P \to \infty$; its velocity approaches c for large P. The velocity of any member of the family is shown, both numerically and analytically, to be $\partial E/\partial P$, the derivative being taken along the family (Roberts, 1972). The results (in nondimensional units) are given in Table 4.3. It is interesting to note that the radius of the ring $\tilde{\omega}$ vanishes on the lower branch very near the cusp. The cusp of the dispersion curve is at $E_M = 50.7$, $P_M = 69.6$.

The question of the ultimate fate of quantized vortex rings as they shrink has a long history. Onsager was often quoted as calling a roton 'the ghost of a vanished vortex ring'. His idea was reviewed in detail in the light of knowledge at the time by Donnelly (1974). Feynman (1955) had a similar idea: a large vortex ring at absolute zero will move forward in such a way that the Magnus force on the ring caused by its forward motion just balances the tension (energy per unit length) of the vortex. At finite temperatures, excitations will cause the ring to decay according to the prescription we have given in Section 3.5. Feynman took the view that when the ring has shrunk to atomic size, quantum pressure effects would enter to oppose the shrinking of the ring. 'In a roton we imagine that the forces tending to contract the ring are already opposed by a kind of stiffness of the ring. It is already as small as possible. No drift motion results. In fact forward drift would expand it and raise the energy, while reverse drift would try to compress it to smaller size, again raising

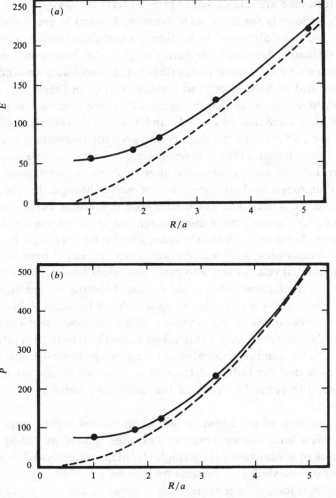

Figure 4.1 Plots of (*a*) the energy *E* and (*b*) momentum *P* of the small quantum rings of Jones and Roberts (1982). The units are dimensionless. The units of *E* are $\rho\kappa^2 a/4\pi^2$ and of *P* are $\rho\kappa a^2/2\pi$ and apply only at $T = 0$. The dashed lines represent Equations (4.2) and (4.3). Comparison of this figure with Figure 1.9 shows that small quantum rings deviate from the large R/a formulae more rapidly than classical rings. The circled points are the calculated results (after Muirhead, Vinen and Donnelly (1984)).

Table 4.3. *Results of Jones and Roberts (1982) for axisymmetric disturbances in the condensate*

U	E	P	m	ω̃
0.4	129.0	233.0	22.6	3.36
0.5	80.7	123.5	13.0	2.31
0.55	66.5	96.5	10.6	1.82
0.60	56.4	78.9	8.97	1.06
0.63	52.3	72.2	8.37	
0.66	50.7	69.6	8.20	
0.68	53.7	74.1	8.80	
0.69	60.0	83.2	9.92	

Here U is the velocity and m is the dipole moment. (The units for this table are $\rho\kappa^2 a/4\pi^2$ for energy, $\rho\kappa a^2/2\pi$ for momentum, $\kappa/2(2)^{1/2}\pi a$ for U, $\kappa a^2/8\pi^2$ for m and $a/2\pi$ for $\tilde{\omega}$.)

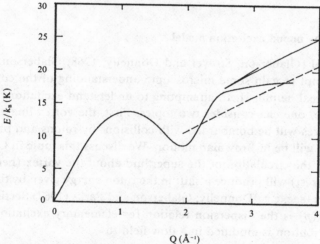

Q (Å^{-1})

Figure 4.2 The axisymmetric solitary wave solutions obtained by Jones and Roberts (1982) compared to the dispersion curve for helium II determined by Donnelly, Donnelly and Hills (1981). The location of the thin core formulae for a Bose condensate (4.2)–(4.4) on the diagram is shown by a dashed line. The core parameter for this plot was taken to be 1.25 Å which puts the energy of the cusp near 2Δ.

the energy. The lowest energy is at zero drift velocity. We may notice in passing that they can drift only perpendicular to their plane, that is, along, or opposite, the direction of the momentum. This agrees with a property derived for rotons from their energy–momentum relation, that the group velocity $\delta\varepsilon/\delta p$ is in the direction of the momentum (or opposite).'

The location of the cusp in the Jones–Roberts calculations can be varied at will by selecting some value of a. Theoretical work on the nucleation of vortex rings by ions by Muirhead *et al.* (1984) and its experimental verification by McClintock's group, all described in Chapter 8, suggests that the cusp may be located as shown in Figure 4.2. This arises because Muirhead *et al.* predicted the barrier opposing nucleation using the calculations shown in Figure 4.1 and a value of a of 1.25 Å. With this choice of core parameter the cusp corresponds to an energy of 17.4 K, or very nearly 2Δ, the energy of two rotons, and the momentum somewhat less than $2P_0$ for rotons. If this idea is correct one can speculate that the fate of a shrinking quantized vortex ring is to decay into two rotons with angle of about 100° between them emitted in the forward direction.

4.3   The bound excitation model

This model (Glaberson, Strayer and Donnelly, 1968; Glaberson, 1969) is an attempt to gain some microscopic understanding of the core of a vortex in real helium II. In attempting to understand excitations near a vortex line, one can consider two topics. First, the vortex line at finite temperatures will be bombarded with collisions by rotons and phonons and hence will be in Brownian motion. We discuss this topic in Chapter 6. Second, the circulation of the superfluid about the vortex (here considered at rest) will produce a shift in the roton energy given by the $\mathbf{p} \cdot \mathbf{v_s}$ interaction (see, e.g., Donnelly, Glaberson and Parks (1967) Section 12). Thus, if $\varepsilon(p)$ is the dispersion relation for elementary excitations, the Bose distribution is modified in a flow field to

$$n(p) = h^{-3}(\exp\{[\varepsilon(p) - \mathbf{p} \cdot (\mathbf{v_n} - \mathbf{v_s})]/k_B T\} - 1)^{-1} \qquad (4.5)$$

The normal fluid is considered to be at rest, on the average, near a vortex core and the superfluid has the velocity $v_s = \kappa/2\pi r$. Values of v_s become so large near a vortex core that the energy of a roton (with Landau parameters Δ, p_0, and μ)

$$\varepsilon(p) = \Delta + (p - p_0)^2/2\mu k_B T \qquad (4.6)$$

becomes depressed, and the number density N_r increases until ultimately the Landau critical velocity v_L is reached and rotons can then be created spontaneously. The situation is shown in Figure 4.3 where the quantities N_r, Δ, and ρ_s/ρ are plotted as a function of distance from the vortex

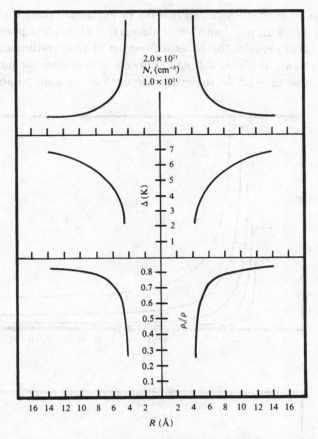

Figure 4.3 Behavior of roton density N_r, roton energy gap Δ, and ρ_s/ρ near the core of a vortex line at $T = 1.6$ K on the bound excitation model (after Glaberson *et al.* (1968)).

core. A careful discussion of this problem was undertaken by Glaberson (1969) who considered the effects of roton-roton interactions which arise when N_r grows and the effects of the uncertainty principle when the roton is localized near the core. Neither of these effects qualitatively changes the expected variation of roton density near the line. At distances shorter than that where the Landau critical velocity is reached, it is assumed that the core is normal, with properties perhaps like helium I. It was assumed that the total density and temperature are constant everywhere. One highly significant result is that the radius of the normal core *increases* with pressure in qualitative accord with experiment (see

Table 4.2) one might have assumed that the radius would scale with the total density, i.e., with $\rho^{-1/3}$ and hence decrease with applied pressure. Glaberson's (1969) results for N_r as a function of temperature and of pressure are shown in Figure 4.4. At the lowest temperature the normal core has a radius of $\sim$2.5 Å, somewhat larger than the core parameter

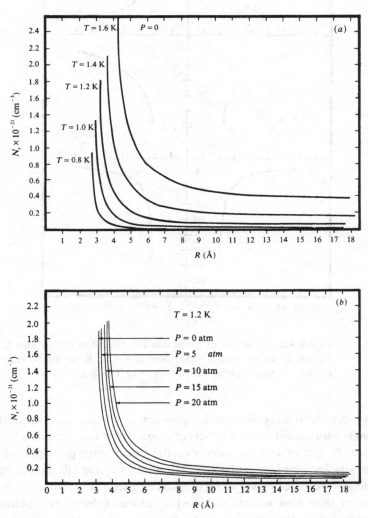

Figure 4.4   (*a*) Behavior of the roton density N_r near a vortex line as a function of temperature at $P = 0$. (*b*) Behavior of the roton density N_r near a vortex line as a function of pressure at $T = 1.2$ K (after Glaberson (1969)).

~0.81 Å quoted in Table 4.1. The radius of the normal core is seen to increase with temperature, and the region of excess roton density (the 'tail') spreads out with increasing temperature. Near T_λ the core radius diverges, and is given approximately by

$$a \to 3.2/(T_\lambda - T)^{1/2}(\text{Å}) \tag{4.7}$$

It is generally believed that the coherence length ξ (Section 2.8) is a measure of the core parameter a and that $\xi \sim \rho_s^{-1}$, and hence according to experiment diverges approximately as $(T_\lambda - T)^{-2/3}$ near T_λ, in conflict with the results of the bound roton model.

4.4   The Hills–Roberts theory

A number of experiments on helium II show evidence that the superfluid density is depleted near solid boundaries. In particular, experiments on the propagation of third sound in thin, unsaturated films of helium II show that the superfluid behaves as if it had a lower areal density than that given by the product of the superfluid density and the film thickness at the same temperature. Rudnick and his collaborators (Rudnick and Fraser, 1970; Scholtz, McLean and Rudnick, 1974) (as well as other groups) have associated this reduction in density with 'healing', the notion that the superfluid density decreases near a boundary introduced in Section 2.8. Hills and Roberts (1977a, b; 1978a, b) have advanced a two-fluid theory which incorporates healing as well as relaxation (the processes which prevent the superfluid fraction from changing instantaneously when the thermodynamic state is altered). Neither healing nor relaxation is incorporated in the Landau two-fluid theory (Section 2.1). The Hills–Roberts theory rests on accepted macroscopic balance laws for mass, momentum and energy together with a postulate for entropy growth. Their theory is entirely hydrodynamical, valid over the entire temperature and pressure range of helium II. The superfluid density is regarded as an independent thermodynamic variable and the development allows for a constitutive dependence on superfluid density gradients. Boundary conditions on the superfluid density are not given *a priori* and some assumption needs to be made. Usually one assumes that ρ_s vanishes at rigid boundaries and at vortex cores. The resulting theory becomes completely determinate once a free energy $A(\rho, T, \rho_s)$ is known.

The details of this calculation are beyond the scope of a book on vortices, but have been reviewed by Glaberson and Donnelly (1986). The Hills–Roberts theory is a two-fluid model, so that while the superfluid density vanishes on a wall in a manner suggested by (2.82), the normal fluid density rises to keep the total fluid density constant.

There are various definitions of 'healing length' possible. Hills and Roberts adopt the idea of a 'displacement thickness' δ, which is defined to be that distance for which $f\delta$ is the superfluid mass (per unit area of wall) 'displaced' from the wall through healing, so that the hypothetical density distribution

$$\rho_s = \begin{cases} 0, & z < \delta \\ f, & z > \delta \end{cases} \left. \right\} \quad (4.8)$$

has the same net superfluid mass as the actual solution. Here f is defined as the bulk superfluid density far from the wall. For the density distribution (4.8), $\delta = D$, a displacement thickness tabulated in Table 4.4.

The Hills–Roberts theory can be used to obtain a vortex core structure, again with the boundary condition that the superfluid density vanishes on the central axis of the vortex. Here the theory of the healing length is qualitatively altered by the superfluid circulation, it is necessary to introduce the concept of a 'dynamical healing length' (see Section 2.8). In this model, the superfluid density goes to zero on the core over a distance smaller than the displacement length of Table 4.4, and is not unlike the Ginzburg–Pitaevskii (1958) solution shown in Figure 2.19. In the Hills–Roberts theory (as in the Ginzburg–Pitaevskii theory), the pressure does not diverge negatively as it does in a classical vortex (see Equation (1.39)). The superfluid density and pressure variations for a Hills–Roberts vortex are shown in Figure 4.5. The vortex core parameter on the Hills–Roberts theory is tabulated in Table 4.5

The vortex core parameter in Table 4.5 can be compared with the experimental data of Tables 4.1 and 4.2. Referring to the result for $\rho = 0.14520$ g/cm^3 appropriate to zero pressure, and interpolating to $T = 0.3$ K, we find $a = 0.946$ Å, which is to be compared with the Rayfield–Reif (1964) result $a = (0.81 \pm 0.08)$ Å. The agreement is satisfactory – perhaps remarkably so – when one considers the difficulties in obtaining absolute magnitudes from experiment as described above in this section. The first row of Table 4.5 shows that the core parameter a increases slowly with temperature, the increase being about 6% between 0.4 and 0.6 K. The data of Table 4.2 show an increase of about 9% in a/a_0 over the same range – thus the data appear to be rising somewhat slower than predicted but the absolute magnitudes and trend are satisfactory.

The low temperature pressure dependence of the core parameter a can be deduced from the first row of Table 4.5, a/a_0 is predicted to increase

Table 4.4. Displacement lengths δ at a plane wall (Å)[a]

Density (g/cm³)	Temperature (K)									
	0.2	0.4	0.6	0.8	1.0	1.2	1.4	1.6	1.8	2.0
0.14520	2.130	2.165	2.241	2.370	2.576	2.890	3.286	3.851		
0.14795	2.162	2.200	2.279	2.411	2.623	2.950	3.363	3.956	4.965	7.300
0.15070	2.207	2.247	2.331	2.472	2.692	3.023	3.454	4.077	5.159	7.829
0.15345	2.261	2.304	1.294	2.565	2.775	3.108	3.561	4.222	5.391	8.535
0.15620	2.320	2.365	2.463	2.625	2.868	3.206	3.683	4.396	5.695	9.552
0.15895	2.380	2.429	2.535	2.710	2.968	3.319	3.824	4.604	6.087	11.18
0.16170	2.440	2.493	2.607	2.796	3.072	3.445	3.986	4.854	6.605	
0.16445	2.502	2.560	2.681	2.886	3.184	3.586	4.174	5.156	7.304	
0.16720	2.568	2.630	2.763	2.983	3.306	3.750	4.397	5.525	8.259	
0.16995	2.641	2.709	2.852	3.095	3.668	3.937	4.664	5.993	9.726	
0.17270							4.989	6.629	12.622	

[a] Courtesy of R. N. Hills.

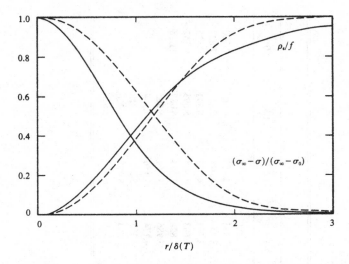

Figure 4.5 The ratio of the superfluid density ρ_s to the bulk density f near a vortex line on the Hills–Roberts theory at two temperatures: $T = 0.2$ K (dashed lines), $T = 2$ K (solid lines). The ratio of local to bulk superfluid density is shown, as is the nondimensional pressure difference $(\sigma_\infty - \sigma)/(\sigma_\infty - \sigma_0)$ with σ_∞ and σ_0 denoting the values of σ at distances far from and on the axis of the vortex. The abscissa is the nondimensional radial distance from the core (after Hills and Roberts (1988)).

by ∼26% going from 0 to 20 atm at $T = 0.4$ K. The data of Table 4.2 show an observed increase of 24% going from 0 to 20.1 atm. Going from 0 to ∼15 atm a/a_0 is predicted to increase by 19% and is observed to increase by 21% going from 0 to 14.6 atm. On the whole, the predictions are in satisfactory accord with low temperature data obtained from vortex ring experiments and emphasize that a does not scale with the average interatomic spacing $\rho^{-1/3}$.

We have remarked that vortex ring core determinations cannot be extended much above 0.6 K because of drag. We would still like to know the vortex core parameter all the way to T_λ for a variety of reasons, including nucleation theories. One method is to find an alternative way to measure δ. Roberts, Hills and Donnelly (1979) attempted to compare their calculations at $p = 0$ to the results of third sound experiments on unsaturated films. Third sound in thin films travels at velocities considerably slower than would be inferred from the film thickness and the expression for the velocity of third sound. This has been interpreted as evidence for healing behavior near the substrate and possibly near the

Table 4.5. *Vortex core energy parameter a (Å)*[a]

Density (g/cm³)	Temperature (K)									
	0.2	0.4	0.6	0.8	1.0	1.2	1.4	1.6	1.8	2.0
0.14520	0.931	0.961	1.014	1.091	1.205	1.372	1.577	1.863		
0.14795	0.945	0.978	1.031	1.110	1.229	1.401	1.614	1.915	2.420	3.851
0.15070	0.966	0.999	1.056	1.140	1.263	1.437	1.659	1.975	2.516	3.842
0.15345	0.991	1.024	1.086	1.175	1.304	1.479	1.711	2.046	2.631	4.191
0.15620	1.017	1.052	1.118	1.214	1.349	1.527	1.772	2.132	2.781	4.695
0.15895	1.043	1.082	1.152	1.255	1.398	1.582	1.841	2.235	2.975	5.506
0.16170	1.070	1.113	1.186	1.297	1.450	1.645	1.921	2.359	3.231	
0.16445	1.098	1.143	1.222	1.341	1.504	1.715	2.014	2.508	3.577	
0.16720	1.124	1.175	1.261	1.388	1.565	1.794	2.124	2.691	4.051	
0.16995	1.158	1.211	1.303	1.443	1.634	1.886	2.257	2.923	4.776	
0.17270							2.417	3.238	6.215	

[a] Courtesy of R. N. Hills.

free surface. The authors showed that quantitative agreement with predicted displacement lengths was best assuming healing at both the
substrate and free surfaces.

Another technique is the measurement of fourth sound in porous
media. When the powder size is sufficiently small, healing behavior

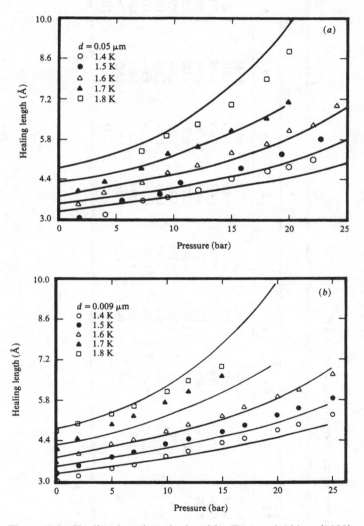

Figure 4.6 Healing lengths calculated by Tam and Ahlers (1982)
from the fourth sound velocity data of (*a*) Tam and Ahlers (1982)
and (*b*) Heiserman *et al.* (1976). The solid lines represent the
displacement length calculation of R. N. Hills (Table 4.4).

influences the index of refraction of the porous medium which can be studied as a function of both temperature and pressure. Tam and Ahlers (1982) have made such a study for a number of fourth sound cavity packings and have interpreted their results in terms of a displacement length. They have also reanalyzed some data of Heiserman, Hulin, Maynard and Rudnick (1976). Two sets of data for 0.05 μm and 0.009 μm data are shown in Figure 4.6. The data cover the range 1.4–1.8 K and 0–25 bar and are compared to the results of Table 4.4. The agreement with the calculations of Roberts *et al.* (1979) is seen to be quite good except possibly at the highest T and p where the healing length becomes large. This is the only evidence we have to date that vortex structure is understood above 1 K.

The situation is even less satisfactory as temperatures approach T_λ. There, measurements of coherence and healing lengths are matters of considerable current interest. What little is known is summarized briefly by Glaberson and Donnelly (1986) Section 2.6.

4.5   Interactions of ions and of vortices

Much of what we know about vortices is information obtained from experiments using ions as probes. Negative ions are electrons in relatively large $(R \sim 16 \text{ Å})$ bubbles cut out of the liquid owing to the repulsive electron–helium atom interaction. Positive ions are smaller (~ 6 Å) solid objects held together by electrostrictive forces. In the vicinity of a vortex line an ion does not experience any azimuthal drag from the superfluid, but it does experience the pressure gradient shown in Figure 2.5. The pressure gradient near a vortex is given by Equation (1.39) and hence a small impurity of volume V will experience an inward force due to the pressure gradient given by $-V \mathrm{d}p/\mathrm{d}r$, and thus varies inversely as the cube of the distance from the vortex line. This estimate is modified if the flow about the impurity is properly allowed for. A discussion of this modification has been given by Donnelly *et al.* (1967) Section 24. A general discussion of the determination of the equation of motion of a sphere immersed in a superfluid containing an arbitrary configuration of quantized vortex lines is given by Paintin, Vinen and Muirhead (1990). The equation of motion is in a form suitable for use with numerical computations of the way in which vortex lines interact with a moving sphere, and therefore is particularly useful in calculations of the behavior of ions and quantized vortices in various configurations. Clearly either species of ion (and indeed ^{3}He impurities) will be pushed into a vortex core by such a large pressure gradient. Another way to appreciate the magnitude of the force is to consider the potential energies involved. When an ion is situated symmetrically on the core, it supplants a substantial volume of high velocity circulating superfluid: recall that the ionic radii are both considerably greater than the core size, which is of order 1 Å. Since the ion displaces fluid with high kinetic energy, one refers to the resulting lowering of the energy of the system as a 'substitution energy.' When an ion of radius R is on the vortex it can be shown that this classical substitution energy is given by (Donnelly *et al.*, 1967; Donnelly and Roberts, 1969)

$$u(0) = -2\pi\rho_s(\hbar/m)^2 R[1 - (1 + a^2/R^2)^{1/2} \sinh^{-1}(R/a)]$$

and off axis

$$u(r) = -2\pi\rho_s(\hbar/m)^2 R[1 - (r^2/R^2 - 1)^{1/2} \sin^{-1}(R/r)]$$

$$(4.9)$$

for $r \gg a$. The resulting potential energy for a negative ion is shown in Figure 4.7(a). The use of the concept of healing in (4.9) is essential. Assuming a hollow core the potential of an ion exhibits unphysical behavior when the ion touches the core. The general expression $u(r)$ was obtained using a healing model of the vortex core (Donnelly and Roberts, 1969, p. 532). We shall comment later on the applicability of (4.9) in the case of a solvated ^{3}He atom.

The very earliest rotating experiments with ions and quantized vortices (Careri, McCormick and Scaramuzzi, 1962) showed that negative ions are captured by quantized vortex lines and positive ions are not. These experiments were done at 1.37 K. If a beam of negative ions is sent across a rotating bucket of helium II, the ion current I diminishes according to

$$dI/I = -n_0\sigma \, dx \qquad (4.10)$$

so that by (2.7)

$$I = I_0 \exp(-2\Omega\sigma x/\kappa) \qquad (4.11)$$

where I_0 is the current in the ion beam for $\Omega = 0$ and σ is the capture diameter for a single vortex line. The magnitude of σ is $\sim 10^{-5}$ cm and decreases with increasing field. Furthermore, above 1.7 K no trapping is observed at all. We shall show that these magnitudes, which at first were very mysterious, have a simple explanation in stochastic theory.

It was suggested by Donnelly (1965) that the behavior of ions in liquid helium might be discussed in terms of Brownian motion in potential wells (Chandrasekhar, 1943). The potential wells are produced by vortices in the superfluid as indicated above, and the Brownian motion of the ions (at temperatures well above 1 K), buffeted by rotons and phonons,

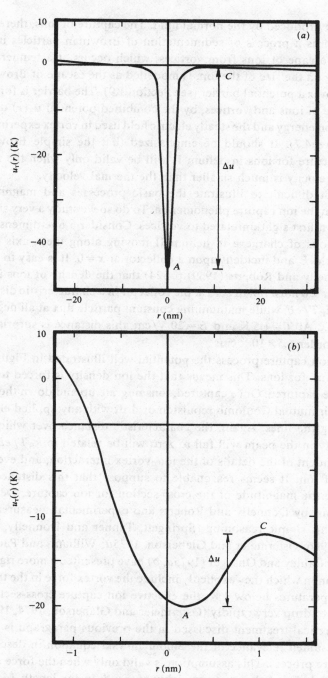

Figure 4.7 Potential energy wells for (*a*) the negative ion and (*b*) the positive ion (after Parks and Donnelly (1966)).

is an effect produced by the normal fluid. The capture of ions, therefore, is pictured as a process of sedimentation of Brownian particles into a well. The escape of ions from vortices, which occurs at a temperature depending on the size of the ion, is modelled as the escape of Brownian particles over a potential barrier (see Section 1.9). The barrier is formed, in the case of ions and vortices, by the combined potential $u_T(r)$ due to substitution energy and the steady electric field used in vortex experiments (see Figure 4.7). It should be emphasized that the simple Brownian motion picture for ions in helium II will be valid only when the mean ion drift velocity is much smaller than the thermal velocity.

It is not difficult to illustrate the basic processes and magnitudes involved in the ion capture phenomenon. To do so we study a very simple problem, at first sight unrelated to vortices. Consider a one-dimensional beam of ions of charge e in helium II moving along the x-axis in an electric field E and incident upon a collector at $x = 0$. It is easy to show (see Donnelly and Roberts (1969) p. 524) that the density of ions in the beam for $x < 0$ must fall to zero at the collector in a characteristic distance of order $k_B T / eE$, while maintaining constant particle flux at all negative values of x. At $T = 1.5$ K and $E = 20$ V/cm, this distance is surprisingly large: of order 6.5×10^{-6} cm.

In the ion capture process the potential well illustrated in Figure 4.7 forms a sink for ions. This means that the ion density is forced to zero as ions are captured. Once captured, ions migrate up and down the lines under their mutual Coulomb repulsion or drift with any applied electric field along the lines. Again, the characteristic distance over which the ion density in the beam will fall to zero will be related to $k_B T / eE$ and is independent of the details of the ion–vortex interaction, and even the species of ion. It seems reasonable to suppose that this distance will determine the magnitude of the cross-section for ion capture. Detailed calculations by Donnelly and Roberts and experimental measurements confirm this simple reasoning (Springett, Tanner and Donnelly, 1965; Tanner, 1966; Ostermeier and Glaberson, 1975a; Williams and Packard, 1978). McCauley and Onsager (1975a, b) have presented a more rigorous calculation in which they explicitly include the vortex force in the theory.

At temperatures below 1 K, the effective ion capture cross-section is observed to drop very rapidly (Ostermeier and Glaberson, 1974, 1975a). The theoretical treatment discussed in the previous paragraph is based on the assumed relevance of the Smoluchowski equation in describing the capture process. This assumption is valid only when the force acting on the ion does not change appreciably over a diffusion length L_D. This condition is obeyed as long as the ion remains a distance r,

$$r \gg L_D = (M_i \mu / e)(2k_B T / M_i)^{1/2} \tag{4.12}$$

from the vortex core. Here M_i, μ and e are the ion mass, mobility, and charge respectively. If an ion finds itself thermalized within a distance r_T of the vortex, where r_T is the radius within which the vortex potential is less than $-k_B T$, the ion will be effectively trapped. The validity of the Smoluchowski equation is therefore ensured by having $r_T \gg L_D$. This inequality is grossly violated at temperatures below ~ 1 K. Ostermeier and Glaberson (1974, 1975a) have performed a 'Monte Carlo' calculation in which ions were permitted to move ballistically in the ion–vortex potential, suffering random collisions with rotons. Individual ion–roton scattering events were taken into account by assuming isotropic scattering and adjusting the frequency of collision to give the correct mobility. This approach yielded excellent agreement with both the temperature dependence and the electric field dependence of the capture cross-section.

The thermal activation of ions out of their potential well and over the barriers of Figure 4.7 depends very much on the depth of the well, and this depth in turn depends on the substitution energy and hence the size of the ion through R and R/a. Again a simple one-dimensional solution illustrates the physics involved. The probability of escape of a Brownian particle over a potential barrier is given by the expression (1.114) with V now called U:

$$P = (\omega_A \omega_C / 2\pi\beta) \exp(-U/k_B T) \tag{4.13}$$

where ω_A and ω_C are curvature parameters describing the characteristics of the well and the barrier, β is a diffusivity, U is the potential difference between well and barrier. The exponential term dominates the total probability and hence reflects the primary role of $U/k_B T$ as the most influential property of the escape process: U, of course, depends strongly on the radius of the ion. In fact, this characteristic of the escape enabled Parks and Donnelly (1966) to estimate the radii of the positive and negative ions from observations of the temperature above which ions cannot remain trapped for appreciable times (a temperature they call the 'lifetime edge'). Shortly afterwards Springett and Donnelly (1966) used the same phenomenon to show how the radius of the negative ion changes with applied pressure. Other related investigations for both species of ion are reported by Douglass (1964), Ostermeier and Glaberson (1975b), Williams, Deconde, and Packard (1975) and Williams and Packard (1978).

Measurements of the escape of ions from vortex rings in very high electric fields (see Figure 4.7) (Cade, 1965; Johnson and Glaberson, 1974) yield consistent results but, because the ion–vortex potential well is strongly distorted by the electric field, the data are more difficult to interpret. More serious is the use of the crude assumption that the vortex line remains perfectly stationary during the escape process, and the neglect of a proper quantum mechanical treatment of the problem for particles of atomic dimensions. We shall return to this latter problem in the case of trapped ^{3}He below. The data which emerge from these measurements indicate the rough size of the trapped ion, and involve the core parameter a but do not provide a determination of the detailed core structure.

4.6   The mobility of ions along vortices

Ions trapped on vortex lines, pushed along the lines by an electric field, experience considerably more drag than is experienced by free ions moving through the bulk liquid. In a very simple sense, this implies that the vortex cores are more 'normal' than the bulk.

The experimental cell used by Ostermeier and Glaberson (1976) is shown in Figure 4.8. The vortex lines are charged with ions in a trapping region at the bottom of the rotating cell. A small amount of charge is then gated into the uniform electric field region and the time-of-flight to the collector at the top is determined. Figure 4.9 is a plot of the inverse low field mobility as a function of inverse temperature for negative and positive ions. The data at temperatures above 0.8 K are in good agreement with earlier results (Douglass, 1964; Glaberson *et al.*, 1968). The solid lines labelled 3 and 4 in the figure are the results of calculations of drag on the ion arising from the scattering of thermally excited vortex waves (Fetter and Iguchi, 1970). Curve 1 is a calculation of ion drag associated with the scattering of bound rotons (Glaberson, 1969). At higher temperatures, there is order of magnitude agreement with all of these approaches but there is serious disagreement at low temperatures. Addition of a small amount of ^{3}He impurities to the system restores agreement with the vortex wave theories at low temperature. It may be that, in the absence of ^{3}He, the vortex line is not sufficiently damped so that some instability plays a role.

The high electric field mobility data of Ostermeier and Glaberson (1975*d*, 1976) have an interesting feature which may be related to the vortex core structure. Figure 4.10 shows the ion velocity as a function

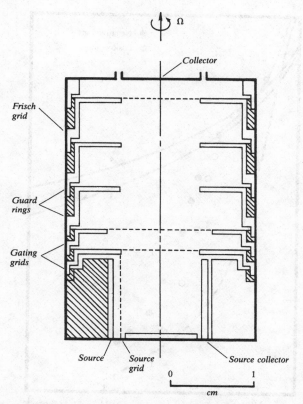

Figure 4.8 A schematic of the experimental cell (after Ostermeier and Glaberson (1976)). The cross-hatched areas indicate insulating surfaces. All other areas and grid are gold-plated stainless steel.

of electric field. The ion velocity saturates in high fields at some nearly temperature-independent velocity that depends on the sign of the ion. The values are ~1100 cm/s for the negative ion and ~1600 cm/s for the positive ion. Making an explicit assumption regarding the superfluid distribution in the vortex core one can obtain the curvature of the harmonic potential that binds the ion to the vortex. The authors suggest that the frequencies corresponding to the harmonic potential well determine characteristic ion velocities beyond which the ions cannot easily be pushed. It is suggested that resonant generation of helical vortex waves (see Section 6.1.1), and therefore substantial drag, occur when two conditions are satisfied:

$$\omega(k_r) - k_r v_L = \mathrm{Re}\ \Omega \tag{4.14}$$

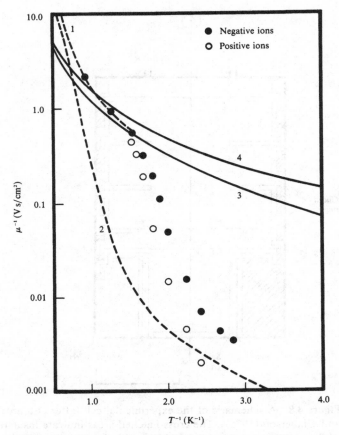

Figure 4.9 Inverse ion mobility in low electric fields as a function of inverse temperature for positive and negative ions. The lines are discussed in the text (after Ostermeier and Glaberson (1976)).

$$v_L = \partial \omega(k_r)/\partial k_r \qquad (4.15)$$

where $\omega(k_r)$ is the frequency of a vortex wave of wave vector k_r and Re (Ω) is the real part of the natural frequency of the ion in its potential well. The first condition states that, at the limiting ion velocity v_L, the frequency of a vortex wave of wave vector k_r (in the frame of reference of the moving ion), is the same as the natural ion frequency. The second condition is that the group velocity of the vortex wave is the same as the ion velocity. These two conditions yield a unique ion velocity that depends on Ω. Assuming that $\rho_s(r)$ in the vicinity of the core is that of a weakly interacting Bose gas, the equations yield values for v_L, for the two ionic

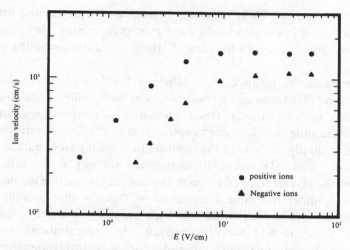

Figure 4.10 The electric field dependence of trapped ion drift velocity at $T = 0.415$ K (after Ostermeier and Glaberson (1976)).

species, about a factor of 3 too large. The authors suggest a number of possible explanations for the discrepancy including uncertainties with respect to the calculation of the ion–vortex interaction potential.

4.7 ^{3}He condensation onto vortex cores

Rent and Fisher (1969) and Ohmi and coworkers (Ohmi, Tsuneto and Usui, 1969; Ohmi and Usui, 1969) predicted that, in ^{3}He–^{4}He solutions, ^{3}He condensation or phase separation of the ^{3}He-rich phase into vortex cores should occur at sufficiently low temperatures. A number of experiments provided indirect evidence for the existence of such a condensation. Ostermeier and Glaberson (1975a), in measuring the capture cross-sections of ions by vortex lines, observed no measureable trapping for the bare positive ion in a 1% solution. According to stochastic theory, significant capture should have occurred at temperatures below about 0.6 K, the positive ion thermal lifetime edge. It was suggested by the authors that their observations might be accounted for by the condensation of ^{3}He onto the core with a resulting increase in the core parameter and thus a reduction of the thermal lifetime edge to below their lowest accessible temperature. A similar mechanism was also proposed by Williams et al. (1975) who observed a lack of positive ion trapping in a similar solution down to 0.1 K. Williams and Packard (1978) measured the thermal trapping lifetime of positive ions on vortex lines in dilute

^{3}He-^{4}He solutions. Their observations indicated a decreasing lifetime and decreasing ion–vortex binding energy with increasing ^{3}He concentration, also consistent with the idea of ^{3}He condensation onto the vortex cores.

Ostermeier, Yarmchuck and Glaberson (1975) and Ostermeier and Glaberson (1976) obtained measurements of the mobility of ions trapped on vortex lines in various ^{3}He-^{4}He solutions at temperatures down to 0.3 K. Assuming that the vortex contribution to the drag experienced by the ion is simply additive to the contributions arising from bulk thermal excitations and ^{3}He atoms, the important quantity is the difference between the inverse trapped ion mobility and the inverse free ion mobility measured under the same circumstances. Plots of this quantity as a function of inverse temperature for both positive and negative ions are shown in Figure 4.11 for various fixed ^{3}He concentrations. At temperatures above some concentration-dependent critical temperature, the data fall along a common curve, in good qualitative and reasonable quantitative agreement with the vortex wave drag theory of Fetter and Iguchi (1970). Comparable agreement with the data can be obtained using the other drag theories.

The sharp increase in the vortex contribution to the ion drag, for both the negative and positive ions, is identified with the onset of ^{3}He condensation onto the vortex core. Figure 4.12 contains a plot of the ambient ^{3}He concentration as a function of the critical temperature at which the vortex drag deviates from its universal high temperature behavior. The line in the figure is the result of a calculation which predicts the onset of condensation of ^{3}He atoms onto the vortex cores, in excellent agreement with the data. In the presence of a vortex line, it is argued that the radial dependence of the ^{3}He number density is given by

$$n(r, T) = n_\infty \exp\left[-U(r, T)/k_B T\right] \qquad (4.16)$$

where n_∞ is the ambient ^{3}He number density and $U(r, T)$ is an effective hydrodynamic potential related to the kinetic energy of the superfluid displaced by a ^{3}He atom. ^{3}He condensation is associated with the number density on the core $n(0, T)$ becoming equal to the critical density measured at phase separation in bulk solutions. The quantity $U(r, T)$ is obtained using the formalism of Parks and Donnelly (1966) for the binding energy of a sphere of radius R to the vortex line. The effective hydrodynamic radius of an ^{3}He atom, R_3, is determined from its measured hydrodynamic mass m_3^*:

$$m_3^* = m_3 + \tfrac{2}{3}\pi R_3^3 \rho \qquad (4.17)$$

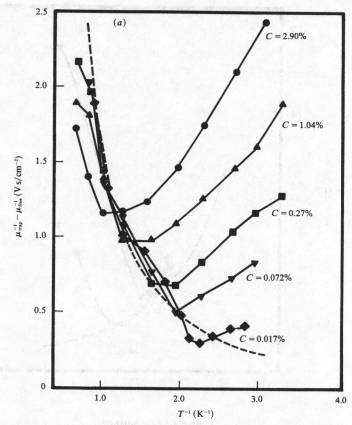

Figure 4.11 The vortex contribution to the trapped ion inverse mobility as a function of inverse temperature. The dashed lines are approximate fits of the Fetter-Iguchi thermal vortex wave theory to the higher temperature data: (a) negative ions; (b) positive ions (after Ostermeier and Glaberson (1975c, 1976)).

where m_3 is the bare ^{3}He atomic mass. Numerous objections can be raised against this very simple macroscopic treatment of the problem, but the good agreement between the data and the calculation, with no adjustable parameters, suggests that the essential physical ideas are correct.

Carrying this simple theory even further, Ostermeier and Glaberson (1976) have attempted to explain their observed ion drag at temperatures below the critical temperature for ^{3}He condensation. Figure 4.13 is a plot of their measured vortex contribution to the ion drag as a function of

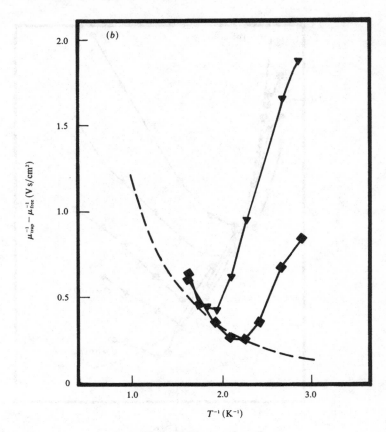

Figure 4.11—continued.

reduced temperature below T_c. In their analysis, they picture a phase separation of the ^{3}He–^{4}He solution into a ^{4}He-rich phase and a ^{3}He-rich phase about the vortex line. The radius of the ^{3}He-rich phase grows continuously from zero as the temperature is lowered below T_c. Calculations of ion mobility, in the circumstances considered, are very involved (Huang and Dahm, 1976) and to a large extent speculative. Nevertheless, the calculated ion inverse mobilities, shown in Figure 4.14, are in satisfactory agreement with the data. The calculated ^{3}He-rich core radii, developed in the calculation, are shown in Figure 4.15.

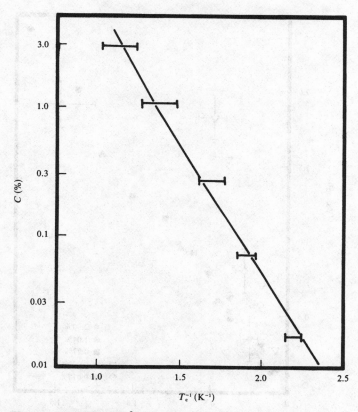

Figure 4.12 Ambient ³He concentration, *C*, as a function of the critical temperature at which ³He condensation onto vortex lines occurs (after Ostermeier and Glaberson (1975c, 1976)).

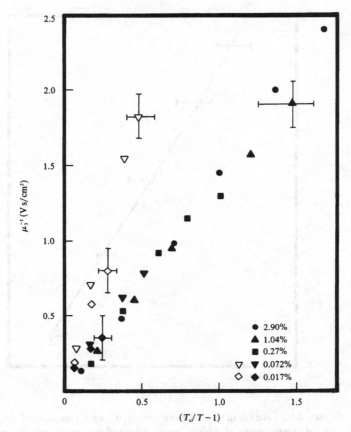

Figure 4.13 The condensed ^{3}He contribution to ion drag along vortices, for positive (open symbol) and negative (closed symbol) ions, as a function of reduced inverse temperature (after Ostermeier and Glaberson (1975*c*, 1976)).

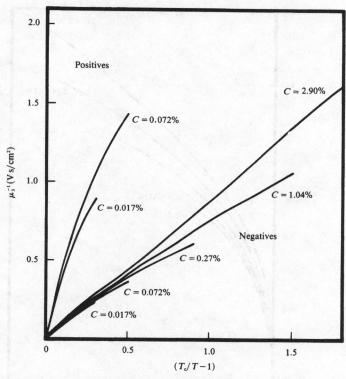

Figure 4.14 The calculated condensed ³He contribution to ion drag along vortices, for positive and negative ions, as a function of reduced inverse temperature (after Ostermeier and Glaberson (1975c, 1976)).

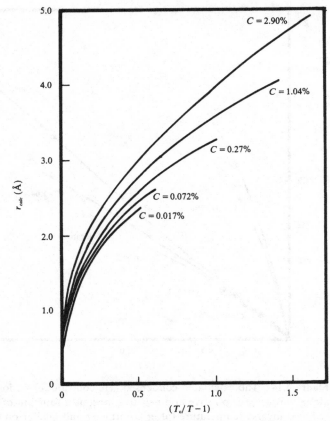

Figure 4.15   The calculated ^{3}He-rich vortex core radius as a function of reduced inverse temperature for various ambient ^{3}He concentrations (after Ostermeier and Glaberson (1975c, 1976)).

5 Vortex arrays

We saw in Chapter 2 how one could begin to understand how vortices are arranged in a simple experiment, namely uniformly rotating helium II. We have some ideas on the dynamics of vortices from Chapter 3 and their internal structure from Chapter 4. Now we turn our attention to a series of sophisticated problems in rotating helium and apply our knowledge to understanding how these work in considerable detail.

5.1 Vortex arrays in a rotating bucket

We have considered the appearance of the first vortex in a rotating bucket in Chapter 2, as well as the general behavior at high rotation rates. In a more sophisticated investigation Tkachenko (1966) showed that in the limit of very high rotation speed the equilibrium configuration of vortices approximates a regular triangular array in a region of space not too far from the axis of the cylinder. Under these circumstances, the energy of the fluid can be written as

$$E = \int \tfrac{1}{2}\rho_s v_{\text{total}}^2 \, d^2 r = \int \tfrac{1}{2}\rho_s(\bar{v}_s + v_{\text{local}})^2 \, d^2 r$$

$$= \int \tfrac{1}{2}\rho_s \bar{v}_s^2 \, d^2 r + \int \tfrac{1}{2}\rho_s v_{\text{local}}^2 \, d^2 r + \int \rho_s \bar{v}_s \cdot v_{\text{local}} \, d^2 r \qquad (5.1)$$

For a uniform array of vortices, having an areal density n_v, it is easy to show that

$$\bar{v}_s = \Omega_v \times r \qquad (5.2)$$

where $\Omega_v = (n_v/2)\kappa$. It follows that

$$E = \tfrac{1}{2}I\Omega_v^2 + \Omega_v \cdot \int \rho_s r \times v_{\text{local}} \, d^2 r + \int \tfrac{1}{2}\rho_s v_{\text{local}}^2 \, d^2 r \qquad (5.3)$$

where $I = \int \rho_s r^2 \, d^2 r$ is the classical moment of inertia per unit length of the fluid. The last two integrals can be identified with the angular momentum and energy associated with the vortices:

$$E = \tfrac{1}{2} I \Omega_v^2 + \Omega \cdot \int n_v L_v \, d^2 r + \int n_v E_v \, d^2 r \tag{5.4}$$

where E_v, the energy per unit length of a vortex, is

$$E_v \sim (\rho_s \kappa^2 / 4\pi) \ln (b/a) \tag{5.5}$$

and L_v, the angular momentum per unit length of a vortex, is

$$L_v = (\kappa \rho_s b^2 / 16) l_z \tag{5.6}$$

and where the mean interline spacing is $b = (\kappa / 3^{1/2} \Omega_v)^{1/2}$. The corresponding expression for the angular momentum of the fluid is

$$L = I \Omega_v + \int n_v L_v \, d^2 r \tag{5.7}$$

The free energy is then

$$E - L\Omega = I \Omega_v (\tfrac{1}{2} \Omega_v - \Omega) + \int d^2 r \, n_v [E_v + L_v (\Omega_v - \Omega)] \tag{5.8}$$

and since for a large cylinder the second term is much smaller than the first, it follows that

$$\Omega_v = \Omega \tag{5.9}$$

On the average, therefore, the superfluid mimics solid body rotation at the rotation speed of the container. This same result follows from the requirement that in equilibrium there can be no dissipation so that the vortices must move with the normal fluid velocity.

An interesting effect comes about because of the incompatibility of triangular symmetry and the circular symmetry of the rotation field. Campbell and Ziff (1978, 1979) have pointed out that significant distortions of the triangular lattice occur at substantial distances from the outside edge of the array. They discuss the circular distortion in terms of a destabilizing velocity, i.e., the difference between the velocity induced at a vortex in a triangular array and that corresponding to solid body rotation. The destabilizing velocity typically varies as $(r/R)^5$ and does not seem to decrease as the vortex density is increased.

Several of the detailed vortex distributions predicted by Campbell and Ziff are shown in Figure 5.1 for 18 vortex lines. Only one of these, 18_1, is the lowest free energy configuration, the other configurations being only local free energy minima (numbers 2, 3, and 7) or free energy saddle points (numbers 4, 5, and 6). Figure 5.2 shows two predicted patterns

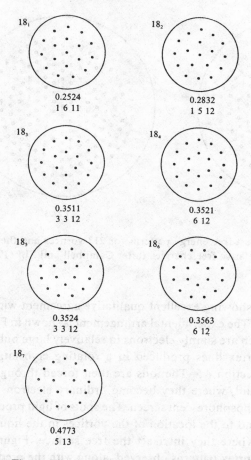

Figure 5.1 Seven stationary patterns of 18 vortices. Immediately below each pattern is the corresponding value of Δf_0 - the free energy of the pattern relative to what it would have been as a component of an infinite array. Also indicated are the number of vortices in each circular ring in the patterns (after Campbell and Ziff (1979)).

for 217 vortices wherein the circular distortion can be readily observed. In the calculations just discussed, the effects of the boundary are ignored and are presumed to be unimportant except very near the boundary.

5.2   Photographs of vortex arrays

Williams and Packard (1974, 1978), Yarmchuck, Gordon and Packard (1979) and Yarmchuck and Packard (1982) have obtained 'photographs'

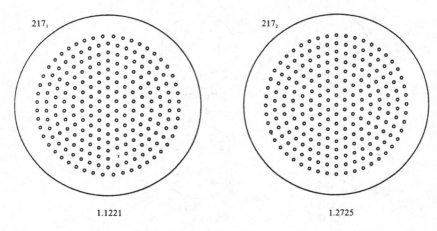

Figure 5.2   Lowest free energy patterns for 217 vortices and their corresponding relative free energies (after Campbell and Ziff (1979)).

of vortex distributions showing excellent qualitative agreement with the predicted distributions. The experimental arrangement is shown in Figure 5.3. Negative ions, which are simply electrons in relatively large bubbles, are trapped on the vortex lines produced in a rotating cryostat, in a manner discussed in Section 4.5. The ions are then forced through the free surface of the liquid, where they become ordinary electrons, and are accelerated onto a phosphorescent screen. The spots of light produced on the screen correspond to the location of the vortices in the liquid by indicating the points where they intersect the free surface. Figure 5.4 shows two different 6-vortex patterns observed, along with the predicted patterns. The 'S' pattern corresponds to an absolute minimum free energy whereas the 'MS' pattern corresponds to a metastable configuration. The agreement is reasonably good but it should be mentioned that there are some noticeable discrepancies. The patterns are invariably slightly distorted from those predicted. Furthermore, whereas the interline spacing is frequently within ~5% of its predicted value, some of the data yield values differing by as much as 30%. Many observations of the mean separation between the vortices in a two-line pattern yield values averaging about 10% smaller than predicted. It is suggested that vortex pinning at the bottom or side surfaces of the experimental cell may play a role. The vortices are then not perfectly rectilinear and some discrepancies of the sort observed are to be expected. Interesting oscillation modes observed in the vortex distributions will be discussed later.

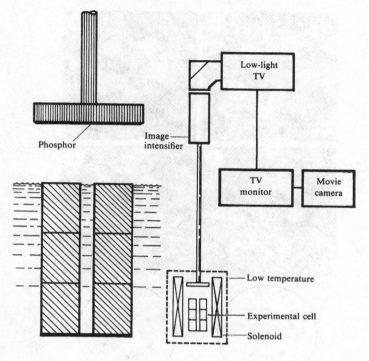

Tritium source

Figure 5.3 A block diagram of the apparatus. The cylindrical vessel is a 2 mm diameter hole drilled in a stack of three carbon composition resistors. It is 22.7 mm in height. Voltage differences applied across each resistor section produce axial electric fields for the manipulation of the ions. A tritiated titanium foil forms the bottom surface of the vessel and serves as the ion source. A 700 V potential difference is applied between the phosphor screen and the top of the vessel for acceleration of the electrons. The solenoid produces a magnetic field of 0.5 T, which prevents defocusing of the accelerated electrons (after Yarmchuck *et al.* (1979), Yarmchuck and Packard (1982)).

5.3   Vortex arrays in a rotating annulus, stability of the flow

5.3.1   *Entry of vortices*

We have discussed the entry of vortices in a rotating annulus in Section 2.6. We now examine the more general problem of arbitrary rotations of the boundaries of the annulus.

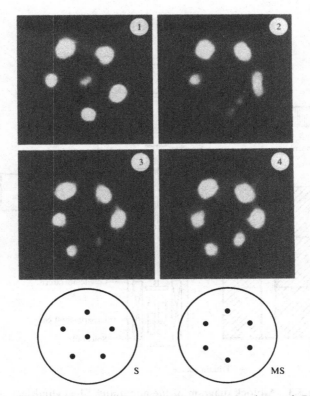

Figure 5.4 Two different 6-vortex patterns observed. Below these are the predicted stable (S) and metastable (MS) patterns (after Yarmchuck and Packard (1982)).

A direct experimental test of Feynman's rule (2.17) was provided by Bendt (1966, 1967*a*). Feynman's rule states that the vortex line density is given by the ratio of the vorticity in the normal fluid to the quantum of circulation. Bendt constructed a highly precise pair of concentric cylinders of radii $R_1 = 2.722$ cm, $R_2 = 3.333$ cm with a gap $d = 0.6105$ cm, which was reduced to $d = 0.608 \pm 0.002$ cm with aquadag coatings used for generation and detection of second sound (Figure 5.5).

Recall from Section 1.8 that the laminar flow between concentric rotating cylinders (see Figure 1.20) is given by Equation (1.91) with the coefficients A and B given by (1.92) and (1.93). The vorticity for this flow $\omega = 2A$ (1.95) can be readily adjusted by varying Ω_1 and Ω_2.

Bendt's cylinders were carefully chosen so that $(R_2/R_1)^2 = 1.50$. Thus by arranging to rotate either cylinder and by constructing a special gearing

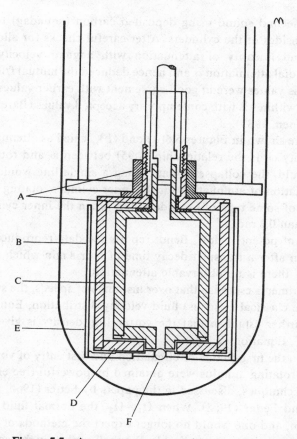

Figure 5.5 Apparatus constructed by Bendt (1966) for examining the flow of helium II between rotating cylinders by means of the attenuation of second sound. The bearings A are made of Kel-F, and the rotating cylinders are B and C; D marks the holes drilled in the lower plate to release heat generated by the bias currents to the aquadag transducers painted on the cylinders; E is an aluminum shield and F a ball bearing.

which makes $\Omega_1/\Omega_2 = 1.50$ he was able to produce four flows of different vorticity ω:

(1) solid body rotation, $\Omega_1 = \Omega_2$, $\omega = 2\Omega_1$
(2) $\Omega_2 = 0$, $\omega = 4\Omega_1$
(3) $\Omega_1 = 0$, $\omega = 6\Omega_2$
(4) potential flow, $\Omega_1/\Omega_2 = 3/2$, $\omega = 0$

Bendt determined the relative vortex line density by measuring the

attenuation of second sound using deposited carbon (aquadag) transducers on the insides of the cylinders. After careful checks for allowed
power input and linearity of attenuation with angular velocity, he
measured the radial attenuation α and hence deduced the mutual friction
coefficient B. His values were in good agreement with earlier values and
indeed agree to within 9% with contemporary accepted values (Barenghi,
Donelly and Vinen, 1983).

The results are shown in Figures 5.6(a) and (b), scaled as attenuation
α versus vorticity ω. If the relationship (1.95) between ω and rotation
rates did not hold, the collapse of data along a single line would not
occur. The deviation of attenuation for the inner cylinder rotating indicates the onset of some type of secondary flow when the inner cylinder
rotates faster than 0.4 rad/s.

For the case of potential flow, Bendt reports the data reproduced in
Figure 5.7 taken after a transient decay time of 6 or 8 min which show
that for $\Omega_1 < \Omega_s$ there is no observable attenuation.

Bendt's experiments establish that over his range of speeds, the superfluid adopts the classical (viscous) fluid velocity distribution, Equation
(1.91). They further establish that the vortex line density is given by
Feynman's rule, Equation (2.17).

The results on the first quanta of circulation and first entry of vortices
in a uniformly rotating annulus were obtained by powerful free energy
minimization techniques, discussed in the papers by Fetter (1966, 1967)
and Stauffer and Fetter (1968). When $\Omega_1 \neq \Omega_2$, the normal fluid is in
shearing motion, and one would no longer expect the methods of equilibrium thermodynamics to be applicable. We shall explore in this section
some ideas on vortex arrays that might appear at low rates of rotation
in the presence of shear.

We take our starting point from the discussion of the rotating annulus
by Donnelly and Fetter (1966), referred to briefly in Section 2.6. Although
the discussion concerned equilibrium flow in an annulus, strictly speaking
there is relative motion of the normal and superfluids. To appreciate this
one has only to examine Figure 5.8. The velocity distributions of Figure
5.8 are drawn on the following assumptions. The normal fluid velocity
is distributed according to Equation (1.91) which is the sum of solid
body rotation and potential flow terms. By analogy with the ideas of
Donnelly and Fetter (1966), the superfluid acquires circulation $\Gamma =
2\pi\Omega_1 R^2$ to match the velocity of the normal fluid midway between
cylinders, and is otherwise distributed in potential flow, $v_s = \Gamma/2\pi r$ within
the annulus. Thus even in the solid body rotation treated by Donnelly

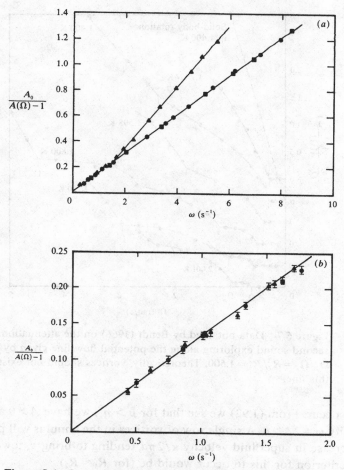

Figure 5.6 Second sound attenuation results obtained by Bendt (1966) with the apparatus illustrated in Figure 5.5: (*a*) results showing attenuation of second sound under the conditions indicated; (*b*) details showing all results below the critical velocity when plotted against the vorticity ω of the normal component. This is a direct test of Feynman's rule (2.17).

and Fetter ($\Omega_1 = \Omega_2$), there is relative motion between normal and super-fluid components, despite the fact that the normal fluid is free of shear.

If we follow this hint from Donnelly and Fetter (1966), we might speculate that, in general, vortices will appear when their presence can minimize the relative velocity between the two fluids, and ignore the shear motion of the normal fluid. A simple argument suggests how this

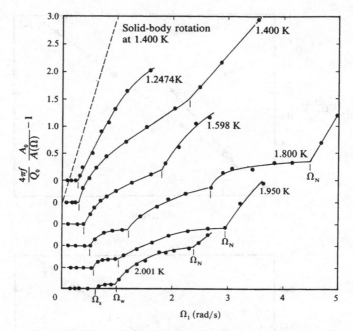

Figure 5.7 Data obtained by Bendt (1966) on the attenuation of second sound exploring along the potential flow line given by $\Omega_1/\Omega_2 = R_2^2/R_1^2 = 1.500$. Theoretically, vortices should not exist on this line.

might occur. From (1.92) we see that for $\mu > \eta^2$, we have $A > 0$ and the vorticity $\omega = 2A > 0$. A single row of vortices in the annulus will provide an increase in superfluid velocity $\kappa/2\pi d$ tending to bring v_s toward v_n. So a criterion for this to occur would be (for $R_1 \approx R_2$)

$$\Omega_2 R_2 - v_s = \Omega_2 R_2 - \Gamma/2\pi R_2 \geq \kappa/\pi d$$

or

$$D_2 - D_1 \geq R_2/\pi d \quad (A > 0)$$

(5.10)

where $D_i = \Omega_i R_i^2/\kappa$ is a dimensionless quantum parameter. For $\mu < \eta^2$, we have $A < 0$ and vorticity $\omega = 2A < 0$, a single row of vortices in the annulus will provide a decrease in superfluid velocity $\kappa/\pi d$ tending to bring v_s down toward v_n. Then the criterion would become

$$v_s - \kappa/\pi d \geq \Omega_2 R_2$$

or

$$D_1 - D_2 \geq R_2/\pi d \quad (A < 0)$$

(5.11)

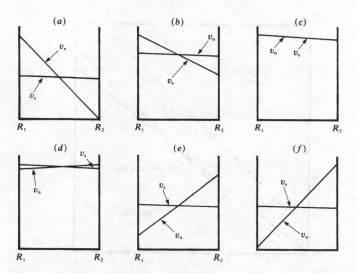

Figure 5.8 Distribution of the flow of the two fluids in the annulus between concentric cylinders in the absence of vortex lines, $\eta = 0.95$. The distribution of the normal component is given by (1.83) and the superfluid is in potential flow arranged to match the normal fluid velocity in the middle of the gap. Distributions are given for six values of μ: (*a*) $\mu = 0$ (inner cylinder rotating); (*b*) $\mu = 0.5$; (*c*) $\mu = 0.9025$ (potential flow); (*d*) $\mu = 1$ (solid-body rotation); (*e*) $\mu = 5$; (*f*) $\mu = 100$. $\mu = \infty$ corresponds to rotation of the outer cylinder (after Donnelly and LaMar (1988)).

Building on the qualitative argument just given, Swanson and Donnelly (1987) have shown that the minimum free energy for vortices in a general flow between rotating cylinders is obtained by minimizing the average of $\rho_s(v_s - v_n)^2$. They find that a single row of vortices appears when $|A| > \kappa L/\pi d^2$ where the logarithmic factor $L = \ln(2d/\pi a)$ and a second row when $|A| > 1.85\kappa L/\pi d^2$. They argue that this relation is valid anywhere in the (D_1, D_2)-plane, i.e., even for counterrotating cylinders. In our notation their criterion reads

$$|D_2 - D_1| > (R_2 + R_1)\ln(2d/\pi a)/\pi d \tag{5.12}$$

Equation (5.12) defines the boundaries for the entry of vortices in the annulus. These are shown by the dashed lines in Figure 5.9. Thus the primary (i.e., unperturbed) state in the lower triangle contains positive vortices, and the primary state in the upper triangular contains negative vortices. Along the dotted potential flow line and indeed between dashed lines, no vortices enter the annulus.

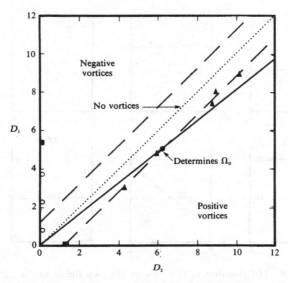

Figure 5.9 Measurements of the first appearance of attenuation of second sound in helium II Taylor–Couette flow. The dashed lines around the dotted potential flow line represent the boundaries of the vortex-free region. The data represent measurements by the following authors: solid triangles – Bendt and Donnelly (1967) and Bendt (1967b); open circles – Donnelly (1959); open diamond – Snyder (1974); solid square – Wolf, Perrin, Hullin and Elleaume (1981); and solid circle – Heikkila and Hollis Hallett (1955). The solid line corresponds to solid-body rotation $\Omega_1 = \Omega_2$ or $D_1 = D_2 \eta^2$. The intersection of the solid-body rotation and the criterion for entry of positive vortices reproduces Ω_0 of Section 2.6.

Solid-body rotation is indicated by the solid line in Figure 5.9, i.e., $D_1 = D_2 \eta^2$. When this criterion is combined with (5.12), we find the condition for entry of vortices in an annulus in solid body rotation is

$$\Omega_2 \geq (\kappa / \pi d^2) \ln (2d / \pi a) = \Omega_0 \qquad (5.13)$$

which recovers Equation (2.39) for Ω_0.

5.3.2   Stability of the flow

Chandrasekhar and Donnelly (1957) were the first to consider the stability of the flow of helium II between rotating cylinders in the presence of two-fluid flow plus vortices. They noted that if the Landau equations of motion are used then the superfluid would obey the Rayleigh stability criterion (see Section 1.8), and in particular would be unstable for any rotation of the inner cylinder. The normal fluid would act as a classical viscous fluid with viscosity η, density ρ_n and kinematic viscosity $\nu_n = \eta/\rho_n$, and thus would be unstable beyond the classical critical velocity modified to include ρ_n. The authors argue that when vortices and mutual friction are included the situation is drastically changed. The presence of vortex lines introduces a coupling due to mutual friction between the two fluids. The superfluid instability is raised above its value of zero for rotation of the inner cylinder and the normal instability is raised above the classical critical value.

The theory of Chandrasekhar and Donnelly omits a term in the equations of motion for a rotating fluid which was not appreciated at the time, namely the effect on stability of the tension (energy per unit length) of the vortex lines (see Section 3.2). The stability problem becomes very complicated with the full equations, but some insight may be obtained by considering a derivation equivalent to the Rayleigh criterion (see Chandrasekhar (1961) Sections 66, 67).

Mamaladze and Matinyan (1963) recognized that if the flow of helium II between concentric cylinders involves an array of vortices, the tension in these vortices themselves would provide a restoring force for a displaced element of fluid. They then argued that the Rayleigh criterion would be modified in the direction of extra stability.

Their calculations proceeded in the same spirit as those of Lord Rayleigh. The situation is considered at absolute zero where there is no viscosity. They assumed the vortex lines in the annulus are dense enough to allow continuum calculations, and assumed further that the distribution of the superfluid velocity is that of a viscous fluid, Equation (1.91). They used the equation of motion of the rotating superfluid with vortex lines (Equation (3.26) with $\mathbf{F}_{ns} = T = 0$). Making small perturbations about the equilibrium state, they derived a stability criterion which they suggested should replace Rayleigh's for pure superflow.

Barenghi and Jones (1987, 1988) have reconsidered the stability problem of helium II between rotating cylinders. They have begun their work by examining the effect of vortex tension in the pure superfluid. Barenghi and Jones note first that Mamaladze and Matinyan failed to realize that A can be positive or negative depending on Ω_1 and Ω_2 (see Figure 5.9). Thus their equations for the perturbed velocities are not correct. In addition, Barenghi and Jones have investigated nonaxisymmetric perturbations: Mamaladze and Matinyan, following Chandrasekhar and Donnelly assumed that the first modes to become unstable are axisymmetric. Barenghi and Jones find that when the inner cylinder rotates, the nonaxisymmetric modes become unstable at a lower velocity than the axisymmetric modes, and these long wavelength axial modes reduce the critical velocity to zero. Further they find that nonaxisymmetric modes of long wavelength can be unstable when the *outer* cylinder is rotating.

The situation at finite temperatures is the subject of a paper by Barenghi and Jones (1988). Here one must consider perturbations of the full HVBK equations of motion (Equations (2.3), (3.25) and (3.26)). They find that vortex tension produces instability for nonaxisymmetric modes of long wavelength. The wavelength is so long that the results indicate that only above 2 K will critical wavelengths be shorter than the length of most apparatuses.

The experimental situation has been reviewed by Donnelly and LaMar (1988). They show that in helium II the stability with rotation of the inner cylinder or outer cylinder is not much different, in contrast to classical results (Section 1.8, Figure 1.21) and in qualitative agreement with the results of Barenghi and Jones. New experiments are clearly called for to expand on the theoretical work.

5.4   Thermorotation effects

We have described in Section 2.8 the concept of phase slip introduced by Anderson. Individual vortex lines crossing a flow such as in Figure 2.20 produce a chemical potential difference given by (2.105). When a dense array of vortices moves, the chemical potential difference is given equivalently by (2.106). Yarmchuck and Glaberson (1978, 1979) have carried out a series of experiments which illustrate these principles. The concept of their experiment is shown in Figure 5.10. A pair of horizontal parallel glass plates is arranged to form a closed channel of rectangular cross-section closed at one end with a heater nearby, and open at the other end to the liquid helium bath. The channel is rotated about the vertical axis. The heater produces a counterflow of the two fluids given by (2.6): $j = \rho_n v_n + \rho_s v_s = 0$, and since the superfluid has no entropy, the heat flux $q = \rho S T v_n$. Rotation produces an array of vortices with vorticity $\omega = 2\Omega$. Detectors mounted with the rotating channel allowed a direct measurement of the chemical potential gradient parallel to the counterflow and the temperature gradient perpendicular to the counterflow. The direct measurement of chemical potential gradients was pioneered by these authors.

In the absence of superfluid acceleration (2.106) gives rise to a vortex flow induced chemical potential gradient given by

$$\nabla \mu_v = -(v_L - v_s) \times \omega \tag{5.14}$$

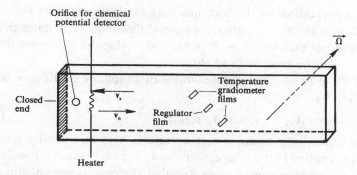

Figure 5.10 A schematic drawing of the counterflow channel of Yarmchuck and Glaberson (1978, 1979). The center superconducting film is used to regulate the ambient bath temperature and the outer two are used in a bridge as temperature difference detectors. The chemical potential detector is described by the authors.

where v_L is written relative to v_s which is in motion in this experiment. In this case, all of the chemical potential gradient is associated with vortex motion

$$\nabla \mu_v = -\nabla p/\rho + S \nabla T \tag{5.15}$$

whereas, if acceleration is allowed, the system can respond to the chemical potential gradient by accelerating:

$$\nabla \mu = \nabla \mu_v - dv_s/dt \tag{5.16}$$

Equation (5.16) expresses the fact that a chemical potential gradient gives rise to a time varying phase gradient, which, in turn, can arise from either accelerating fluid ($\dot{v}_s = \nabla \varphi$) or moving vortices. In a system in steady rotation at frequency Ω the discussion of Section 1.4 shows that

$$\nabla \mu = \nabla \mu_v - 2\Omega \times v_s \tag{5.17}$$

where v_s is now measured in the rotating coordinate system and the centripetal force is absorbed in the pressure gradient. In order to relate this last expression to measurable quantities, it remains to determine v_L in terms of experimentally controllable normal and superfluid velocities. From Equations (3.6) and (3.7) and the condition $f_M + f_D = 0$ we write

$$-\rho_s \kappa s' \times (v_L - v_s) = \gamma_0 (v_n - v_L) + \gamma_0' s' \times (v_n - v_L) \tag{5.18}$$

where v_n and v_s are the normal fluid velocities averaged over a region containing many straight vortex lines, κ is the vortex circulation, s' is the direction of local vorticity presumed perpendicular to $(v_L - v_s)$, and γ_0 and γ_0' are parameters defined in Chapter 3. Equation (5.18) is a general expression for the frictional drag experienced by the vortex line as a consequence of a transverse normal fluid flow, and states that the net force per unit length on a vortex line - Magnus force plus friction force - must be zero, as in (3.16).

For the case of thermal counterflow in an infinite medium, v_n and v_s are given by

$$\mathbf{v}_n = v_0 \mathbf{1}_x, \qquad \mathbf{v}_s = -(\rho_n/\rho_s) v_0 \mathbf{1}_x \tag{5.19}$$

where it has been assumed that a uniform heat flux density given by $\rho S T v_0$ is applied in the x-direction and $\mathbf{1}_x$ is a unit vector. Choosing the z-axis as the direction of $\mathbf{s}'$ we find that the component equations for the vortex line velocity are given by

$$\left.\begin{aligned}
\gamma_0(v_0 - v_{Lx}) + \gamma_0' v_{Ly} &= \rho_s \kappa v_{Ly} \\
-\gamma_0 v_{Ly} + \gamma_0'(v_0 - v_{Lx}) &= -\rho_s \kappa [v_{Lx} + (\rho_n/\rho_s) v_0]
\end{aligned}\right\} \tag{5.20}$$

where $v_{Lx} \mathbf{1}_x + v_{Ly} \mathbf{1}_y = \mathbf{v}_L$. These equations can be reduced to expressions for v_{Lx} and v_{Ly} separately:

$$\left.\begin{aligned}
v_{Lx} &= \frac{v_0}{\gamma_0^2 + (\gamma_0' - \rho_s \kappa)^2} [\gamma_0^2 + (\gamma_0' - \rho_s \kappa)(\gamma_0' + \rho_n \kappa)] \\
v_{Ly} &= \frac{v_0}{\gamma_0^2 + (\gamma_0' - \rho_s \kappa)^2} \gamma_0 \rho \kappa
\end{aligned}\right\} \tag{5.21}$$

Using Equations (5.14) and (5.18) and these expressions for the vortex line velocity, we can calculate the contribution to the chemical potential gradient due to vortex line motion

$$\nabla \mu_v = 2\Omega v_0 \left[\frac{\gamma_0 \rho \kappa}{\gamma_0^2 + (\gamma_0' - \rho_s \kappa)^2} \mathbf{1}_x + \left(\frac{\rho}{\rho_s} + \frac{\rho \kappa (\gamma_0' - \rho_s \kappa)}{\gamma_0^2 + (\gamma_0' - \rho_s \kappa)^2} \right) \mathbf{1}_y \right] \tag{5.22}$$

where ω has been taken as 2Ω. In order to simplify this expression, the parameters γ_0 and γ_0' will be replaced by their equivalent forms in terms of the macroscopic mutual friction parameters B and B'. Using Table 3.1 we obtain the much simpler form

$$\nabla \mu_v = (\rho_n/\rho_s) \Omega v_0 [B \mathbf{1}_x - B' \mathbf{1}_y] \tag{5.23}$$

and Equation (5.17) yields

$$\nabla \mu = (\rho_n/\rho_s) \Omega v_0 [B \mathbf{1}_x + (2 - B') \mathbf{1}_y] \tag{5.24}$$

Neglecting the pressure gradient owing to rotation (5.24) gives the temperature gradient:

$$\nabla T = -(\rho_n \Omega v_0 / \rho_s S)[B \mathbf{1}_x + (2 - B') \mathbf{1}_y y], \qquad \nabla p = 0 \tag{5.25}$$

Under experimentally accessible conditions, the magnitude of this temperature gradient does not exceed several microdegrees per centimeter.

Equation (5.25) indicates roughly what is to be expected in a rotating counterflow experiment if the counterflow channel is very large. The effects of the channel surfaces on vortex line motion, however, are not included, and in order to deal with these effects, the macroscopic HVBK equations must include terms due to vortex curvature. Effects of curvature cannot be completely neglected in a channel of finite height because in such a system the normal fluid velocity is not constant throughout space and therefore will affect the lines differently at different points along their lengths. For a solution of the full hydrodynamic equations for flow in a rotating channel of finite height and including the effects of vortex line pinning on the channel surfaces, the reader is referred to Appendix A of Yarmchuck and Glaberson (1979).

The experimental arrangement used by Yarmchuck and Glaberson (1978, 1979) is shown schematically in Figure 5.10. Temperature gradients parallel to the channel axis and perpendicular to it, as well as chemical potential gradients parallel to the channel axis were obtained as a function of heater power and rotation speed.

The parallel component of the temperature gradient is shown in Figure 5.11 as a function of heater power at several rotation speeds at $T = 1.3$ K. For heater powers below the critical power associated with the onset of turbulent counterflow, the temperature gradient is proportional to the heater power and increases with increasing rotation speed. Plots of the linear regime slopes versus rotation speed, for both parallel and perpendicular components of the temperature gradient, are shown in Figure 5.12. The solid lines are fits to the data using a full solution to the equations as outlined above, for a channel of finite height but infinite aspect ratio. The only fitting parameters are the mutual friction coefficients B and B' and the values obtained are in excellent agreement with those determined from second sound damping experiments.

A plot of the chemical potential gradient $\nabla\mu$ versus heater power Q for $\Omega = 0$ and $\Omega = 10$ rad/s is shown in Figure 5.13. Also shown for comparison is the temperature gradient. There exists a critical heater power Q_{c1}, not observable in the temperature measurements, below which $\nabla\mu$ is zero. Q_{c1} was associated with the power at which the vortex array 'depins' and begins to move in response to the counterflow. Below Q_{c1} the vortices are pinned to protuberances in the channel walls and accommodate the counterflow by deforming. If the boundary condition on $\nabla \times \mathbf{v_s}$ in the calculation is adjusted, so as to produce no longitudinal

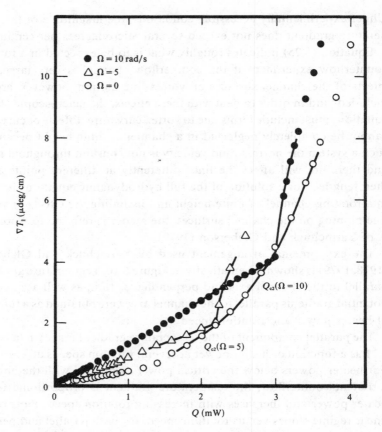

Figure 5.11 A plot of the parallel component of the temperature gradient as a function of heater power for several rotation speeds (after Yarmchuck and Glaberson (1978, 1979)).

chemical potential gradient, it is found that, indeed, little influence on ∇T is predicted – the pressure gradient does, of course, become large. A systematic study of vortex pinning is discussed in the next section.

5.5   Vortex pinning

5.5.1   *Neutron stars*

The dynamics of quantized vortices has influenced the study of the dynamics of pulsars, which, according to one view, are rotating neutron stars. When the thermonuclear fuel in a star has been largely used up it

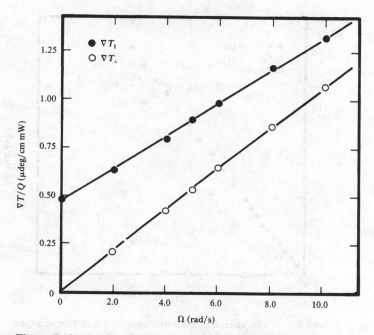

Figure 5.12 A plot of $\nabla T_{\parallel}/Q$ and $\nabla T_{\perp}/Q$ in the linear regime as a function of rotation speed. The solid lines are discussed in the text (after Yarmchuck and Glaberson (1978, 1979)).

seems that three final stages can be envisioned: first, a white dwarf star, in which gravity is balanced by the degeneracy pressure of a dense electron gas; second, a neutron star in which collapse continues beyond the white dwarf structure until gravity is balanced by neutron degeneracy pressure; third, gravitational collapse continues beyond the neutron star stage until a black hole is produced.

The structure of a neutron star is suggested in Figure 5.14, and is thought of as consisting of several distinct components that may rotate at different frequencies. The bulk of the star consists of a 'neutron liquid' at a density of $\sim 10^{14}$ g/cm^3. Its mass is about one solar mass ($\sim 10^{33}$ g) and so its radius ~ 10 km is very small compared to the original star. Since the original star rotated and had a magnetic field of ~ 100 G, conservation of angular momentum and flux, as a first approximation, suggests that the neutron star will be rotating at $\sim 10^3$ s^{-1} and have a field of $\sim 10^{12}$ G, which is not necessarily aligned with the axis of rotation. The surface of such a rotating, highly conducting star, possessing a large magnetic field, will experience large electric fields which will tear charged

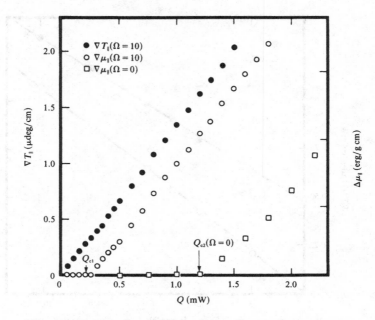

Figure 5.13 The chemical potential gradient and temperature gradient as a function of heater power for $\Omega = 0$ and $\Omega = 10$ rad/s (after Yarmchuck and Glaberson (1978, 1979)).

particles from it, produce a corotating magnetosphere, and lead to emission of electromagnetic radiation by a 'lighthouse beam' effect.

Some hundreds of pulsars have now been detected which have periods ranging from 1.6 ms to a few seconds. The pulsar period is identified with the rotational period of the neutron star, and hence a mechanism for the observed period seems consistent with ideas of neutron star structure.

Neutrons at a density of $\sim 10^{14}$ g/cm^3 comprise a fermion system with a degeneracy temperature $\sim 10^{11}$ K, much larger than the estimated temperature $\sim 10^8$ K. So the neutrons are effectively near absolute zero. If the neutrons near the Fermi surface are paired, the stage is set for condensation and superfluidity of the neutrons. The protons can also form a condensate and a superfluid. Since the major mass of the star consists of neutrons, it appears that most of the rotational energy resides in the superfluid. The circulation in a neutron vortex is $h/2m_n$ where m_n is the neutron mass and $2m_n$ the mass of the pair, thus Equation (2.17) gives a density of vortices of $\sim 2 \times 10^5$ cm^{-2} for the Crab pulsar ($\Omega \sim$ 191 s^{-1}). Such estimates suggest that the neutron superfluid has properties

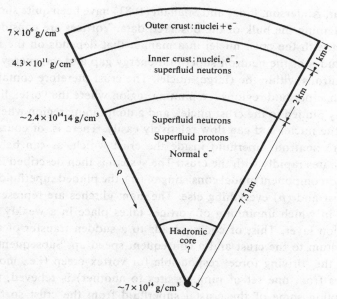

Figure 5.14 Sketch of the structure of a neutron star. The magnitudes of the various radii and densities ρ can vary 50% depending on estimates of the parameters in the equation of state (after Anderson, Alpar, Pines and Shaham (1982)).

similar to helium II undergoing solid body rotation at temperatures near 1 mK (see Pines (1971), Tilley and Tilley (1986)).

The rotational frequency of pulsars decreases very slowly at a constant rate suggesting that the neutron superfluid is gradually losing its rotational energy. This may be due to effects such as the scattering of protons and neutrons off the vortex cores. In 1969, two of the youngest and most rapidly rotating pulsars, the Crab and Vela, exhibited jumps – glitches – in rotational speeds $\Delta\Omega/\Omega \sim 10^{-8}$ and $\sim 10^{-6}$ respectively. These events were followed by periods during which the deceleration was greater than before the event, with deceleration resuming to normal with relaxation times of the order of ~ 7 days and ~ 1 year. These speed-up events offer the possibility of gaining some understanding of the interior.

There have been numerous attempts at explaining the glitch phenomena, and in most of them (e.g., Packard (1972), Ruderman (1969, 1976), Baym and Pines (1969, 1971), Anderson, Pines, Ruderman and Shaham (1978), Krasnov (1977), Campbell (1979)) a substantial role is played by quantized vortices.

Alpar, Anderson, Pines and Shaham (1981) have been quite successful in explaining the bulk of the observed data. Vortices are considered to interact with the crust nuclei in a manner that depends on the relative magnitudes of the neutron superfluid energy gap and the energy gap of the neutrons within the (large) nuclei. The crust therefore contains two neutron superfluid regions – a pinning region where the vortex lines are strongly pinned to the crust nuclei, and a nonpinning region where lines avoid the nuclei and can flow relatively easily. There is, of course, also the core neutron superfluid inside the crust which, as can be shown, equilibrates rapidly with the crust. The system is then described in terms of a two-component model consisting of (a), the pinned superfluid within the crust and (b) everything else. The giant glitches are represented as events in which unpinning of vortices takes place in a weakly pinned transition layer. This, of course, leads to a sudden transfer of angular momentum to the crust and a consequent speed-up. Subsequent to the glitch, the 'driving force' responsible for vortex creep (i.e., motion of vortices from one set of pinning sites to another) is relieved, thereby decoupling some of the crustal superfluid from the crust so that the effective moment of inertia is reduced and the deceleration rate increases. The depinning event itself is triggered by a buildup to some critical value of the differential rotation rate between the crust and the pinned super-fluid. Macroglitches in the Crab pulsar are pictured as being triggered by some external event, the younger and hotter pulsar having a faster steady state vortex creep rate so that creep alone can relieve the differential rotation stress without spontaneous discontinuous unpinning events.

5.5.2   *Experiments on vortex pinning*

Hegde and Glaberson (1980) performed a series of experiments in order to investigate systematically the pinning of vortices to surface protuberances. It had been generally believed (see, e.g., Tsakadze (1978)) that vortex pinning could be described in terms of viscous-like flow of vortices along surfaces. The observation described above (see Figure 5.13), that no chemical potential gradient at all could be observed for a vortex array subject to counterflow intensity less than some critical value, suggests that static pinning is important.

The experimental arrangement was similar to that of Figure 5.10. A glass channel of large aspect ratio was rotated about an axis perpendicular to the channel axis and to the large-area walls. The channel was closed at one end, near which a resistive heater was placed, and open at the other end to a pumped liquid helium bath. A small hole in one of the channel walls near the heater allowed a connection to a chemical potential detector. The absence of a chemical potential gradient was taken as an indication of the absence of vortex motion.

The large-area channel walls, which are perpendicular to the vortices in the absence of thermal counterflow, were of various kinds: (*a*) cleaved mica, (*b*) polished plate glass, (*c*) glass coated with a dense distribution of either 1 μm, 10 μm or 100 μm diameter glass microspheres. The areal density of the microspheres was less than that of close-packed spheres but much larger (except in the case of the 100 μm diameter spheres) than that of the rotation-induced vortices.

A plot of the critical heat flux in the channel, q_{c1}, versus the square root of the rotation speed $\Omega^{1/2}$, is shown in Figure 5.15 for the temperature $T = 1.3$ K: q_{c1} clearly becomes independent of Ω for all channel surfaces at high rotation speed. Of course, as the rotation speed is reduced to zero, q_{c1} becomes equal to the critical heat flux corresponding to the transition to turbulence. q_{c1} is plotted against temperature at a fixed high rotation speed of $\Omega = 6$ rad/s in Figure 5.16. The data evidently divide

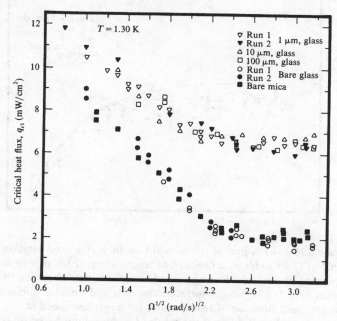

Figure 5.15 A plot of the critical heat flux for the onset of vortex motion as a function of the square root of the rotation speed, at the temperature $T = 1.30$ K (after Hegde and Glaberson (1980)).

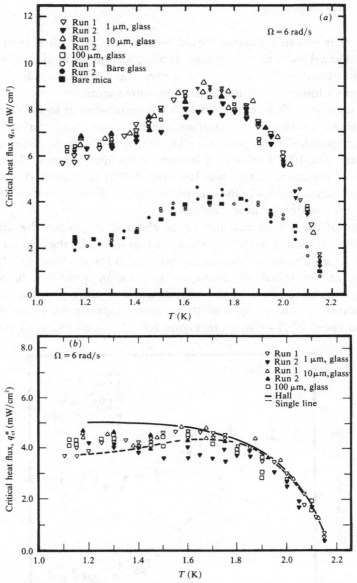

Figure 5.16 (*a*) A plot of the critical heat flux, at a fixed rotation speed of $\Omega = 6$ rad/s as a function of temperature. (*b*) A plot of the depinning critical heat flux q_{cl}^*, equal to the difference between the critical heat flux in 'rough' surfaced channels and 'smooth' surfaced channels, as a function of temperature, for a rotation speed of $\Omega = 6$ rad/s. The solid line is the result of a calculation based on the HVBK equations and assuming the boundary condition that the vorticity is parallel to the surface at the surface. The dashed line is the result of an isolated line calculation using a similar boundary condition (after Hegde and Glaberson (1980)).

into two distinct groups – 'rough' surfaced channels and 'smooth' surfaced channels – within each of which they are completely coincident.
That the data for cleaved mica and for bare polished glass are completely
coincident, in spite of the fact that their surface characteristics on a
microscopic scale are very different, suggests that q_{c1} in these systems
has nothing at all to do with surface roughness. The authors suggest that
there are two distinct mechanisms involved in determining q_{c1}, only one
of which is associated with surface roughness, and that the two mechanisms produce additive contributions to q_{c1}. In Figure 5.16(*b*) we show
a plot of q_{c1}^*, the difference between q_{c1} for the 'rough' channels and
averaged values of q_{c1} for the 'smooth' channels, as a function of temperature. The solid line is a result of a calculation to be discussed below.
This result, having no adjustable parameters, is clearly in excellent
agreement with the data.

It was assumed that all the 'rough' systems studied involve perfectly
pinned vortex lines. By this was meant that the lines will remain pinned
as long as they can terminate on their pins at an angle greater than zero,
with respect to the surface. Yarmchuck and Glaberson solved the linearized Hall–Vinen–Bekharevich–Khalatnikov (HVBK) hydrodynamic
equations for counterflow in a rotating channel of infinite aspect ratio
(no side walls) and infinite length. This solution is clearly inadequate in
our situation where the local vorticity can, and indeed will, greatly exceed
2Ω. The full nonlinear equations were numerically solved by Hegde and
Glaberson (1980), and the calculated value of the heat flux at which the
vorticity at the channel surface is parallel to the surface is shown as the
solid line in Figure 5.16(*b*). This is given approximately by the expression

$$q_{c1} \sim \frac{2\nu\rho ST}{d[G_2^2 + (\rho/\rho_s - G_1)^2]^{1/2}} \tag{5.26}$$

where ρ is the fluid density, ρ_s the superfluid density, S the specific
entropy, T the temperature, and d the channel height; G_1 and G_2 are
related to the mutual friction coefficients by (Table 3.1, Equation (3.23))

$$G_1 = \frac{e}{\alpha^2 + e^2}, \qquad G_2 = \frac{\alpha}{\alpha^2 + e^2} \tag{5.27}$$

where $\nu \sim (\kappa/4\pi) \ln(b/a)$, and b is the interline spacing. Because B,
and particularly B', are not known very accurately, there is $\sim 10\%$ uncertainty in the theoretical values of q_{c1}.

The HVBK equations ought to be valid in situations where all characteristic lengths in the problem are much greater than the intervortex line
spacing. This requirement is well satisfied in the interior of the channel.

Near the channel surfaces, however, the radii of curvature of the vortex lines go to zero faster than the interline spacing and the HVBK equations are not strictly valid. In this region, the behavior of isolated vortex lines would be a better representation of the situation. It turns out that isolated-line behavior is qualitatively similar to the average hydrodynamic behavior so that these considerations are probably not very important.

To obtain a feeling for isolated-line behavior, Hegde and Glaberson (1980) solved the (nonlinear) equation (see Section 3.1)

$$\rho_s \kappa s' \times v_{sl} = \gamma_0 v_n + \gamma_0' \kappa s' \times v_n \tag{5.28}$$

where v_{sl} and v_n are the local superfluid and normal fluid velocities, s' is a unit vector along κ, and r_0 and r_0' are constants related to B and B', as discussed in Chapter 3. This equation states that the net force on a stationary vortex line, Magnus force plus friction force, must be zero. v_n is taken as that for Poiseuille flow and $v_{sl} = v_{s0} + v_i$, where $v_{s0} = -\rho v_n / \rho_s$ (i.e., 'plug' flow) and v_i is the Arms and Hama self-induced velocity given by Equation (1.76). The values of the calculated heat flux for which the vortices terminate parallel to the surface are shown as the dashed curve in Figure 5.16. Although this curve agrees about as well with the data as the HVBK curve, the interactions between the vortices are indeed large and the effects of the image in the surface have been ignored so that the good agreement is probably fortuitous.

The nature of the critical heat flux in the 'smooth' channels is not understood. The HVBK equations for a channel of finite width and length have not been solved and the highly speculative possibility remains that the vortices are capable of distributing themselves so as to produce no vortex motion, below some critical heat flux, even in the absence of pinning. In the presence of pinning, once q_{c1}^* is exceeded the vortices would be capable of redistributing themselves so as to maintain immobility of the vortices in the steady state. Only when the heat flux exceeds the sum of q_{c1}^* and q_{c0}, the critical heat flux in the smooth channels, would steady state vortex motion occur. The qualitative difference between the temperature dependence of q_{c1} and q_{c1}^* in Figure 5.16 lends support to the assumption that the mechanisms involved produce additive contributions to q_{c1}.

The picture of vortex pinning developed here is that, at least in a situation where pinning sites are dense compared to the vortex distribution, the vortices remain pinned as long as they conceivably can. Onset of depinning is characterized in terms of a macroscopic boundary condition – the curl of the (average) superfluid velocity field is parallel to the surface at the surface. The details of the interaction between the vortex line and its pinning site do not appear to be important here, but might be important in situations where the vortex distribution is dense compared to that of the pins or where the distance between vortices is not much smaller than the channel height.

5.5.3 Simulation studies of vortex pinning

A calculation of the interaction between an isolated vortex line and a hemispherical pinning site was done by Schwarz (1981). The evolution of a vortex line terminating on a pinning site and subject to a transverse superflow was followed numerically. The analysis is the same as that used by Hegde and Glaberson (1980) in their single-line calculation, except in two respects. Hegde and Glaberson solved directly for the stationary line shape whereas Schwarz determined the time evolution of the shape. The line, as expected, 'spirals' into its stationary configuration when pinned. More importantly, Schwarz solved the boundary value

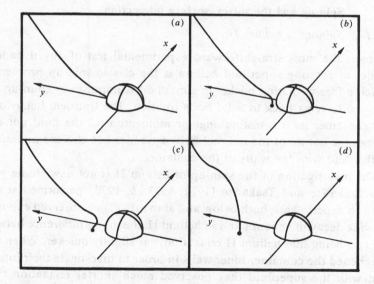

Figure 5.17 Behavior of a pinned vortex as the superfluid velocity is increased from slightly below the depinning critical velocity to slightly above it. Parameters for this calculation are $D = 10^{-3}$ cm, $b = 10^{-4}$ cm, and $\alpha = 0.1$. The stationary configuration (a) corresponding to $v_s = 0.64$ cm/s becomes unstable when v_s is increased to 0.67 cm/s, the vortex reconnecting to the plane and moving off as in (b)–(d) (after Schwarz (1981)).

problem so that the effects of the image vortex in the surface were explicitly included.

The calculation confirmed the picture of depinning suggested above. Below some critical superflow velocity a stationary line shape exists. As the superflow velocity is increased, the line comes into the 'pin' at increasingly smaller angles with respect to the plane surface. Finally, the line passes too close to the plane surface and the effect of the image vortex in that surface dominates. A stationary shape no longer exists and the line 'breaks' away from its pin. Figure 5.17 shows various calculated line shapes. Figure 5.17(*a*) shows the stationary line shape at a superflow velocity just below the critical velocity and Figures 5.17(*b*)–(*d*) show the time evolution of a vortex line under slightly supercritical conditions. This treatment of depinning is closely related to the extrinsic critical velocity model proposed by Glaberson and Donnelly (1966) but, for the reasons discussed earlier, is not particularly well suited for a quantitative analysis of a situation involving dense vortex arrays.

5.6   Spin-up and the vortex–surface interaction

5.6.1   *Spin-up in helium II*

Perhaps the most straightforward experimental test of any dynamical model of rotating superfluid helium is the classic spin-up problem in which a freely rotating bucket of superfluid is impulsively spun-up and allowed to relax back to solid body rotation. The transient behavior of the container as it transfers angular momentum to the fluid not only probes the nature of the internal fluid dynamics but also the interaction of the fluid with the walls of the container.

The investigation of the spin-up of helium II is not new. Some years ago Tsakadze and Tsakadze (1972, 1973*a,b*, 1975) performed several spin-up experiments, both below and above T_λ. They observed exponential-like decay in helium I and in helium II, the major difference between the two being that helium II relaxation was slightly quicker. When they roughened the container inner walls in order to investigate their interaction with the superfluid they observed much shorter relaxation times ($\sim \frac{1}{5}\tau_{\text{smooth}}$) which were independent of temperature through T_λ. This curious result has been considered to be a consequence of normal fluid turbulence (Alpar, 1978).

More recently Campbell and Krasnov (1982) have attempted to explain the results of some early experiments of Reppy and Lane (1961, 1965) and Reppy, Depatie and Lane (1960) in which helium II was spun-up from rest. In these experiments it was found that after an initial normal fluid relaxation the superfluid component induced a relaxation characterized by

$$\Omega_c = A \left[1 - \frac{B}{1 + B + C \exp{(-t/\tau)}} \right] \tag{5.29}$$

where Ω_c is the container angular velocity, qualitatively very different from the typical exponential decay of a classical fluid. Campbell and Krasnov have produced a model of these experiments which incorporates a viscous vortex–boundary interaction in which the vortex drag force is simply proportional to the relative vortex–surface velocity. By varying the strength of the interaction they were able to fit Reppy and Lane's data quite well, giving further evidence to the widely held assumption that a moving vortex line is indeed subject to a viscous force at the fluid–boundary interface. The quality of their model's predictions does not, however, make a compelling case for a viscous interaction. They considered only spin-up from rest, a case in which necessarily large fractional changes in vorticity occur. Clearly, in such cases vortex nucleation at the outer walls of the container becomes an important, not the primary, superfluid relaxation mechanism. The importance of remnant vorticity (see Section 5.8) has not yet been considered.

5.6.2 Experiments on the vortex–surface interaction

Some experimental evidence for a 'static-friction' type of vortex–boundary interaction, in which the vortices appear to move along a surface exerting a force equal to the vortex line tension, has been presented by Adams, Cieplak and Glaberson (1985).

A schematic diagram of their experimental apparatus is shown in Figure 5.18. The helium cell consisted a hollow lead-coated magnesium cylinder, A, 5 cm high and 2.8 cm in radius. The cylinder contained a set of thin aluminum disks and spacers which formed eight cylindrical cells, B, each with a typical height-to-radius ratio of 0.12. The cylinder was sealed with a magnesium cap into which a small hole had been drilled, thus minimizing film flow out of the container during a run. After submerging the container in helium II for a sufficient time for it to fill through the cap hole, the inner jar, C, was emptied via a fountain pump, D. The container was levitated by means of a superconducting magnet, E, surrounding its base. Axial stability was provided by a second superconducting magnet, F, positioned over the top portion of the container.

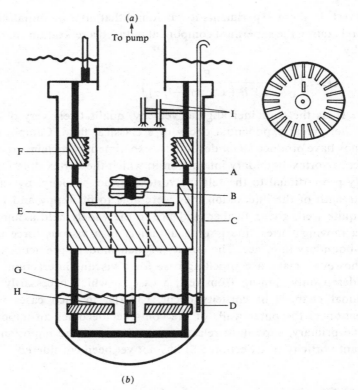

Figure 5.18 (*a*) Schematic diagram of the experimental apparatus. (*b*) The arrangement of photodetector marks, used in monitoring the container's angular velocity, on the top of the cell (after Adams, Cieplak and Glaberson (1985)).

The container was accelerated by a noncontacting induction motor consisting of a thin copper sleeve, G, surrounded by four superconducting drive coils, H.

The experimental procedure involved impulsively spinning-up (and spinning-down) the container of helium II from $\Omega_c = \omega_0$ to $\Omega_c = \omega_1 = \omega_0 + \Delta\omega$. The container's angular velocity, $\Omega_c(t)$, was then monitored as a function of time. The experiments were performed at $T = 2.1$ K and $T = 1.3$ K with both smooth disks and roughened disks, coated with #320 aluminum oxide powder.

The character of the $T = 1.3$ K relaxations depended dramatically on the disk surface roughness. Figure 5.19 shows a typical response of the smooth-surfaced cell to an impulsive torque and Figures 5.20 and 5.21

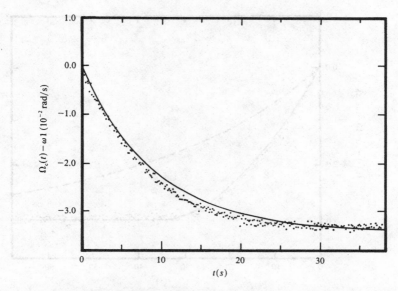

Figure 5.19 The angular velocity of the smooth disk cell after an impulsive spin-up torque ($\omega_0 = 8.54$ rad/s, $\Delta\omega = 0.283$ rad/s). The solid line represents the prediction of the model for $\xi = 0.005$ and $f_p = 0$ (after Adams *et al.* (1985)).

are typical responses for the rough-surfaced cell. Note the exponential-like behavior of the former and the nearly linear response of the latter. Substantial departure from exponential behavior was characteristic of all the low temperature rough disk relaxations for which $\omega_0 > 1.5$ rad/s. The remarkable linearity of the decay curve, observed to be symmetric with respect to spin-up and spin-down, indicates that the superfluid applied a strong, relatively constant, internal torque throughout most of the relaxation.

The data are interpreted in terms of a vortex-boundary force which is taken to be the sum of a 'static' and a viscous boundary force,

$$\mathbf{F}_b = \begin{cases} -(\mathbf{f}_D + \mathbf{f}_M)|_{V_L = V_b} & \text{for } |\mathbf{f}_D + \mathbf{f}_M| < f_p \\ -f_p \hat{V}_{Lb} - \xi \rho_s \kappa (\mathbf{V}_L - \mathbf{V}_b) & \text{for } |\mathbf{f}_D + \mathbf{f}_M| > f_p \end{cases} \quad (5.30)$$

where f_p is the minimum force per unit length required to break loose a pinned vortex, $\mathbf{f}_M$ is the Magnus force on a unit length of line Equation (3.6), $\mathbf{f}_D$ is the drag force on a unit length of line Equation (3.7), ξ is the dimensionless viscous interaction coefficient, $\mathbf{V}_L$ and $\mathbf{V}_b$ are the vortex line and boundary velocities and $\hat{V}_{Lb}$ is a unit vector in the direction of $\mathbf{V}_L - \mathbf{V}_b$. Though f_p and ξ are treated as adjustable parameters, $f_p L$ is

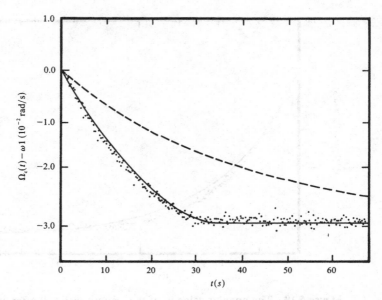

Figure 5.20 The angular velocity of the rough disk cell after an impulsive spin-up torque ($\omega_0 = 2.21$ rad/s, $\Delta\omega = 0.283$ rad/s). The solid curve is the prediction for $\xi = 0.005$ and $f_p = 0.53 \times 10^{-6}$ dyn/cm. The dashed curve is the prediction for $\xi = 0.005$ and $f_p = 0$ (after Adams *et al.* (1985)).

assumed to be of the order

$$T_L = (\rho_s \kappa^2 / 4\pi) \ln (b/a) \tag{5.31}$$

(see (2.16)) where T_L is the line 'tension' of an isolated vortex, a is the vortex core parameter, b is the intervortex line spacing and L is the vortex line length. This boundary force is, of course, in addition to the mutual friction and Magnus forces acting on the vortex lines throughout the fluid.

Simulations of the experiments were made in which ξ and f_p were independently adjusted to fit the low temperature data. The exponential-like behavior of the smooth disk data suggests that viscous drag was the dominant superfluid–container interaction. The smooth disk decays were therefore fit by setting $f_p = 0$ and varying ξ. This is equivalent to the approach taken by Campbell and Krasnov. The solid line shown in Figure 5.19 is the numerical fit. The fact that the data fall along a somewhat straighter line than that of the fit is probably a consequence of not having perfectly smooth disks. Reasonable fits were obtained with ξ ranging from 0.005 to 0.007 for all of the smooth disk data, ξ increasing with

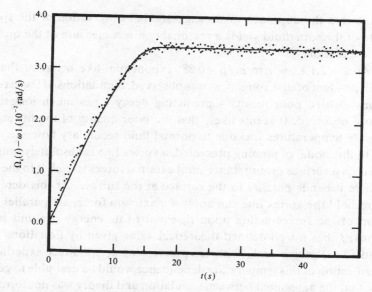

Figure 5.21 The angular velocity of the rough disk cell after an impulsive spin-down torque ($\omega_0 = 3.35$ rad/s, $\Delta\omega = -0.283$ rad/s). The solid curve is the prediction for $\xi = 0.005$ and $f_p = 0.53 \times 10^{-6}$ dyn/cm (after Adams *et al.* (1985)).

decreasing rotation speed. We shall see that these values of ξ are consistent with treating the drag as arising from mutual friction in the Ekman layer.

The rough disk relaxations were fit by simply 'turning on' the 'static' interaction. It was assumed that ξ was independent of surface roughness, thus leaving f_p as the only adjustable parameter. Excellent fits, such as shown as the solid lines in Figures 5.20 and 5.21, to all of the 'rough' disk data for which $\omega_0 > 1.5$ rad/s were obtained with $f_p = 0.53 \times 10^{-6}$ dyn/cm and were relatively insensitive to the value of ξ. This value of f_p corresponds to a maximum line 'tension', assuming a vortex line cannot apply a force to the boundary greater than its 'tension', of

$$T_L \approx \tfrac{1}{2} L f_p$$

$$= 1.59 \times 10^{-7} \text{ dyn} \tag{5.32}$$

in good agreement with the theoretical value,

$$T_L = \frac{\rho_s \kappa^2}{4\pi} \ln(b/a)$$

$$= 1.5 \times 10^{-7} \text{ dyn} \tag{5.33}$$

Note that the approach taken suggests that observation of the spin-up rate of the superfluid yields a reasonably direct measure of the quantum of circulation.

At $T = 2.1$ K, where $\rho_n/\rho = 0.88$, exponential-like behavior that was independent of disk roughness was observed. Simulations of these relaxations yielded poor results – predicting decay times much longer than those observed. It seems likely that the poor quality of the fits at these higher temperatures was due to normal fluid secondary flow.

In the model of pinning presented, a vortex line is absolutely immobilized on a surface protuberance until external forces stretch it to the point where it bends parallel to the surface at the surface. At this depinning threshold the vortex line can apply a maximum force, antiparallel to the sum of the forces acting upon it, equal to its energy per unit length. Thus, f_p has a well-defined theoretical value given by Equations (5.32) and (5.33) which scale with temperature as ρ_s. A direct experimental verification of this temperature dependence would be desirable to guarantee that the agreement between simulation and theory was not fortuitous.

A possible explanation for the necessity of including a viscous interaction in the model lies in the inadequacy of the normal fluid equations solved. By neglecting the axial and radial components of the normal fluid velocity Adams *et al.* (1985) have explicitly assumed that the normal fluid relaxes via viscous diffusion. However, it is well known (see Section 1.4) that secondary flow is the primary relaxation mechanism in all contained (Newtonian) spin-up flows (Greenspan, 1968). Secondary flow is characterized by a quasi-steady Ekman layer at each disk surface through which fluid is pumped radially by centrifugal action as shown in Figure 1.9. Adams *et al.* believe that it is this viscous layer, unaccounted for in the simulations, that requires the *ad hoc* addition of a viscous vortex–boundary interaction to their model.

For a typical spin-up experiment at $T = 1.3$ K in which $\omega_0 = 3$ rad/s, $\eta = 16 \times 10^{-6}$ P, and $\rho_n = 0.007$ g/cm^3, the thickness of the Ekman layer on each disk is (see Tritton (1988) Section 16.5)

$$d_E \approx (\eta/\rho_n\omega)^{1/2}$$
$$\approx 0.03 \text{ cm} \tag{5.34}$$

The total normal fluid friction force on a vortex line due to its motion through the top and bottom Ekman layers (assuming that in the layers $V_n \approx V_b$) is given by analogy to (3.7):

$$F_E = 2d_E\gamma_0(V_L - V_b) \tag{5.35}$$

After equating F_E to the viscous interaction of the model and solving for ξ, they obtain

$$\xi = 2d_E \gamma_0 / L\rho_s \kappa$$

$$\approx 3.3 \times 10^{-3} \tag{5.36}$$

a value surprisingly close to that used in the simulations. Equation (5.36) is also qualitatively consistent with the observed ω_0 dependence of ξ. It thus appears that the viscous vortex–boundary interaction is indeed associated with the mutual friction exerted by the vortices moving through the Ekman layers.

At low temperatures the normal fluid friction parameter, γ_0, is approximated by $\rho_s \kappa \alpha$ (see Table 3.1, noting $\alpha' \sim 0$)

$$\gamma_0 \approx \rho_s \rho_n \kappa B / 2\rho \qquad (B \sim 1.5) \tag{5.37}$$

which when substituted into the expression for ξ gives,

$$\xi \approx (B/\rho L)(\rho_n \eta / \omega_0)^{1/2} \tag{5.38}$$

If ξ does indeed represent vortex drag through normal fluid Ekman layers it should scale with temperature as $\rho_n^{1/2}$ at low temperatures.

The value of ξ used to fit the data is four orders of magnitude smaller than typical values used by Campbell and Krasnov. It can easily be shown that at low temperatures the smooth disk ($f_p = 0$) relaxation time, τ_s, is proportional to $(\xi^2 + 1)/\xi$, so that $\tau_s(\xi = 0.005) \simeq \tau_s(\xi = 200)$. Although the smooth data could be fit with either $\xi = 0.005$ or $\xi = 200$, the rough data restricted ξ to small values.

Note finally that the vortex boundary interaction is likely to have a profound effect on turbulent flow through channels, particularly when the channel walls are rough (see, e.g., Yamauchi and Yamada (1985)).

5.7   Remnant vortices in helium II

The superfluid at rest, being the background fluid for helium II, is ideally described by a single-particle wavefunction, uniform over the whole container, except on the walls. This means, in particular, that the phase is coherent everywhere. It is difficult to imagine how such order can exist in practice when helium II is created by cooling violently boiling helium I through the lambda transition. The same problem arises in imagining how a large melt can solidify to a perfect crystal without dislocations. Dislocations in phase, of course, are vortices in helium II, and one expects by these considerations to find vortices in any sample of helium II, no matter how carefully prepared.

Experiments by Awschalom and Schwarz (1984) suggest that there are residual vortices upon cooling through the lambda transition. Figure 5.22 shows a conjectural sketch of how such vortices might look, pinned to protuberances of various sizes on the boundaries of the channel. The effect of such vortices on critical velocities is a matter of current study. Their behavior in packed powders and small channels is discussed in Section 8.4.

It is argued by these authors that the density of remnant vortex lines is limited by the nonlocal interaction between the lines: if they are too close together, they will move each other around to cause a decrease in the number of lines present. In this way, one obtains an estimate

$$L_R \lesssim 2 \ln (D/a)/D^2 \tag{5.39}$$

for the remnant vortex line length density in a channel of size D. Note that this increases with decreasing D. It has been pointed out by Schwarz that as one goes into any corner, the distance D across which the vortices are pinned becomes smaller and smaller, with a corresponding increase in L_R. This suggests the amusing possibility that every interior corner is permanently decorated with a dense cobweb of pinned vortices.

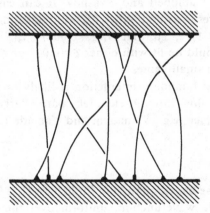

Figure 5.22 Sketch of pinned vortices in a channel pinned to protuberances of different sizes.

6   Vortex waves

Vortex waves are a very important phenomenon in the understanding of quantized vortex lines. The observation of vortex waves by Hall described in Section 2.7 was of crucial importance in the establishment of our subject insofar as it appeared to support the idea of vortex waves on individual vortex filaments and even led to a reasonable estimate (for the time) of the vortex core parameter a. In this chapter we discuss modern evidence for helical waves on quantized vortices, then go on to discuss Tkachenko waves, motions of the entire vortex array, and the possibility of solitary waves.

Some of the material in this chapter such as Section 6.2.3 is highly specialized and could well be omitted in a first reading. A comprehensive review of vortex oscillations and rotating superfluids has been prepared by Sonin (1987). Much of the material presented here is drawn from the review by Glaberson and Donnelly (1986).

6.1   Waves on isolated vortices

6.1.1   Helical waves of constant amplitude

We have discussed helical waves of constant amplitude on classical vortices in Section 1.7. The dispersion formulae can be carried over directly to helium II by simply substituting the quantum of circulation κ in (1.79) for the classical circulation Γ:

$$\omega^{\pm} = \frac{\kappa}{2\pi a^2}\left(1 \pm \left\{1 + ka\left[\frac{K_0(ka)}{K_1(ka)}\right]\right\}\right) \tag{6.1a}$$

where K_n is a modified Bessel function of order n. The Pocklington (1895) relationship for waves on a hollow ring with $n = kR$ the mode number and $k = 2\pi/\lambda$, λ being the wavelength, is also applicable to quantized vortex rings. In both cases the core is considered to be thin compared to other dimensions.

The dispersion formula (6.1a) has two branches, a fast (positive) wave ω^+ and a slow (negative) one ω^-, corresponding to + and − respectively in (6.1a), see Figure 6.1. The formula generally used in the literature of helium vortices is an approximation to ω^- for long wavelengths (see Equation (1.80)):

$$\bar{\omega} \approx -(\kappa k^2/4\pi)[\ln(2/ka) - \gamma]$$

$$\approx -(\kappa k^2/4\pi)[\ln(1/ka) - 0.116] \tag{6.1b}$$

Similar formulae have been shown to apply to vortex waves in an imperfect Bose gas by Pitaevskii (1961) and by Grant (1971). The danger of using (6.1b), common in the liquid helium literature, is shown in Figure 6.2.

The physical meaning of the positive (fast) branch is discussed in Section 1.7.1. The group velocity for the slow branch is shown in Figure 6.3 at the same temperature as Figure 6.1. The maximum group velocity is about twice the velocity of second sound at the same temperatures.

The question of the existence of axisymmetric (varicose) waves on quantized vortices has not been addressed in the low temperature literature. There are difficulties in finding a classical solution appropriate to quantized vortices because of our limited knowledge of the core structure of quantized vortices (see Section 1.7.1). In addition, for helium II, there is the problem of sound generation from the wave motion, which leads to damping of the waves.

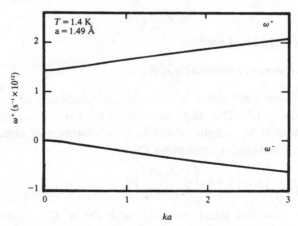

Figure 6.1 The dispersion curves for helical vortex waves in helium II (after Glaberson and Donnelly (1986)).

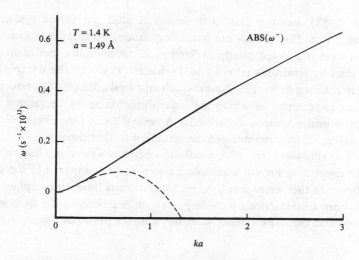

Figure 6.2 The dispersion curve for the slow branch compared to the long wave approximation (dashed line). The error in using the approximation beyond $ka \ll 1$ is evident (after Glaberson and Donnelly (1986)).

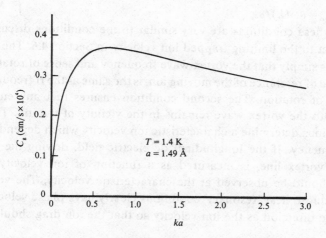

Figure 6.3 The group velocity C_g for slow helical waves (after Glaberson and Donnelly (1986)).

Hall (1958) was the first to demonstrate that the vortex system in rotating helium II could sustain oscillatory modes. His pioneering investigations were discussed briefly in Section 2.7. A similar experiment was performed by Andronikashvili and Tsakadze (1960) on the damping of a disk in rotating helium II. Nadirashvili and Tsakadze (1968) performed a further experiment in which the logarithmic damping decrement was measured under various conditions. A review by Andronikashvili and Mamaladze (1966) summarized the situation to that time.

In all of this work, the disk oscillation periods were quite low so that the corresponding helical wavelengths were large compared to the distance between the vortex lines. As we shall discuss later, this implies that vortex–vortex interactions were important in determining the dynamical behavior and the experiments were not a direct probe of simple Kelvin waves.

6.1.2   *Experimental determination of the dispersion for helical waves*

Halley and Cheung (1969) and Halley and Ostermeier (1977) suggested that an rf electric field, transverse to a vortex line charged with ions moving along the line, would couple strongly to vortex waves under suitable conditions. They argued that it is reasonable to expect strong coupling, i.e., 'resonant' generation of vortex waves, when the following two conditions are satisfied:

$$\left.\begin{aligned}\omega_{\rm rf} &= \omega(k) - kv_{\rm ion}\\v_{\rm ion} &= \partial\omega(k)/\partial k\end{aligned}\right\} \quad (6.2)$$

Note that these conditions are very similar to the conditions discussed with respect to the limiting trapped ion velocity in Section 4.6. The first condition is simply that the vortex-wave frequency and sense of rotation, in the frame of reference of the moving ion, is the same as the rf frequency and sense of rotation. The second condition ensures that any energy pumped into the vortex wave remains in the vicinity of the ion. These two conditions determine a characteristic ion velocity which depends on the rf frequency. If the longitudinal dc electric field, driving the ions along the vortex line, is measured as a function of ion velocity, an anomaly should be observed at the characteristic velocity. The vortex waves produced near 'resonant' conditions clearly have phase velocities in the same direction as the ion velocity so that the ion drag should be enhanced.

The only propagating vortex-wave modes at reasonable frequency are those which are polarized in a sense opposite to the circulation sense of the vortex (Figure 1.15). Because the vortex-wave group velocity is larger than its phase velocity, the sign of ω_{rf} is opposite to that of $\omega(k)$ and conditions (6.2) can only be satisfied simultaneously for an rf electric field polarization in the same sense as the vortex circulation. The ion-velocity anomaly should therefore be observed for only one rf field polarization for a given sense of rotation of the apparatus.

Ashton and Glaberson (1979) were able to observe the predicted effect. The experiment involves the use of a rotating ^{3}He refrigerator in which a sample of ^{4}He could be cooled to 0.3 K while rotating at 10 rad/s. This rotation speed yields a uniform distribution of vortex lines oriented along the rotation axis with a density of 2×10^4 cm^{-2}. A schematic of the experimental cell is shown in Figure 6.4. In order to observe the predicted anomaly properly, it was necessary to ensure that both the longitudinal dc electric field and the transverse rf electric field were reasonably uniform. This was accomplished by having the drift field defined by four stacks of electrodes, each of which consisted of eight electrodes stepped in dc potential. Rf potentials were ac coupled to the electrodes and applied in a circularly polarized mode. Circular polarization not only helped distinguish real from spurious effects (because of the intrinsic polarization of the vortex waves), but also helped achieve rf-field uniformity over the cross-section of the drift region.

A plot of ion velocity versus dc electric field, for the cases of no rf field and for the rf field polarized in the clockwise and counterclockwise senses, is shown in Figure 6.5 (in this case the ion cell was rotated in the counterclockwise sense). Note that the ion velocity is anisotropic with respect to rf field polarization. Reversing the apparatus-rotation sense, and hence the vortex-circulation sense, reproduces the data with the roles of the two rf-field polarizations reversed. An anomalous kink and plateau in the velocity versus dc electric field curve, for counterclockwise rf-field polarization, is observed near the characteristic velocity determined by (6.2). A simultaneous solution of Equations (6.2) yields $v_{ion} = 3.0$ m/s for the rf frequency used. The small discrepancy between this value and the plateau velocity observed can be explained in terms of the ac-field inhomogeneity – the ions move much faster near the top and bottom of the drift region where the ac-field amplitude is small. At the characteristic velocity, the wavelength of the resonantly generated vortex waves (2000 Å) is two orders of magnitude larger than the ion radius and three orders of magnitude larger than the vortex-core radius. It is still in the long wave region, $ka \sim 3 \times 10^{-3}$. The kink is not perfectly sharp, of course, because of finite vortex-wave damping as well as residual

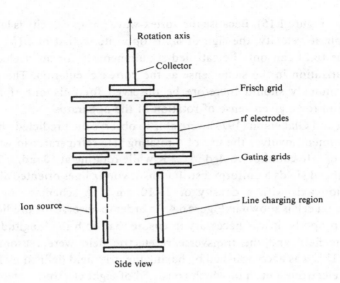

Side view

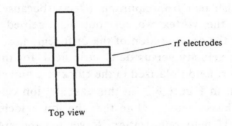

Top view

Figure 6.4 A schematic drawing of the experimental cell for observing vortex waves (after Ashton and Glaberson (1979)).

field inhomogeneities. This observation confirms the most important prediction of Halley and Ostermeier and constitutes a measurement of the vortex-wave dispersion relation (6.1b) at rf frequencies.

6.1.3 Solitary waves; sideband instability of Kelvin waves

We have noted theoretical and experimental evidence in Chapter 1 of the propagation of solitary waves on thin line vortices. The Hasimoto soliton is described in Section 1.7.2.

The Hasimoto soliton can also propagate on quantized vortices. Since Hasimoto's derivation assumed $L/a =$ constant, the first variant was to restore the 'far field' eliminated by Hasimoto's use of the localized

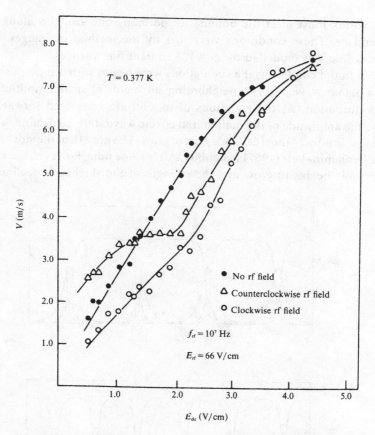

V (m/s) vs E_{dc} (V/cm)

T = 0.377 K

● No rf field
△ Counterclockwise rf field
○ Clockwise rf field

$f_{rf} = 10^7$ Hz

$E_{rf} = 66$ V/cm

Figure 6.5 A plot of ion velocity as a function of dc electric field for zero rf field and with the rf field polarized in the clockwise and counterclockwise senses (after Ashton and Glaberson (1979)).

induction approximation. It turns out that the changes in behavior on adding the far field are not much different than those engendered by using a variable cut off length L.

Samuels and Donnelly (1990) have carried out simulation studies of vortex waves on a fast computer. They were surprised to find that Kelvin waves (see Section 1.7.1) at $T = 0$ K are unstable above a threshold amplitude. They used the Biot-Savart law and the Arms and Hama localized induction approximation as discussed in Section 1.6 for their calculations.

The system studied consisted of a vortex line extended between two parallel planes 10^{-3} cm apart. The boundary conditions chosen were that the vortex must meet the boundaries normally and can slip along the boundary. These conditions were met by the method of images. The vortex line was modelled by $N = 128$ straight line vortices.

For initial conditions the simulations were begun with a superposition of a planar wave and two neighboring sidebands of small amplitude as a perturbation. As the equations of motion are integrated forward in time the amplitude of the main initial cosine wave starts to decline, while the amplitudes of the sidebands start to grow. The growth of the sidebands is a Benjamin–Feir (1967) instability. After some time, however, the main harmonic begins to grow and the strength of the sidebands declines as

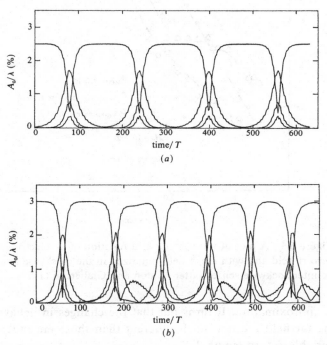

Figure 6.6 Sideband instability and recurrence phenomenon for a quantized vortex line. At $t = 0$ we impose a cosine wave of $n = 11$ half waves of amplitude $A_0 = 0.025\ \lambda$ with two sidebands (10 and 12 half waves with small amplitude). (a) Here we plot the amplitude of the main harmonic ($n = 11$) and the lower harmonics (10, 9, 8 in order of decreasing amplitude) of the three closest sidebands as a function of time. A plot of the upper harmonics ($n = 12, 13, 14$) looks very similar. (b) The same initial conditions as (a) except that the initial amplitude of the cosine wave is raised to produce what is called 'confined chaos'.

shown in Figure 6.6(a). The recurrence continues in the sense of the Fermi, Pasta and Ulam (1965) phenomenon. If the initial amplitude is increased, the simple recurrence is not observed. Instead a state of 'confined chaos' is observed as shown in Figure 6.6(b).

This phenomenon has not yet been studied in a driven system such as would be appropriate to the Ashton–Glaberson experiment of Section 6.1.2. If a similar instability exists, that may account for this very broad resonance illustrated in Figure 6.5.

6.2 Collective effects of Tkachenko waves on infinite vortex arrays

Helical waves propagate along isolated vortex lines with dispersion given by (6.1b). If we move into a frame of reference rotating with angular velocity Ω_0 in the same sense as the velocity field of the vortex line, the vortex will appear to rotate faster, i.e., at an angular velocity given by

$$\omega = \Omega_0 + (\kappa k^2/4\pi)[\ln (1/ka) + 0.116] \tag{6.3}$$

We now consider the effect on the spectrum of allowing for a uniform distribution of vortices. It is obvious that this extension of the theory is necessary; a vortex line moves with the local net velocity at its core, no matter what the source, so that it will respond to the fields generated by other vortices in addition to any self-induced field. Thus a complicated interaction can occur between vortices that, in general, should not be neglected, although in certain limits the behavior is relatively simple.

6.2.1 Rajagopal's calculations

This problem was solved in an approximate fashion by Rajagopal (1964). He made several simplifying assumptions:

(1) There is a mean spacing between vortices in a uniform, but not necessarily ordered, array, given by a parameter b.
(2) There are no vortices within a radius b of the particular vortex under consideration. Beyond this the vortex density is constant and nonzero.
(3) An integral over these velocity fields is sufficiently accurate so that a summation over discrete vortices is not required.

He then found the angular velocity of this central vortex's oscillation in a frame rotating at Ω_0 to be

$$\omega = (2 - J)\Omega_0 + (\kappa k^2/4\pi)[\ln (1/ka) + 0.116] \tag{6.4}$$

where

$$J = 1 - \int_{kb}^{\infty} K_0(s)s \, ds = \int_0^{kb} K_0(s)s \, ds \tag{6.5}$$

and K_0 is a modified Bessel function. The J integral results from the approximate summation of the velocity contributions from all of the other vortices.

Two limits are of interest here and are concerned with the extent that the vortices influence one another; i.e., the degree to which the modes are collective. If the mean spacing b becomes very large relative to the wavelength λ $(=2\pi/k)$, then the vortex oscillations become independent and the frequency must approach the Kelvin result. Indeed, as $kb \to \infty$, $J \to 1$, retrieving (6.3) which is just Kelvin's result in a frame rotating at speed Ω_0. Conversely, in the collective limit, as $kb \to 0$, $J \to 0$ in such a way that

$$\omega \sim 2\Omega_0 + (\kappa k^2/4\pi) \ln (b/a) \tag{6.6}$$

The situation is as follows:

$$\omega = F(\Omega_0)\Omega_0 + \nu k^2 \tag{6.7}$$

where ν is a weak function of k and where $F(\Omega_0)$ is 2 for a dense array of vortices (relative to the wavelength) and goes smoothly to 1 in the limit of independent oscillations. In general, of course, the value should be somewhere in between.

It is appropriate to notice the correspondence limit of the dispersion relation just found. For a uniform, but not necessarily ordered, array, $b \propto (1/n_0)^{1/2} = (\kappa/2\Omega_0)^{1/2}$. Since $\kappa = h/m$, if we let $h \to 0$, then $b \to 0$, which implies that for any k, $F(\Omega_0) \to 2$. Also, $\nu \to 0$, since $\nu \propto k$, so that we are left with the result $\omega = 2\Omega_0$, which is just the dispersion relation for classical inertial waves with wave vector along z. The classical result for general $\mathbf{k}$ is (Section 1.4.1)

$$\omega = 2\Omega_0 \hat{k} \cdot \hat{z} \tag{6.8}$$

If $\mathbf{k}$ is perpendicular to z these modes do not propagate.

The HVBK macroscopic two-fluid equations (Equations (2.3), (3.25), (3.26)) deal with velocity fields which are averaged over distances large compared to b. Quite naturally, these equations yield a dispersion relation for oscillation modes, Equation (6.6), characteristic only of the extreme collective limit. On the other hand, the macroscopic equations are well suited to a determination of the interaction between the normal fluid and the vortex system and the consequent wave damping.

Rajagopal's calculation was important in pointing out the need for considering collective effects in the vortex wave spectrum. However, an assumption made in his work was that the vortices oscillate in phase, which eliminated the important possibility that vortex waves could exist with wave vectors oriented perpendicular to the vortices themselves. Such modes arise naturally from lattice stability calculations, for stability of a vortex lattice is just the condition required for the existence of oscillatory behavior of the vortices, neglecting bending. Tkachenko (1966a, b, 1969) was the first to attempt a stability calculation of this kind for an infinite array of classical vortices; he found that triangular lattices were stable, and that the normal modes of such a lattice (plane waves) consisted of elliptical motions of the vortices about their equilibrium positions. Other studies of stability yield similar results. It should be emphasized, however, that these results are valid only for an infinite system. When a finite set of vortices is considered, it is found that the triangular lattice is often not the lowest free-energy configuration. The effect this might have on the spectrum will be discussed later.

6.2.2 A simple calculation of motion of vortices in a Tkachenko wave

The lattice sums involved in the calculation of the dispersion relation are complicated. Therefore it is instructive to attempt a simple derivation of an approximate expression. Consider a triangular array of rectilinear vortices aligned along the z-axis. Assume the presence of a long wavelength plane Tkachenko standing wave mode having nodal lines parallel to the y-axis (see Figure 6.7). Consider vortices that are near an antinodal line. Each vortex executes elliptical motion about its equilibrium position, where the ellipse has semiminor and semimajor axes $\delta_L(x)$ and $\delta_T(x)$ respectively. The frequency with which the ellipse is traversed is clearly of order

$$\omega \sim v_{L,max}/\delta_L \sim v_{T,max}/\delta_T \tag{6.9}$$

where $v_{L,max}$ and $v_{T,max}$ are the maximum longitudinal and transverse components of the vortex velocity.

We begin with a calculation of $v_{T,max}$. This is the velocity that the vortex has when the displacements of all of the vortices are purely longitudinal. Under these circumstances $\delta_T(x) = 0$ and $\delta_L(x) = \delta_{LO} \cos(lx)$ where l is the wavevector of the wave. Corresponding to

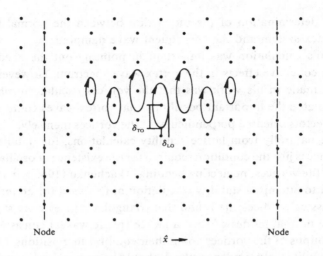

Figure 6.7 A Tkachenko standing wave mode. The dots are the undisturbed triangular vortex lattice points. δ_{TO} and δ_{LO} are the semimajor and semiminor axes of the elliptical paths executed by the vortices (after Glaberson and Donnelly (1986)).

these displacements is a vortex density oscillation along the x-axis

$$\Delta n(x) = n(x) - n_0 \sim l\delta_{LO}n_0 \sin(lx) \tag{6.10}$$

where $n_0 = 2\Omega/\kappa$ is the equilibrium density. It is only the excess vortex density, $\Delta n(x)$, that is important here since, in the rotating coordinate system, a uniform vortex distribution induces no velocity at a vortex position. A column of vortices having a density λ lines/cm, each line having a circulation κ, acts as a vortex sheet and induces a velocity (parallel to the column) $v \sim \pm \lambda\kappa/2$ at all points in the fluid not too close to the vortices. Fortunately, the net vortex density change for any region between adjacent nodal lines is zero. We need therefore only be concerned with the excess vortices within a distance $(\pi/2)l$ from the vortex of interest. It follows almost immediately that

$$v_{T,max} \sim \kappa n_0 \delta_{LO} \tag{6.11}$$

Now consider the situation when the displacements of all of the vortices are purely transverse to the wavevector – $\delta_L(x) = 0$ and $\delta_T(x) = \delta_{TO} \cos(lx)$. There is no vortex density oscillation and no transverse component of the vortex velocity. A column of vortices induces a velocity component perpendicular to itself which oscillates as a function of displacement along the column with an amplitude that falls off exponentially with distance from the column. The characteristic decay length is of order b, the interline spacing. It follows that we can obtain a reasonable estimate for the longitudinal component of the velocity of a vortex by considering only the effects of vortex columns immediately adjacent to the one containing the vortex of interest. A short calculation will show that at a displacement ε from a symmetry position along the vortex column and at a distance b from the column, the longitudinal velocity (i.e., the velocity along the wave vector or perpendicular to the column) is of order $\kappa\varepsilon/b^2$. For a long wavelength, $\varepsilon \sim (lb)^2\delta_T$, so that

$$v_{\mathrm{L,max}} \sim \kappa l^2 \delta_{\mathrm{TO}} \tag{6.12}$$

Note that, although all of the vortices in a particular column have the same longitudinal velocity, the average longitudinal superfluid velocity must be zero if the fluid is assumed incompressible. In a Tkachenko wave, the transverse vortex velocity is close to the average transverse superfluid velocity whereas the longitudinal vortex velocity is very different from the average transverse superfluid velocity.

Combining Equations (6.11) and (6.12) we have

$$\omega^2 \sim \frac{v_{\mathrm{L,max}}}{\delta_{\mathrm{LO}}} \frac{v_{\mathrm{T,max}}}{\delta_{\mathrm{TO}}} \sim \kappa^2 l^2 n_0 \tag{6.13}$$

so that $\omega(l) = C_{\mathrm{T}} l$ where $C_{\mathrm{T}} \sim \kappa n_0^{1/2} \sim (\Omega\kappa)^{1/2}$. Furthermore, the eccentricity of the elliptical motion is given by

$$\frac{\delta_{\mathrm{TO}}}{\delta_{\mathrm{LO}}} \sim \frac{(\Omega_0/\kappa)^{1/2}}{l} \tag{6.14}$$

As can be seen, the vortices rotate about their equilibrium positions in a sense opposite to that of the superflow about each of the lines.

6.2.3 *Combined Tkachenko and bending waves*

It is now necessary to consider the fact that in laboratory systems it is difficult, if not impossible, to attain the conditions required for the application of the previous analysis – the reason being that vortices are easily bent. If the vortices are contained in a finite chamber, they will attempt to clamp or 'pin' to the walls hydrodynamically, so that any motion of a vortex requires, if it is to remain firmly attached, that the vortex line bend. This, in turn, induces a velocity at its core that will result in the excitation of a wave whose wave vector lies along the axis of the undisturbed vortex. Thus a theory is required that combines both the Tkachenko and the Kelvin contributions to the dynamics.

The combined effects have been studied by Williams and Fetter (1977) and Sonin (1976). The two results are essentially identical, and we will draw extensively from both papers. The starting point is simply the Biot–Savart formula, Equation (1.74), summed over all the vortices except the one under consideration. If we work in a frame rotating at speed Ω_0 and recognize that the velocity induced at site i will result in a small displacement $u_i(z_i, t)$ of the vortex away from its equilibrium position r_i, then

$$
\begin{aligned}
\frac{\partial \mathbf{u}_i}{\partial t} = &-\Omega_0 \hat{z} \times \mathbf{u}_i + \frac{\kappa}{4\pi} \sum' \int_{-\infty}^{\infty} \left\{ \hat{z} \times \left(\frac{\mathbf{r}_{ij}}{|R_{ij}^0|^3} - \frac{4\pi \Omega_0 \mathbf{r}_i}{\kappa} \right) \right. \\
&+ \hat{z} \times \frac{(\mathbf{u}_i - \mathbf{u}_j) - (z_i - z_j)\mathbf{u}_j'}{|R_{ij}^0|^3} \\
&\left. - \hat{z} \times \frac{3 r_{ij}[R_{ij}^0 \cdot (\mathbf{u}_i - \mathbf{u}_j)]}{|R_{ij}^0|^5} + \frac{\mathbf{u}_{ij}' \times \mathbf{r}_{ij}}{|R_{ij}^0|^3} \right\} \, dz_j
\end{aligned}
\tag{6.15}
$$

where $\mathbf{r}_{ij} \equiv \mathbf{r}_i - \mathbf{r}_j$, $R_{ij}^0 \equiv \mathbf{r}_{ij} + (z_i - z_j)\hat{z}$, and $\mathbf{u}_{ij}' \equiv \partial \mathbf{u}_i / \partial z$. In order for the equilibrium configuration to rotate at the same speed as the frame, we impose the condition on the density of vortices $n_0 = 2\Omega/\kappa$, which forces the first term to drop out.

We assume that the vortices are of length L, impose periodic boundary conditions in the z-direction, and require translational invariance in the z-direction and discrete translational invariance in the (x, y)-plane. We can then expand the u_i in plane waves,

$$
\mathbf{u}_i(z_i) = (NL)^{-1/2} \sum_k \sum_l (\exp(-i\mathbf{l} \cdot \mathbf{r}_i) \exp(-ikz_i)]\mathbf{u}_{kl}
\tag{6.16}
$$

where N is the total number of vortices in the system, $k = 2\pi m/L$, $m = 0$, $\pm 1, \pm 2, \ldots$, and $\mathbf{l}$ is a reciprocal lattice vector in the first Brillouin zone. This is substituted into Equation (6.15), the integrals are performed, and the Fourier-transformed equations become, after invoking the assumed triangular symmetry of the lattice,

$$
\left(\frac{\partial \mathbf{u}_{kl}}{\partial t} \right)_x = (\Omega_g' - \eta + \xi)(\mathbf{u}_{kl})_y + \alpha(\mathbf{u}_{kl})_x
\tag{6.17}
$$

$$
\left(\frac{\partial \mathbf{u}_{kl}}{\partial t} \right)_y = -(\Omega_g' - \eta - \xi)(\mathbf{u}_{kl})_x - \alpha(\mathbf{u}_{kl})_y
\tag{6.18}
$$

$$
\left(\frac{\partial \mathbf{u}_{kl}}{\partial t} \right)_z = 2i[\nu_y(\mathbf{u}_{kl})_x - \nu_x(\mathbf{u}_{kl})_y]
\tag{6.19}
$$

where the Greek letters represent the following lattice sums:

$$\Omega_g' = \Omega_0 + \frac{\kappa k^2}{4\pi} \sum_j' K_0(kr_j) \tag{6.20}$$

$$\eta = \frac{\kappa k^2}{4\pi} \sum_j' [1 - \exp{(i\mathbf{l} \cdot \mathbf{r}_j)}] K_0(kr_j) \tag{6.21}$$

$$\xi = \frac{\kappa k^2}{4\pi} \sum' [1 - \exp{(i\mathbf{l} \cdot \mathbf{r}_j)}] \frac{y_j^2 - x_j^2}{r_j^2} K_2(kr_j) \tag{6.22}$$

$$\alpha = \frac{\kappa k^2}{4\pi} \sum_j' [1 - \exp{(i\mathbf{l} \cdot \mathbf{r}_j)}] \frac{2x_j y_j}{r_j^2} K_2(kr_j) \tag{6.23}$$

$$\nu_x = \frac{\kappa k^2}{4\pi} \sum_j' [1 - \exp{(i\mathbf{l} \cdot \mathbf{r}_j)}] \frac{x_j}{r_j} K_1(kr_j) \tag{6.24}$$

$$\nu_y = \frac{\kappa k^2}{4\pi} \sum_j' [1 - \exp{(i\mathbf{l} \cdot \mathbf{r}_j)}] \frac{y_j}{r_j} K_1(kr_j) \tag{6.25}$$

Self-induced motion is accounted for by modifying the first lattice sum as follows:

$$\Omega_g = \Omega_g' + \frac{\kappa k^2}{4\pi} \left[\ln\left(\frac{2}{ka}\right) - \gamma \right] \tag{6.26}$$

(Equation (6.1b)).

A long wavelength approximation to the spectrum can be obtained as follows. We change the coordinate system so that the x and y components transform to components in the direction of l and transverse to it. The resulting equations, with somewhat different lattice sums than those above, are

$$\frac{\partial u_l}{\partial t} \exp{(i\mathbf{Q} \cdot \mathbf{R}_j)} \equiv v_l = [\Omega_0 + \beta)u_l + \alpha u_q] \exp{(i\mathbf{Q} \cdot \mathbf{R}_j)} \tag{6.27}$$

$$\frac{\partial u_t}{\partial t} \exp{(i\mathbf{Q} \cdot \mathbf{R}_j)} \equiv v_t = [-(\Omega_0 - \gamma)u_q - \alpha u_t] \exp{(i\mathbf{Q} \cdot \mathbf{R}_j)} \tag{6.28}$$

where $\mathbf{Q}$ is the total wave vector and $\mathbf{R}_j$ is the three-dimensional position vector of the vortex line. In the approximation $lb \ll 1$, and assuming $b = [\kappa/(\Omega_0 3^{1/2})]^{1/2}$ in a triangular lattice, the lattice sums reduce to

$$\alpha \approx 0 \tag{6.29}$$

$$\beta \approx \Omega_0 \frac{k^2 - l^2}{k^2 + l^2} + \frac{\kappa k^2}{4\pi} \ln\left(\frac{r}{a}\right) + \frac{\kappa l^2}{16\pi} \tag{6.30}$$

$$\gamma \approx -\Omega_0 - \frac{\kappa k^2}{4\pi} \ln\left(\frac{r}{a}\right) + \frac{\kappa l^2}{16\pi} \tag{6.31}$$

where r is the smaller of the two lengths b and $1/k$. These sums neglect the effects of the nature of the core. Solving for the eigenvalues of Equations (6.27) and (6.28), we have

$$\omega^2 = (\Omega + \beta)(\Omega - \gamma) - \alpha^2 \tag{6.32}$$

Equation (6.32) is greatly simplified in two limits. If $l \to 0$, ω becomes $2\Omega_0 + (\kappa k^2/4\pi) \ln (r/a)$, thus recovering the Kelvin result. A second limit is that of $k \to 0$, the case of pure Tkachenko waves. Then

$$\omega^2 = (\Omega_0 + \beta)(\Omega_0 - \gamma)$$

$$= \left(2\Omega_0 - \frac{\kappa l^2}{16\pi}\right) \frac{\kappa l^2}{16\pi} \sim (\Omega_0 \kappa/8\pi) l^2 \tag{6.33}$$

This result follows when $2\Omega_0 \gg \kappa l^2/16\pi$, i.e., when there is a dense lattice and long wavelength $2\pi/l$. From this we can see that pure Tkachenko waves are nondispersive and travel at a speed

$$C_T = (\Omega_0 \kappa/8\pi)^{1/2} \tag{6.34}$$

Returning to the original equations, we find that the vortices in such a mode will move on elliptical paths, with the major axis perpendicular to l. As $|l|$ decreases, the eccentricity increases, so that for very long wavelengths the motion reduces to pure transverse displacements of the line relative to l. The ratio of maximum transverse displacement to maximum longitudinal displacement is

$$\frac{u_{t,\max}}{u_{l,\max}} = \left(\frac{\Omega_0 - \gamma}{\Omega_0 + \beta}\right)^{1/2} = \frac{2\Omega_0}{l} \left(\frac{8\pi}{\Omega_0 \kappa}\right)^{1/2} = \frac{2\Omega_0}{C_T l} \tag{6.35}$$

Pure Tkachenko modes have velocity and frequency proportional to $\kappa^{1/2}$. Presumably, an array of discrete classical vortices of constant strength could also exhibit Tkachenko waves. The classical experiment has not yet been tried but an apparatus to provide an array of vortices is described in Section 1.5.

6.2.4   *Macroscopic hydrodynamic equations*

It is of interest to write down a set of macroscopic hydrodynamic equations, in which an appropriate averaging is done over a length scale much larger than the interline spacing, from which one can obtain the normal modes of the system. The HVBK equations were such an early attempt (see Section 3.2). We reproduce them here including a term associated with the bending of the vortex lines, as viewed in a coordinate system rotating with angular velocity $\boldsymbol{\Omega}$.

$$\rho_s \frac{d\mathbf{v}_s}{dt} = -\frac{\rho_s}{\rho}\nabla p + \rho_s S\nabla T - 2\rho_s \boldsymbol{\Omega}\times\mathbf{v}_s + \frac{\rho_s}{2}\nabla(|\boldsymbol{\Omega}\times\mathbf{r}|^2)$$

$$+ \rho_s \nu(\boldsymbol{\omega}\cdot\nabla)\mathbf{s}' - \mathbf{F}_{ns}$$

$$\rho_n \frac{d\mathbf{v}_n}{dt} = -\frac{\rho_n}{\rho}\nabla p - \rho_s S\nabla T - 2\rho_n \boldsymbol{\Omega}\times\mathbf{v}_n$$

$$+ \frac{\rho_n}{2}\nabla(|\boldsymbol{\Omega}\times\mathbf{r}|^2) + \eta_n[\nabla^2\mathbf{v}_n + \tfrac{1}{3}\nabla\cdot\mathbf{v}_n)] + \mathbf{F}_{ns} \tag{6.36}$$

$$\mathbf{F}_{ns} = \rho_s \omega \alpha \mathbf{s}'\times[\mathbf{s}'\times(\mathbf{v}_n - \mathbf{v}_s)] + \rho_s \omega \alpha' \mathbf{s}'\times(\mathbf{v}_n - \mathbf{v}_s)$$

$$+ \rho_s \omega \alpha \nu \operatorname{curl}\mathbf{s}' + \rho_s \omega \alpha' \nu(\mathbf{s}'\cdot\nabla)\mathbf{s}' \tag{6.37}$$

where as before α and α' are the mutual friction coefficients, $\boldsymbol{\omega} = \nabla\times\mathbf{v}_s + 2\boldsymbol{\Omega}$, $\mathbf{s}'$ is a unit vector along $\boldsymbol{\omega}$, and $\nu = (\kappa/4\pi)\ln(b/a)$ where b is of order the interline spacing, a is the core parameter, and κ is the vortex circulation.

To these can be added equations expressing the incompressibility of the superfluid and normal fluid components

$$\left.\begin{array}{l} \nabla\cdot\mathbf{v}_s = 0 \\ \nabla\cdot\mathbf{v}_n = 0 \end{array}\right\} \tag{6.38}$$

These last two equations are justified since, at the relevant frequencies, the wavelengths of first and second sound are much larger than any other lengths of interest.

In the superfluid acceleration equation, the terms $2\rho_s\boldsymbol{\Omega}\times\mathbf{v}_s$ and $\tfrac{1}{2}\rho_s\nabla(|\boldsymbol{\Omega}\times\mathbf{r}|^2)$ are the Coriolis and centripetal forces associated with the transformation to the rotating coordinate system. The term important for our consideration is the vortex bending term $\rho_s\nu(\boldsymbol{\omega}\cdot\nabla)\mathbf{s}'$. It comes from assuming a contribution to the energy of the system simply proportional to the length of vortex line present (see Section 3.2). It is straightforward to linearize the equations and determine plane wave solutions of wavevector k. In the absence of mutual friction, the superfluid and normal fluid equations decouple and the normal modes correspond to ordinary damped inertial waves in the normal fluid obeying

$$\omega(k) = \pm 2\Omega(k_z/k) + i(\eta/\rho_n)k^2 \tag{6.39}$$

where $k^2 = k_x^2 + k_y^2 + k_z^2$ and $\boldsymbol{\Omega} = \Omega \mathbf{1}_z$ and to mixed inertial-vortex line waves in the superfluid having the dispersion relation

$$\omega(k) = \pm(k_z/k)[(2\Omega + \nu k_z^2)(2\Omega + \nu k^2)]^{1/2} \tag{6.40}$$

For propagation along the rotation axis, the dispersion relation corresponds to the long wavelength limit in Rajagopal's analysis. It is these modes which were presumably observed in the early vortex wave experiments. As is true for classical inertial waves, these modes do not propagate in a direction perpendicular to the axis of rotation. The HVBK equations do not generate Tkachenko modes because, as stated, it was assumed that the energy of the system depends only on the vortex line density and not on vortex lattice deformations. As we have seen, the vortex lines do not, in general, move with the average superfluid velocity so that a complete description of the system must involve an additional equation of motion for the vortex lattice deformation field.

Volovik and Dotsenko (1980) and Dzyaloshinskii and Volovik (1980) have developed powerful Poisson bracket techniques for dealing with the dynamics of defects in condensed systems and have used those techniques to derive linearized macroscopic equations for a rotating superfluid which yield Tkachenko modes. A more physically intuitive and more easily generalized approach was recently developed by Baym and Chandler (1983). We shall follow their approach.

6.2.5   Equations of Baym and Chandler

It is assumed that, in equilibrium, the vortices form a regular two-dimensional lattice

$$\mathbf{r}_{ij}^0 = i\boldsymbol{\alpha} + j\boldsymbol{\beta} \tag{6.41}$$

where $\boldsymbol{\alpha}$ and $\boldsymbol{\beta}$ are the fundamental translation vectors of the lattice. In nonequilibrium situations, the vortices will be displaced from their equilibrium positions by a two-dimensional deformation vector

$$\boldsymbol{\varepsilon}_{ij} = \mathbf{r}_{ij} - \mathbf{r}_{ij}^0 \tag{6.42}$$

Note that $\boldsymbol{\varepsilon}_{ij}$ is a two-dimensional object so that for rotation about the z-axis, $\boldsymbol{\varepsilon}_{ij}$ has only x and y components. $\boldsymbol{\varepsilon}_{ij}$ is not the velocity of a line element – it is the projection of that velocity in the (x, y)-plane.

We now average the superfluid velocity and the lattice deformation over a region of space large compared to the interline spacing but small compared to any other macroscopic lengths. We are left with macroscopic fields $\mathbf{v}_s(\mathbf{r}, t)$ and $\boldsymbol{\varepsilon}(\mathbf{r}, t)$. The macroscopic vorticity in the fluid, $\boldsymbol{\omega} = \nabla \times \mathbf{v}_s$ has a direction parallel to that of the vortex lines and has a magnitude proportional to the two-dimensional vortex density in a plane perpendicular to that direction. It follows that $\mathbf{v}_s$ and $\boldsymbol{\varepsilon}$ are related to each other through a simple equation of continuity. Let $n_v(\mathbf{r}, t)$ be the number density of vortex lines passing through a plane parallel to the (x, y)-plane. Of course, this is just proportional to the z component of $\boldsymbol{\omega}$:

$$\omega_z(\mathbf{r}, t) = \kappa n_v(\mathbf{r}, t) \tag{6.43}$$

The equation of continuity is

$$\frac{\partial n_v(\mathbf{r}, t)}{\partial t} + \nabla \cdot [n_v(\mathbf{r}, t)\dot{\boldsymbol{\varepsilon}}(\mathbf{r}, t)] = 0 \tag{6.44}$$

which simply states that the vortex density in some region can change only if vorticity flows into or away from that region. Corresponding, but more complicated, equations can be written for the vortex density in the (x, z)- and (y, z)-planes. These equations can be combined into a single vector equation

$$\partial \boldsymbol{\omega}(\mathbf{r}, t)/\partial t + \nabla \times [\boldsymbol{\omega}(\mathbf{r}, t) \times \dot{\boldsymbol{\varepsilon}}(\mathbf{r}, t)] = 0 \tag{6.45}$$

This, in turn, can be integrated to yield

$$\partial \mathbf{v}_s/\partial t + \boldsymbol{\omega} \times \dot{\boldsymbol{\varepsilon}}(\mathbf{r}, t) = -\nabla \phi(\mathbf{r}, t) \tag{6.46}$$

where energy conservation requires that $\phi(\mathbf{r}, t) = \mu(\mathbf{r}, t) + v_s^2(\mathbf{r}, t)/2$ and where μ is the chemical potential. Finally, this can be rewritten as

$$\partial \mathbf{v}_s/\partial t + (\mathbf{v}_s \cdot \nabla)\mathbf{v}_s + \boldsymbol{\omega} \times (\dot{\boldsymbol{\varepsilon}} - \mathbf{v}_s) = -\nabla \mu \tag{6.47}$$

In order to determine the momentum conservation equation, it is first necessary to discuss the (kinetic) energy associated with the elastic vortex deformation. Restricting consideration first to two-dimensional behavior, it can be shown that the energy associated with deformation of a triangular lattice is

$$E_{\text{elastic}} = \int d^3r \frac{\rho_s \kappa \Omega_0}{16\pi} \left[-2(\nabla \cdot \boldsymbol{\varepsilon})^2 + \left(\frac{\partial \varepsilon_x}{\partial x} - \frac{\partial \varepsilon_y}{\partial y} \right)^2 \right.$$
$$\left. + \left(\frac{\partial \varepsilon_x}{\partial y} + \frac{\partial \varepsilon_y}{\partial x} \right)^2 \right] \tag{6.48}$$

The momentum conservation equation is then

$$\partial j_i/\partial t + \partial \Pi_{ik}/\partial r_k = 0 \tag{6.49}$$

where the stress tensor is

$$\Pi_{ik} = \rho v_i v_k + p\delta_{ik} - \gamma_{ik}^{\text{el}} \tag{6.50}$$

and where p is the pressure and γ_{ik}^{el} is given by

$$\gamma_{ik}^{\text{el}} = \frac{\delta E_{\text{el}}}{\delta(\partial \varepsilon_k/\partial r_i)} = \frac{\rho \kappa \Omega_0}{8} \left(\frac{\partial \varepsilon_i}{\partial r_k} + \frac{\partial \varepsilon_k}{\partial r_i} - 3\delta_{ik} \frac{\partial \varepsilon_j}{\partial r_j} \right) \tag{6.51}$$

Writing $\mathbf{j} = \rho \mathbf{v}_s$, the momentum conservation Equation (6.47) becomes

$$\partial \mathbf{v}_s / \partial t + (\mathbf{v}_s \cdot \nabla) \mathbf{v}_s = -\nabla \mu - \boldsymbol{\sigma}_{el} / \rho_s \qquad (6.52)$$

where the elastic forcing term is

$$\boldsymbol{\sigma}_{el} = \frac{\rho \kappa \Omega_0}{8} [2\nabla(\nabla \cdot \boldsymbol{\varepsilon}) - \nabla^2 \boldsymbol{\varepsilon}] \qquad (6.53)$$

Comparison of this equation with Equation (6.47) shows that the elastic force $-\boldsymbol{\sigma}_{el}$ is what drives the line displacement $\mathbf{v}_s - \dot{\boldsymbol{\varepsilon}}$.

We now include the effect of line bending on the stress tensor. For small line deflections, the excess length of line associated with the bending is given by

$$\frac{\delta l}{l_0} = \frac{1}{2} \left[\left(\frac{\partial \varepsilon_x}{\partial z} \right)^2 + \left(\frac{\partial \varepsilon_y}{\partial z} \right)^2 \right] \qquad (6.54)$$

where l_0 is the undeflected line length. It follows immediately that the energy associated with this bending is given by

$$E_b = \int d^3 r \frac{1}{2} \left[\left(\frac{\partial \varepsilon_x}{\partial z} \right)^2 + \left(\frac{\partial \varepsilon_y}{\partial z} \right)^2 \right] n_v \nu \qquad (6.55)$$

where $\nu = (\rho_s \kappa^2 / 4\pi) \ln(b/a)$ is the energy per unit length associated with the line. This leads to an additional term in the stress tensor

$$\gamma_{zk}^b = \frac{\partial E_b}{\partial (\partial \varepsilon_k / \partial z)} = \frac{2\Omega_0 \nu}{\kappa} \qquad (6.56)$$

and therefore to an additional force term

$$\boldsymbol{\sigma}_b = -\frac{\partial}{\partial z} \left(\frac{\Omega_0 \nu}{\kappa} \frac{\partial \boldsymbol{\varepsilon}}{\partial z} \right) \qquad (6.57)$$

which is the same form as the vortex bending term in the HVBK equations.

In the rotating coordinate system, assuming incompressibility of the superfluid, the linearized equations become

$$\nabla \cdot \mathbf{v}_s = 0 \qquad (6.58)$$

$$\frac{\partial \mathbf{v}_s}{\partial t} + 2\boldsymbol{\Omega} \times \dot{\boldsymbol{\varepsilon}} = -\nabla \mu \qquad (6.59)$$

$$\frac{\partial \mathbf{v}_s}{\partial t} + 2\boldsymbol{\Omega} \times \mathbf{v}_s = -\nabla \mu - \frac{\kappa \Omega_0}{8\pi} [2\nabla_\perp (\nabla \cdot \boldsymbol{\varepsilon}) - \nabla_\perp^2 \boldsymbol{\varepsilon}]$$

$$+ \frac{\Omega_0 \nu}{\rho \kappa^2} \frac{\partial^2 \boldsymbol{\varepsilon}}{\partial z^2} \qquad (6.60)$$

where $\nabla_\perp$ refers to the components of the gradient in the (x, y)-plane only.

Certain extreme cases are easily dealt with. For wave motion propagating along the rotation axis, the equations reduce to the HVBK equations (without dissipation, of course) and the corresponding mixed Kelvin-vortex waves are obtained. For propagation perpendicular to the rotation axis, the vortex bending term plays no role and the equations reduce to

$$\frac{\partial^2}{\partial t^2}(\nabla \times \varepsilon)_z + \left(\frac{\kappa}{16\pi}\nabla^2 + 2\Omega_0\right)\frac{\partial}{\partial t}(\nabla \cdot \varepsilon) = 0 \qquad (6.61)$$

$$\frac{\partial}{\partial t}(\nabla \cdot \varepsilon) + \left(\frac{\kappa}{16\pi}\nabla^2\right)(\nabla \times \varepsilon)_z = 0 \qquad (6.62)$$

The solutions, in the long wavelength limit, are Tkachenko waves with a dispersion relation

$$\omega_T^2(l) = (\kappa\Omega_0/8\pi)l^2 \qquad (6.63)$$

For propagation at arbitrary angles with respect to the rotation axes, the dispersion relation determined using the lattice sum approach is obtained.

Baym and Chandler (1986) pointed out an interesting consequence of the inequivalence of the average superfluid velocity field $v_s(r, t)$ and the vortex lattice deformation velocity field $\dot{\varepsilon}(r, t)$. In a frame of reference at rest with respect to the superfluid at some point ($v_s = 0$), the energy per unit length of a vortex line is not precisely equal to the quantity ν. In addition there is a small term which can be written in the form $\frac{1}{2}m^*\dot{\varepsilon}^2$ where the vortex effective mass m^* is approximately the mass of superfluid excluded by the vortex core. There is a corresponding small contribution to the mass current. The effect of these terms is to allow an additional normal mode of the system in which the vortices rotate about their equilibrium positions in the same sense as that of the superflow about the vortices and opposite to that of Tkachenko modes. This mode is not likely to be directly observable because of its very high frequency: $\omega \sim \kappa/a^2 \sim 10^{12}\ \mathrm{s}^{-1}$. This 'inertial' mode is not confined to vortex arrays, but exists also in the presence of a single isolated vortex line, as discussed in Section 3.1.

6.2.6 Experimental investigation of Tkachenko waves

A series of experiments was performed by Tsakadze and Tsakadze (1973a, b, 1975) and Tsakadze (1976) in order to observe Tkachenko oscillations. An experimental cell was employed that allowed for free rotation of various shapes of buckets, including a spherical one so as to simulate a neutron star interior. The buckets were filled with superfluid and suspended magnetically on a long rod. The vessels were driven to high rotation speeds and their period of rotation was monitored. Oscillations of the period were observed at frequencies that were not simply related to expected Tkachenko mode behavior. Following Sonin (1976), they analyzed their data in terms of an empirical viscous vortex slip coefficient, making a rather long extrapolation to the relevant experimental regime, and obtained results which were consistent with Tkachenko waves. The results were, however, not conclusive in establishing their existence or in verifying details of the theory.

Tkachenko waves have probably been observed by Andereck, Chalupa and Glaberson (1980) and Andereck and Glaberson (1982). Their experimental cell is shown in Figure 6.8. A stack of Macor disks was suspended on a stainless-steel torsion fiber from a relatively massive platform which was itself suspended on a fiber from the cryostat. The disks were 0.0127 cm thick, 3.05 cm in diameter and were separated from each other by spacers ranging in thickness from 0.02 cm to 0.27 cm. The entire assembly was immersed in helium and mounted in a rotating ^{3}He refrigerator. The disks were driven into torsional oscillation magnetically and the response was detected electrostatically. The response was monitored using phase-sensitive detection with the phase in quadrature with respect to the drive so that the empty disk response was suppressed. As the drive frequency was swept through a vortex resonance, the effective complex moment of inertia of the disk assembly was altered and response such as shown in Figure 6.9 was observed. At relatively high temperatures, the quality factor of the resonance was found to decrease with increasing temperature, presumably because of mutual friction, but was temperature independent below 1.3 K, where it is believed to be limited by disk spacing inhomogeneity.

In order to ensure that the vortices are pinned at the disk surfaces, the disks were coated with a layer of 10 μm diameter glass beads. Although this affected the large-amplitude response, no significant effect on the 'zero'-amplitude resonance frequencies was observed. The authors observed hysteretic behavior with respect to drive amplitude changes after a change of rotation speed. The hysteresis disappeared after the disks were driven at sufficiently large amplitude. This suggests that the vortices were indeed strongly pinned at the surfaces, which is consistent with the observations of Yarmchuck and Glaberson (1979) and of Hegde and Glaberson (1980), but at variance with the presumptions of Hall (1958) and Tsakadze (1976).

In Figure 6.10 we show the measured resonance frequencies for various disk spacings as a function of rotation speed. Also shown for comparison,

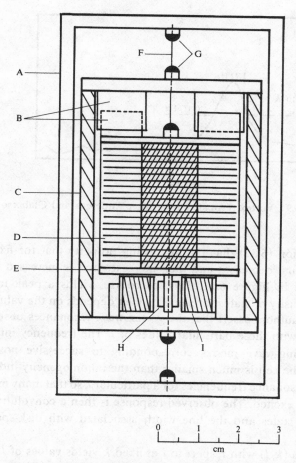

Figure 6.8 The experimental cell used to observe Tkachenko waves by Andereck *et al.* (1980), Andereck and Glaberson (1982). Subassemblies shown are: (A) mounting frame, (B) detection capacitor plates, (C) noise isolation cylinder, (D) disk and spacer stack, (E) epoxy on brass sheath, (F) torsion fiber, (G) epoxy, (H) magnet, and (I) driving coils.

as the dashed line, is the predicted resonance frequency for the lowest purely longitudinal (Kelvin) mode appropriate for the 0.076 cm spacing.

Consider a plane wave vortex displacement of wave vector $\mathbf{q} = \mathbf{k} + \mathbf{l}$, where $\mathbf{k}$ is the component of the wave vector parallel to the rotation axis and $\mathbf{l}$ its perpendicular component. In the experimental situation, the value of k is fixed by the disk-spacing, $k = \pi/\text{disk-spacing}$. As pointed out by Williams and Fetter (1977), the vortex oscillation dispersion

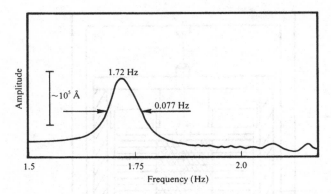

Figure 6.9 Vortex resonance plot (after Andereck and Glaberson (1982)).

relation, Equation (6.32), has the interesting property that for fixed k the frequency first decreases and then increases as l is increased from zero, as shown in Figure 6.11. This, of course, yields a peak in the vortex-wave density of states for some l which depends on the values of k and Ω. The authors assert that the vortex-wave resonances observed are associated with these particular values of l. The frequency interval between standing-wave modes corresponding to successive possible values of l in the cell is much smaller than the inhomogeneity-induced spread of the resonance frequencies for a particular l, so that many modes are necessarily excited. The observed response is then a convolution of the density of states and the line width associated with disk-spacing inhomogeneity.

Minimizing $\omega(k, l)$ with respect to l at fixed k yields values of l such that $lb \sim 1$, so that the long wavelength limit, Equation (6.32), was not strictly reached in the experiment and a detailed sum over the vortex-line lattice was performed to obtain the correct dispersion relation. The detailed calculation yielded results qualitatively similar to that of the continuum calculation (Figure 6.10). Furthermore, since $lR \sim 200 \gg 1$, where R is the disk radius, the cylindrical geometry of the experiment has negligible influence on the resonance frequencies. The predicted values of the resonance frequencies, with no adjustable parameters, are shown as the solid lines in Figure 6.9. There can be no doubt that the data are properly accounted for.

Several points concerning the experiment should be mentioned. Resonances were observed only when a cylindrical sheath was placed outside the stack of disks. Whether the effect of the sheath was to minimize the

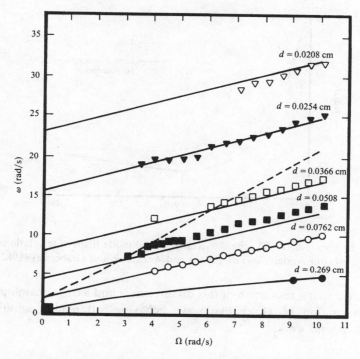

Figure 6.10 Resonance frequencies as a function of rotation speed for various disk spacings. The solid lines are predictions based on the density of states discussed in the text. The dashed line is the prediction of Rajagopal for a disk spacing of 0.076 cm in the absence of a Tkachenko contribution. The shaded area represents the region explored in Hall's oscillating disk experiments (Andereck *et al.* (1980); Andereck and Glaberson (1982)).

influence of side to side mechanical vibrations or to change the resonance cavity characteristics fundamentally was not clear. The theoretical resonance condition does not involve the question of the coupling of the mechanical system to the vortex oscillations. This coupling remains something of a mystery. Finally, the experiment does not appear to be sensitive to a density of states peak associated with the Brillouin zone edge of the triangular lattice. Perhaps the global circular distortion of the lattice, discussed in Section 5.1, smears out this peak. It should also be pointed out that Sonin (1983) has recently suggested an alternative interpretation of the experimental data which does not involve Tkachenko waves at all. This interpretation, in terms of the normal modes inherent in the HVBK equations, yields reasonable agreement with the data. It is

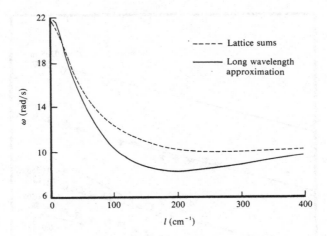

Figure 6.11 Tkachenko–Kelvin mixed mode dispersion relations (after Andereck *et al.* (1980) and Andereck and Glaberson (1982)).

apparent that a resolution of this disagreement and a truly unambiguous demonstration of Tkachenko wave behavior will require additional experimental work.

6.3   Collective effects – finite vortex arrays

Thus far we have been concerned with the normal modes in systems of infinite extent. Campbell (1981) has carried out extensive calculations of the transverse normal modes of finite rectilinear vortex arrays, extending the early work of Lord Kelvin and of Havelock (1931). The free energy of N vortices contained within a circular boundary of radius R, in a frame of reference rotating with angular frequency Ω, is written in the form

$$f/f_0 = - \sum_{j<k} \sum \ln |z_j - z_k|^2 + \tfrac{1}{2} \sum_j \sum_k \ln |1 - z_j z_k^*|^2$$
$$- \frac{2\pi R^2 \Omega}{\kappa} \sum_j (1 - |z_j^2|) + \sum_j \ln \left(\frac{R}{a} \right) \qquad (6.64)$$

where $f_0 = \rho \kappa^2 / 4\pi$ and $z_j = x_j + iy_j$ is the complex position of vortex j. The second term on the right comes from the effect of the boundary and is left out in an unbounded fluid. The last term plays no role in the situation considered here, where the total number of vortices is fixed. Minimizing this quantity with respect to the vortex positions yields the equilibrium quasi-triangular arrays discussed earlier.

The free energy plays the role of a stream function for the vortex velocity. In a situation where dissipation dominates the vortex dynamics, the vortices always move antiparallel to the gradient of the free energy. In the opposite limit, no dissipation at all, the vortices move perpendicular to the gradient with a velocity proportional to its magnitude. We shall mostly consider this latter situation. The equations of motion thus generated are a $2N \times 2N$ matrix equation. The eigenvalues and eigenvectors of the matrix are the normal modes of the system. The vortex trajectories for various normal modes and for certain values of N in an unbounded fluid are shown in Figure 6.12. As expected, the vortices execute elliptical motion about their equilibrium positions in a sense opposite to that of the vortex circulation. There are, of course, $2N$ normal modes for a given stable (or metastable) configuration, only N of which are independent. Some of these are common to all configurations: a rotation mode having zero frequency, a breathing mode whose amplitude vanishes in the limit of zero dissipation, and a displacement mode. The rotation mode is simply a rotation of the array as a unit about the center of symmetry and obviously has no restoring force associated with it. The breathing mode – a simple expansion or contraction of the array – corresponds to a relaxation of the system back to its equilibrium configuration, without oscillation, in a time that depends on the mutual friction coefficients. Ignoring the dragging along of the normal fluid by the vortices, the vortex density relaxation time is given by

$$\tau \sim (2\sigma \cos \theta \sin \theta \, \Omega)^{-1} \tag{6.65}$$

where

$$\sigma = \rho_s \kappa / (\rho_s \kappa - \gamma_0') \tag{6.66}$$

and

$$\theta = \tan^{-1} [\gamma_0 / (\rho_s \kappa - \gamma_0')] \tag{6.67}$$

and where γ_0 and γ_0' are the vortex drag coefficients defined in Chapter 3. Perhaps the most interesting mode is the displacement mode in which the array is displaced as a unit, at some instant of time, along some direction. In the absence of a boundary, this mode is equivalent to a shift of the center of symmetry from the axis of rotation of the frame of reference, and has the frequency Ω. The presence of a boundary decreases the frequency of this mode. The frequency for N not too large is then approximately the same as that of a single vortex of circulation N at the center of the container:

$$\omega \sim \Omega - N\kappa / 2\pi R^2 \tag{6.68}$$

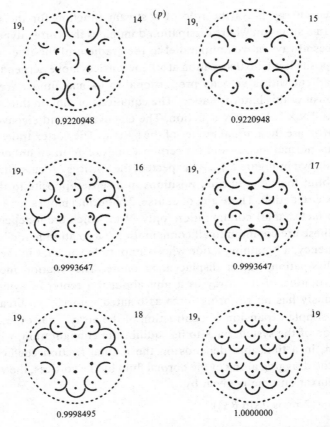

Figure 6.12 Some of the vortex normal modes for an equilibrium distribution of 19 vortices. Only half of each ellipse is shown to illustrate better the correlation of vortex motion. Below each mode is its corresponding oscillation frequency in units of $\sigma\Omega$; θ is taken as zero (after Campbell (1981)).

In the continuum limit, $\Omega \sim N\kappa/2\pi R^2$, this mode softens considerably but, because of the vortex-free region near the boundary, remains finite. This then corresponds to an $m = 1$ 'edge wave' discussed by Campbell and Krasnov (1981). Another important feature of finite vortex arrays in the absence of a boundary is the rigorous absence of angular momentum from any oscillation mode.

In general, relatively long wavelength phenomena predicted by the continuum calculations are confirmed by the detailed finite, but large, vortex array calculations. An important exception involves situations where the perturbation wavelength is comparable to the size of the system. Williams and Fetter (1977) have solved their continuum equations for the normal modes of a vortex distribution in a cylindrical container. They utilize the tractable but physically unreasonable boundary condition that the vortices have no radial velocity at the boundary. For the axisymmetric case, the vortex displacements are elliptical with an amplitude that varies as a Bessel function of the distance from the axis. Campbell has simulated this mode by considering large finite arrays of vortex lines, constraining the outer ring of lines to be fixed in position. He finds that the lowest normal modes of the systems have a frequency about a factor of 2 smaller than predicted by the continuum calculation. Campbell suggests that the global circular distortion, discussed in Section 5.1, is responsible for this effect. The built-in dislocations render the array considerably softer, for very long wavelength perturbations, than is the case in a perfect triangular array. Furthermore, this effect does not seem to depend on the number of vortices in the system. The Tkachenko wave experiments of Andereck *et al* (1980) involve transverse wavelengths much smaller than the size of the system and are therefore not affected by these considerations.

Yarmchuck and Packard (1982), using their vortex photography technique, have observed oscillations in vortex arrays containing small numbers of vortices. Detailed measurements were made of the oscillation periods and damping constants for two-, three- and four-vortex arrays. None of the observed oscillation behavior could be directly associated with any of the normal modes predicted by Campbell (1981). For example, two vortices were observed to execute damped oscillatory azimuthal motion, i.e., oscillation in the angle formed by a line connecting the two vortices and some reference line passing through the center position. There is clearly no restoring force for rectilinear vortices perturbed in this manner. It is suggested that the vortices are, in fact, not rectilinear, being pinned to either the bottom or side surfaces of the experimental cell. The experiment is, of course, sensitive only to positions of the vortices at the free surface. They calculated the response of small vortex arrays $(N = 2, 3, 4)$ to three-dimensional perturbations, and obtained a simple expression for the frequency of azimuthal modes:

$$\omega = k[(\Omega\kappa/2\pi) \ln (b/a) + \tfrac{1}{2}(N-1)]^{1/2} \qquad (6.69)$$

where N is the number of vortices and k is the longitudinal wavevector. Assuming that the vortices were pinned at the bottom surface of the cell, this expression yields frequencies 2–5 times smaller than observed. A number of possible explanations of the discrepancies were put forward, the most likely being that the vortices were pinned at the sides of the cell rather than at the bottom.

In the case of three-vortex arrays, it was found that the vortex orbits, after subtracting out the azimuthal oscillation and the instrumentally-induced oscillation of the center position, were similar to those for displacement waves as predicted by Kelvin and Havelock. As mentioned earlier, this displacement mode corresponds to a displacement of the array, as a unit, off the axis of the cylindrical cell, and has a frequency very close to the rotation frequency.

6.4   A vortex instability

Glaberson, Johnson and Ostermeier (1974) and Ostermeier and Glaberson (1975) have pointed out the existence of a simple hydrodynamic instability involving vortex lines. The instability arises in the presence of counterflow along the lines. Consider the linearized HVBK equations Equation (6.36). Taking $\Omega = \Omega 1_z$ and introducing two scalar potentials, $\phi_1(r, t)$ and $\phi_2(r, t)$ for the gradient terms, solutions to the equations are sought, having the form

$$
\left.\begin{array}{l}
\phi_1 = \phi_{10} \exp\left[i(\mathbf{k} \cdot \mathbf{r} + \omega t)\right], \\
\phi_2 = \phi_{20} \exp\left[i(\mathbf{k} \cdot \mathbf{r} + \omega t)\right] \\
\mathbf{v}_s = \mathbf{v}_{s0} \exp\left[i(\mathbf{k} \cdot \mathbf{r} + \omega t)\right], \\
\mathbf{v}_n = U_0 1_z + \mathbf{v}_{n0} \exp\left[i(\mathbf{k} \cdot \mathbf{r} + \omega t)\right]
\end{array}\right\} \quad (6.70)
$$

In the absence of mutual friction, the normal modes have the form given in Equations (6.39) and (6.40) except that, in Equation (6.39), ω is replaced by $\omega + U_0 k_z$.

The state of marginal stability, when the mutual friction force is included, is determined by the condition Im $\omega = 0$. This determines the critical value of the axial normal fluid velocity for a mode having wavevector k:

$$
U_{0,c} = \mp k^{-1}[(2\omega + \nu\kappa_z^2)(2\Omega + \nu k^2)]^{1/2} \quad (6.71)
$$

The critical frequency is given by Equation (6.40) so that the condition for instability is: a mode of wavevector k is marginally stable when the projection of the normal fluid velocity onto that wavevector is equal to the phase velocity of that mode.

The simplest modes to consider are those propagating along the vortex lines. Generalizing to off-axis modes, for which one should, in principle, include Tkachenko effects, probably does not qualitatively affect the calculation. The stability condition for longitudinal modes is similar to a Landau condition in which the critical velocity is given by

$$U_{0,c} = (\omega/k)_{min} = 2(2\Omega\nu)^{1/2} \tag{6.72}$$

For normal fluid velocities larger than this value, there exist infinitesimal helical deformations of the vortex lines which grow exponentially in time.

We now present a simplified derivation of the critical velocity for axial normal flow along an isolated vortex line. This derivation, although not completely rigorous, lends considerable insight into the nature of the instability. Consider a vortex line oriented along the z-axis and deformed into a helix of wave number k and infinitesimal amplitude δ. A unit vector along the vortex line is given by

$$\mathbf{s}'(z) = -k\delta[\sin(kz)]\hat{\mathbf{x}} + k\delta[\cos(kz)]\hat{\mathbf{y}} + \hat{\mathbf{z}} \tag{6.73}$$

and the superfluid velocity at the line is given approximately by

$$\mathbf{v}_s(z) = \nu k^2 \delta\{[\sin(kz)]\hat{\mathbf{x}} - (\cos(kz)]\hat{\mathbf{y}} + k\delta\hat{\mathbf{z}}\} \tag{6.74}$$

where $\nu = (\kappa/4\pi)\ln(1/ka)$. The normal fluid is assumed to be in solid-body rotation at frequency Ω about the z-axis and at the same time translating along the z-axis with velocity U_0:

$$\mathbf{v}_n(z) = -\Omega\delta[\sin(kz)]\hat{\mathbf{x}} + \Omega\delta[\cos(kz)]\hat{\mathbf{y}} + U_0\hat{\mathbf{z}} \tag{6.75}$$

Writing the velocity of a vortex line element as v_L, the Magnus force per unit length on the line is (Equation (3.6))

$$\mathbf{f}_M = \rho_s \kappa \mathbf{s}' \times (\mathbf{v}_L - \mathbf{v}_s) = \rho_s \kappa \mathbf{s}' \times (\mathbf{v}_L - \mathbf{v}_s) \tag{6.76}$$

The drag force per unit length experienced by a vortex line is (Equation (3.7))

$$\mathbf{f}_D = -\gamma_0 \mathbf{s}' \times [\mathbf{s}' \times (\mathbf{v}_n - \mathbf{v}_L)] + \gamma_0' \mathbf{s}' \times (\mathbf{v}_n - \mathbf{v}_L) \tag{6.77}$$

The motion of the line is determined by requiring that the net force on each line element vanishes:

$$\mathbf{f}_D + \mathbf{f}_M = 0 \tag{6.78}$$

The helical deformation of the line will either grow or decay, depending on whether the radial component of the line velocity is positive or negative. Solving (6.78) for this radial component and setting it to zero yields a 'critical' value for U_0:

$$U_{0,c} = (1/k)(\Omega + \nu k^2) \tag{6.79}$$

This expression is essentially the same as that derived from the HVBK equations, for a wavevector along the rotation axis, except that 2Ω is replaced by Ω. As pointed out before, this reflects the difference between isolated line dispersion ($\omega = \Omega + \nu k^2$) and extreme collectivization of the helical waves ($\omega = 2\Omega + \nu k^2$).

In an experiment reported by Cheng, Cromar, and Donnelly (1973), thermal counterflow was impressed along the axis of rotation in rotating helium. The attenuation of a transverse negative-ion beam – due to trapping of the ions on vortex lines – was found to decrease significantly as a result of the counterflow. These results probably can be explained in terms of the vortex-array instability. In a thermal-counterflow experiment (assuming Poiseuille flow for the normal fluid), one can estimate

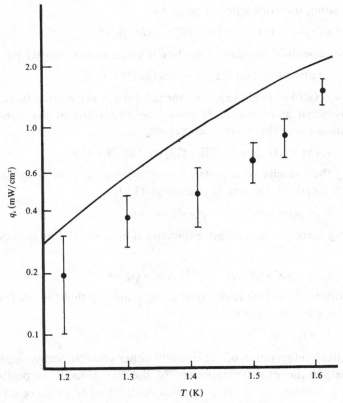

Figure 6.13 Critical heat current for the onset of the vortex instability as a function of temperature. The points are taken from smoothed data of Cheng *et al.* (1973), and the solid line is a plot of the theoretical critical heat current from Equation (6.80) (Glaberson *et al.* (1974)).

the heat current necessary to begin disrupting the vortex array by

$$q_c \sim \tfrac{1}{2}\rho_s ST(2\Omega\nu)^{1/2} \tag{6.80}$$

where ρ_s is the superfluid density, S is the specific entropy, and T is the temperature. The points in Figure 6.13 are the measured heat currents at which 20% of the recoverable ion beam was restored and the solid line is a plot of Equation (6.80), where Ω is taken as 2.5 rad/s. There is good qualitative and fair quantitative agreement between the theoretical and experimental results.

A more direct observation of the instability has been obtained by Swanson, Barenghi and Donnelly (1983). They measured the attenuation of second sound propagating in a direction perpendicular to an array of vortex lines induced by rotation. In the presence of thermal counterflow along the lines an onset of excess attenuation was observed at critical counterflow velocities in very good agreement with the predicted values.

6.5   Thermally induced vortex waves

We remarked in Section 6.1 that the dispersion relation (6.1) applies to vortex rings with the substitution $n = kR$. An interesting question is to find the population of vortex waves found on vortex rings in equilibrium with the remainder of the liquid at a temperature T. Such an investigation has been carried out by Barenghi, Donnelly and Vinen (1985) with rather surprising results. They found that near the lambda transition the free energy to create certain sizes of rings with waves seems to vanish so that spontaneous production of such rings should occur, destroying the superfluidity of helium II. The authors refer to this phenomenon as a 'free energy catastrophe', and speculate on its cause.

Let us consider a sample of helium II at temperature T when no vortex rings are present. Then the free energy is

$$F_1 = F_0 + F_{ex} \tag{6.81}$$

where F_0 is a constant and F_{ex} is the free energy of the excitations. In the case in which a vortex ring of given size and orientation is present we have

$$F_2 = F_0 + F_{ex} + E + F_w \tag{6.82}$$

where the free energy of the ring F consists of the energy E given by (1.34) and the contribution F_w of the thermally excited waves. We assume that F_{ex} is unaffected by the presence of the vortex ring. If the quantity

$$\Delta F_0 = F_2 - F_1 = E + F_w < 0 \tag{6.83}$$

then the system lowers its energy by creating vortices spontaneously. We are led therefore to studying the quantity $\Delta F_0 = E + F_w$ as a function of temperature and ring radius. The free energy of the vortex waves is calculated by quantizing the oscillations of the ring and evaluating the partition function using the Pocklington dispersion curve numerically, applying a Debye cut off in the usual way.

The surprising result is indeed that, for a given ring size R, there is a temperature $(< T_\lambda)$, above which F_w is large and negative enough that the ΔF_0 becomes negative. This can occur over a temperature range extending roughly 0.3 K below T_λ involving rings from $\sim 10^{-6}$ to 1 cm. Numerical calculations are given by the authors. Such a process would destroy superfluidity, so that the conclusion is inconsistent with experiment.

Barenghi *et al.* (1985) then proceed to examine various effects to discover whether they are likely to prevent the catastrophe from occurring. They discuss the effects of the choice of Debye cut off, the effect of finite amplitudes, mode interactions and mutual friction. None of these appears to be large enough to change the results of the calculation significantly. They conclude by observing that elementary excitations tend to be localized near vortex lines owing to the $\mathbf{p} \cdot \mathbf{v}_s$ interaction (see Section 2.5). On a microscopic theory there is no clear distinction between vortex waves and other excitations bound to the vortex. Pitaevskii (1961), for example, considered vortex lines in an imperfect Bose gas showing that among the solutions of the relevant equations (which show the Bogoliubov spectrum in a uniform condensate – phonons for small k, free particles for large k) is the spectrum of vortex waves. This suggests that for real liquid helium the vortex waves should be regarded as a particular form of elementary excitation bound to the vortex. The result is that the free energy cannot be computed by a method which does not take into account the modification of the excitation spectrum of the liquid produced by the presence of vortex lines.

6.6   Effect of mutual friction on vortex waves

One of the calculations carried out by Barenghi *et al.* (1985) is of interest here: namely the effect of mutual friction on vortex waves. Consider the vortex ring as a rectilinear vortex of length $2\pi R$ lying along the x-axis. We denote displacements of the vortex in the x- and y-directions by $\xi(z)$, $\eta(z)$. The balance of forces on a massless vortex filament is, as usual,

$$\mathbf{f}_D + \mathbf{f}_M = 0 \tag{6.84}$$

where $\mathbf{f}_M$ is the Magnus force per unit length, (3.6),

$$\mathbf{f}_M = \rho_s \kappa \times (\mathbf{v}_L - \mathbf{v}_i) \qquad (6.85)$$

and $\mathbf{f}_D$ is the friction force per unit length (3.7)

$$\mathbf{f}_D = -\gamma_0 v_L - \gamma_0' \kappa \times \mathbf{v}_L \qquad (6.86)$$

where $\mathbf{v}_i$ is the self-induced velocity of the line and $\mathbf{v}_L$ is its velocity in the laboratory system. We assume that there are no background velocities in either the normal fluid or the superfluid. A vortex has an energy per unit length given by Equation (1.21) and therefore a tension. If the vortex is bent the tension gives rise to a restoring force, acting at right angles to the vortex and in the plane in which the vortex is bent. The motion is such (see the discussion in Section 6.1) that this restoring force is balanced by a Magnus force acting on the moving line. Thus the term $-\rho_s \kappa \times \mathbf{v}_i$ can be replaced by a tension force T_0. From Equations (6.84)–(6.86) we have

$$\left.\begin{array}{l} -\rho_s \kappa \dot{\eta} - \gamma_0 \dot{\xi} + \gamma_0' \dot{\eta} + T_0 d^2 \xi / dz^2 = 0 \\ \rho_s \kappa \dot{\xi} - \gamma_0 \dot{\eta} - \gamma_0' \dot{\xi} + T_0 d^2 \eta / dz^2 = 0 \end{array}\right\} \qquad (6.87)$$

Assuming a wave-like dependence like $\exp[i(\omega t - kz)]$ for both ξ and η we have

$$\left.\begin{array}{l} i\omega(\gamma_0' - \rho_s \kappa)\eta - (i\omega\gamma_0 + T_0 k^2)\xi = 0 \\ i\omega(\gamma_0' - \rho_s \kappa)\xi + (i\omega\gamma_0 + T_0 k^2)\eta = 0 \end{array}\right\} \qquad (6.88)$$

which we can solve for ω:

$$\omega = \frac{2i\gamma_0 T_0 k^2 \pm \{-4\gamma_0^2 T_0^2 k^4 + 4T_0^2 k^4 [\gamma_0^2 + (\gamma_0' - \rho_s k)^2]\}}{2[\gamma_0^2 + (\gamma_0' - \rho_s k)^2]} \qquad (6.89)$$

Therefore the real and imaginary parts of ω are

$$\text{Re } \omega = \frac{T_0 k^2}{\rho_s \kappa} \frac{\gamma(\rho_s \kappa - \gamma_0')}{\gamma_0 \rho_s \kappa} = \frac{T_0 k^2}{\rho_s \kappa}(1 - \alpha') \qquad (6.90)$$

$$\text{Im } \omega = \frac{T_0 k^2 \gamma}{\rho^2 \kappa^2} = \frac{T_0 k^2 \alpha}{\rho_s \kappa} \qquad (6.91)$$

where the mutual friction coefficient γ, defined in Section 3.1, is given by

$$\gamma = \frac{\gamma_0 \rho_s^2 \kappa^2}{[\rho_0^2 + (\rho_s \kappa - \gamma_0')^2]} \qquad (6.92)$$

and the coefficients α and α' are defined in Table 3.1.

We know that in the limit of long wavelengths we have for the negative branch (see (6.1b))

$$\omega^- \simeq \kappa k^2 L / 4\pi \qquad (6.93)$$

where L is a slowly varying logarithmic term. From Equations (6.90) and (6.93) we identify in this limit

$$T_0 = \rho_s \kappa^2 L / 4\pi \qquad (6.94)$$

which is the energy per unit length, Equation (1.38). Equations (6.90), (6.91) can therefore be written as

$$\text{Re } \omega \approx \omega^- (1 - \alpha') \qquad (6.95)$$

and

$$\text{Im } \omega \approx \omega^- \alpha \qquad (6.96)$$

The effects of mutual friction are therefore a shift in the frequency of the wave (given by (6.95)) together with a damping (given by (6.96)). The authors show that, although the damping is large at the higher temperatures, the resulting uncertainty in the energy of the vortex wave is less than its total energy unless the reduced temperature, $\varepsilon = (T_\lambda - T)/T_\lambda$, is less than about 2×10^{-3}. Furthermore, the relative shift in frequency or relative shift in energy is less than 20% unless ε is less than 1.3×10^{-2}.

7   Superfluid turbulence

7.1   Background

All the vortex motions which have been discussed so far in this book
have been regular – vortex rings and vortex lines, perturbed only by
Magnus forces and friction. We now examine a far more common motion
of helium II, namely that associated with the turbulent motion of quantized vortex lines. This subject has had a long and complicated history
which has been the subject of two reviews (Tough, 1982; Donnelly and
Swanson, 1986). Theoretical developments have been described by
Schwarz (1985, 1988). So much progress has been made in recent
years, that we shall take a quite different approach in the present
chapter, namely to discover what can be said about the subject from
the point of view of vortex dynamics before describing the detailed
experimental situation. Superfluid turbulence is sometimes referred
to as quantum turbulence to emphasize the key role quantum mechanics plays in the phenomenon (see Donnelly and Swanson (1986)).
For our purposes it is sufficient to consider the situation in Figure 7.1
which is the simplest and most common method of production of
superfluid turbulence. A channel with a heater at a closed end, is operated
at a heat flux exceeding a critical value q_c. Second sound attenuation
measurements show that vortex line appears in the superfluid and
that it is homogenous and roughly isotropic: i.e., no regular array
exists such as is produced by rotation only. The vortex arrangement is
usually described as a tangle in the sense described in the quote by
Feynman in Section 2.3.2. A numerical simulation of a tangle is shown
in Figure 7.2. The measurements of vortex line are described as giving
a macroscopic average of the vortex line length per unit volume L;
L has dimensions cm^{-2} and $L^{-1/2}$ clearly defines a characteristic length
of the tangle.

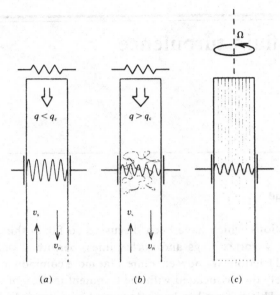

Figure 7.1 Schematic diagram of a thermal counterflow apparatus
for studying quantum turbulence. Second sound transducers are
represented on the side walls with a resonance excited between them:
(*a*) $q < q_c$, (*b*) $q > q_c$, (*c*) rotation only (after Donnelly and Swanson
(1986)).

More sophisticated measurements show that the vortex tangle is not
quite isotropic and that it drifts as a whole toward the heater. Since no
net mass flows in such an experiment,

$$\mathbf{j} = \rho_n \mathbf{v}_n + \rho_s \mathbf{v}_s = 0 \tag{7.1}$$

in either laminar or turbulent flow, the motion occurs by flow of superfluid
toward the heater and counterflow of the normal fluid toward the exit.
At high heat fluxes, a vigorous submerged jet of fluid can be observed
flowing from the exit.

Early experiments in relatively narrow channels reported that after a
critical heat flux, a substantial temperature gradient could be established
down the tube, and the Gorter-Mellink mutual friction term $\mathbf{F}_{ns}$ was
devised to account for the observations (see Section 2.4.2 and Equation
(2.28)). Feynman's speculation on turbulent vortices and Vinen's early
work on identifying mutual friction with vortices are reported briefly in
Section 2.4.2.

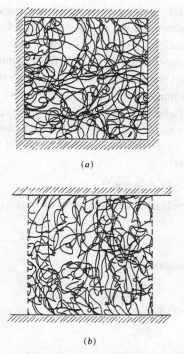

(a)

(b)

Figure 7.2 Simulation of a vortex tangle in a rough channel generated numerically by Schwarz (1988). (*a*) A projection along the flow direction, and (*b*) a projection of a section of channel viewed perpendicular to the flow direction. The tangle is generated by a superflow from right to left in (*b*). The wall and its surface roughness affect the distribution near the walls.

7.2 Application of vortex dynamics to superfluid turbulence

7.2.1 Dimensionless form of vortex dynamics

We have shown in Chapter 3 that if the inertia of the vortex cores is neglected, the motion of an element of line such as is shown in Figure 1.14 is given by Equation (3.17):

$$\mathbf{v}_{L} = \mathbf{v}_{sl} + \alpha \mathbf{s}' \times (\mathbf{v}_n - \mathbf{v}_{sl}) - \alpha' \mathbf{s}' \times [\mathbf{s}' \times (\mathbf{v}_n - \mathbf{v}_{sl})] \qquad (3.17)$$

Here $\mathbf{v}_L = \dot{\mathbf{s}}$ and $\mathbf{v}_{sl}$ consists of the vector sum of the potential flow of the superfluid at large distances from any vortex line, and $\mathbf{v}_i$ the superflow induced by any curvature of the vortex line in the sense discussed in Section 1.6, particularly near Equations (1.74)–(1.77).

We have observed in Section 1.1 that in classical incompressible hydrodynamics two geometrically similar flows are dynamically similar at the same Reynolds number (Equation (1.12)). This powerful concept underlies scaling of flows in all of classical incompressible fluid dynamics. The question then becomes, how can we accomplish scaling in quantum fluid dynamics? It is not difficult to make (3.17) dimensionless: we need only divide by v_{ns} which we define to be the absolute value of the spatial and temporal average of $\mathbf{v}_{ns} = \mathbf{v}_n - \mathbf{v}_s$ in the flow of interest. Such a procedure need not necessarily be restricted to flows obeying (7.1). We then have the dimensionless equation

$$\mathbf{u}_L = \mathbf{u}_i + \alpha \mathbf{s}' \times (\mathbf{u} - \mathbf{u}_i) - \alpha' \mathbf{s}' \times [\mathbf{s}' \times (\mathbf{u} - \mathbf{u}_i)] \tag{7.2}$$

where we have defined the dimensionless vortex line velocity in the frame of the superfluid by

$$\mathbf{u}_L = (\mathbf{v}_L - \mathbf{v}_s)/v_{ns} \tag{7.3}$$

a dimensionless tangle-induced velocity

$$\mathbf{u}_i = \mathbf{v}_i/v_{ns} \tag{7.4}$$

and a dimensionless counterflow velocity

$$\mathbf{u} = \mathbf{v}_{ns}/v_{ns} \tag{7.5}$$

7.2.2   Dynamical similarity

Let us investigate dynamical similarity using Equation (7.2). These ideas have been developed over a period of years by Schwarz (1982), Awschalom, Milliken and Schwarz (1984), Swanson (1985), Swanson and Donnelly (1985), Donnelly and Swanson (1986), Schwarz (1988). The discussion which follows is based upon these references.

Consider two geometrically similar flows whose characteristic dimension is d. Time will scale with d/v_{ns}, all tangle lengths will scale with d and the vortex ratio $\mathbf{u}_i$ must be the same for similar points on the tangles in each flow. The friction parameters α and α' must be the same in each flow, requiring the temperature to be the same. Since, as we have seen in Section 3.3, α and α' are slightly frequency- and velocity-dependent, dynamical similarity will be only approximate as we change velocities. This problem can be neglected in many practical situations. The core parameter a, however, depends only on temperature and cannot scale with d. This fundamental difficulty is ameliorated somewhat by the fact that the ratio d/a appears in a logarithmic term, as we shall now see.

The induced velocity $\mathbf{v}_i$ is given by the Biot–Savart law (1.74) and can be approximated as in (1.75) when the integral is over the whole tangle except for some distance a_{eff} on either side of $\mathbf{s}_0$, where a_{eff} is of order a. Since the radius of curvature $R = |s''|^{-1}$ and $\Gamma = \kappa$ we have

$$v_i = \frac{\kappa}{4\pi R} \ln\left(\frac{R_n}{a_{\text{eff}}}\right) = \frac{\kappa}{4\pi R} l \tag{7.6}$$

where we use R_n in place of L in (1.76b) to avoid confusion with the line density, and $l = \ln(R_n / a_{\text{eff}})$.

Applying (7.6) to two geometrically similar flows will require that $\mathbf{s}' \times \mathbf{s}''$ and thus $\mathbf{v}_i$ and $\mathbf{u}_i$ will be parallel at similar points in the two flows. Thus we need only consider the magnitude of u_i which from (7.4) and (7.6) is

$$|\mathbf{u}_i| = \kappa l / 4\pi R v_{\text{ns}} \tag{7.7}$$

Now let us consider a whole vortex tangle. Let $R_0 = \langle R \rangle$ and $l_0 = \langle l \rangle$ where the brackets denote an average over the whole tangle, and

$$u_i = \kappa l_0 / 4\pi R_0 v_{\text{ns}} \tag{7.8}$$

If $l = l_0$ throughout the tangle, then the requirement that u_i be the same in two flows is replaced by a requirement that u_i be the same. If l varies, equality of u_i will give approximate scaling if the distribution of l is narrow enough. Swanson and Donnelly (1985) show in Appendix B of their paper that the equation of vortex line motion will not differ significantly in channels of different sizes even with a relatively wide distribution.

The ideas of scaling are not particularly useful unless the necessary quantities are experimentally observable. Unfortunately R_0, a natural length in the problem, is not observable. What can be measured is the average vortex line length L per unit volume. Since L has dimensions cm^{-2}, a length scale $L^{-1/2}$ can be deduced, and we can define approximate time averaged proportionality constants c_1 and c_2 by

$$\overline{|s''|} = c_1 L^{1/2} = R_0^{-1}; \qquad \overline{|s''|^2} = c_2^2 L \tag{7.9}$$

which is independent of d for geometrically similar flows. We need in addition to find an expression for l_0, which is a function of the geometry of the tangle. Let us define the tangle geometry parameter g by

$$l_0 = \ln(R_0 / ag) \tag{7.10}$$

When we use the Arms–Hama approximation and the assumption of a narrow curvature distribution, $g = R_0 a_{\text{eff}} / a R_n$. The parameter g is configuration- and core-model-dependent. For a hollow vortex ring described by (1.59) with $\alpha = 2$, $g = e^2/8$. We expect that both c_1 and g will be of order unity.

With the definitions of c_1 and g, we are in a position to relate our scaling parameters to measurable quantities. In the calculations described here most quantities are tangle averages. Let us define the dimensionless applied counterflow velocity

$$V \equiv 4\pi v_{ns} d / \kappa l_0 = L^{1/2} c_1 d / u_i \qquad (7.11)$$

where the second equality is from the definition (7.8) and where now

$$l_0 = \ln (R_0/ag) = \ln (1/agc_1^{1/2}) \approx \ln (1/L^{1/2}a) \qquad (7.12)$$

For dynamical similarity $L^{1/2}d$, c_1, and u_i are independent of d. Thus V is a dimensionless velocity describing tangle properties.

If we define the quantity

$$\beta = \kappa l_0 / 4\pi \qquad (7.13)$$

then V has the form of a Reynold's number (see (1.12b)) with β replacing the kinematic viscosity. Since c and g are of order unity, (7.11) will give useful scaling without a knowledge of c or g as long as $a \ll L^{-1/2}$, i.e., one can use $l \approx \ln (1/L^{1/2}a)$ with relatively good results.

7.2.3   Anisotropy of the vortex tangle

Vortex rings which are in a counterflow will align with their axis parallel to the direction of counterflow and can grow or decay according to Equation (3.43). Simulation studies, which will be described below, show that vortices in a tangle tend to grow by ballooning outwards in a plane perpendicular to v_{ns} with the induced velocity v_i being preferentially oriented parallel to the driving velocity. This leads to anisotropy of the vortex line distribution with vortices concentrated somewhat in planes perpendicular to v_{ns}. One expects the distribution to be rationally symmetric about the v_{ns} direction.

Following a discussion of the anisotropy of the vortex tangle by Swanson and Donnelly (1985), Schwarz (1988) has introduced a useful characterization of anisotropy using unit vectors $\mathbf{1}_{\parallel}$ and $\mathbf{1}_{\perp}$ for directions parallel and perpendicular to the v_{ns} direction.

These are

$$I_{\parallel} = \frac{1}{\Omega L} \int [1 - (\mathbf{s}' \cdot \mathbf{1}_{\parallel})^2] \, d\xi \qquad (7.14)$$

$$I_{\perp} = \frac{1}{\Omega L} \int [-(\mathbf{s}' \cdot \mathbf{1}_{\perp})^2] \, d\xi \qquad (7.15)$$

and

$$I_l \mathbf{1}_\| = \frac{1}{\Omega L^{3/2}} \int \mathbf{s}' \times \mathbf{s}'' \, d\xi \tag{7.16}$$

If the tangle is homogeneous, these measures are independent of scaling or line length. If the tangle were isotropic, the averages $I_\| = I_\perp = 2/3$, $I_l = 0$. Conversely if the tangle consists of curves lying only in planes perpendicular to $\mathbf{v}_{ns}$ then $I_\| = 1$, $I_\perp = 1/2$. The relation

$$I_\|/2 + I_\perp = 1 \tag{7.17}$$

which follows from the symmetry of the problem is useful (see Swanson and Donnelly (1985), p. 394).

When second sound is propagated in a selected direction $\mathbf{1}_r$, the fraction of vortex line intercepted is given by

$$J(\mathbf{1}_r) = \frac{1}{\Omega L} \int [1 - (\mathbf{s}' \cdot \mathbf{1}_r)^2]^{1/2} \, d\xi \tag{7.18}$$

7.2.4 Applications of dynamical similarity

The ideas developed in the previous sections can now be applied to a wide range of experimental problems. A convenient way to do this has been devised by Schwarz (1988). He observed that if (3.17) is used in the spirit of the Arms–Hama approximation (see Section 1.6), then the curve $\mathbf{s} = \mathbf{s}(\xi, t)$ will be given by

$$\dot{\mathbf{s}} = \beta \mathbf{s}' \times \mathbf{s}'' + \mathbf{v}_s + \alpha \mathbf{s}' \times (\mathbf{v}_{ns} - \beta \mathbf{s}' \times \mathbf{s}'')$$
$$- \alpha' \mathbf{s}' \times [\mathbf{s}' \times (\mathbf{v}_{ns} - \beta \mathbf{s}' \times \mathbf{s}'')] \tag{7.19}$$

One can then implement a two-step process. First the factor β can be absorbed into reduced time and velocity scales

$$t_0 = \beta t, \qquad v_0 = v/\beta \tag{7.20}$$

(which now have dimensions cm^2 and cm^{-1}) to give the relationship

$$\partial \mathbf{s}/\partial t_0 = \mathbf{s}' \times \mathbf{s}'' + \mathbf{v}_{s,0} + \alpha \mathbf{s}' \times (\mathbf{v}_{ns,0} - \mathbf{s}' \times \mathbf{s}'')$$
$$- \alpha' \mathbf{s}' \times [\mathbf{s}' \times (\mathbf{v}_{ns,0} - \mathbf{s}' \times \mathbf{s}'')] \tag{7.21}$$

which is invariant under a transformation in which spatial dimensions are multiplied by a scale factor λ, reduced times by λ^2 and reduced velocities by λ^{-1}. If conditions are favorable, then a vortex configuration contained in some reference geometry denoted by an asterisk, with dimensions d^*, subject to reduced velocities $\mathbf{v}_0^*$ and evolving in reduced times t_0^*, will have an evolution in a geometry of scale d given by

$$\mathbf{s}(\xi, t_0) = \mathbf{s}(\lambda \xi^*, \lambda^2 t_0^*) = \lambda \mathbf{s}^*(\xi^*, t_0^*) \tag{7.22}$$

where, e.g.,

$$\lambda = \frac{d}{d^*} = \frac{v_0^*}{v_0} = \left(\frac{t_0}{t_0^*}\right)^{1/2} \tag{7.23}$$

As an example of this procedure, consider the line-length density of the tangle

$$L = (L/\Omega) \int d\xi \tag{7.24}$$

where Ω is the volume of the region involved. Assuming the average line density depends only on the counterflow velocity, the steady state average of line density L^* at some point r^* in a channel of dimensions d^* will scale as

$$
\begin{aligned}
L(\mathbf{r}, d, v_{\mathrm{ns},0}) &= L(\lambda \mathbf{r}^*, \lambda d^*, v_{\mathrm{ns},0}^*/\lambda) \\
&= \lambda^{-2} L^*(\mathbf{r}^*, d^*, v_{\mathrm{ns},0}^*)
\end{aligned} \tag{7.25}
$$

If we believe that the position in the channel is unimportant, then we are dealing with a *homogeneous turbulent state* and expressing λ in terms of velocities from (7.22) we find

$$\frac{L}{L^*} = \frac{1}{\lambda^2} = \frac{v_{\mathrm{ns},0}^2}{v_{\mathrm{ns},0}^{2*}} = \frac{(v_{\mathrm{ns}}/\beta)^2}{(v_{\mathrm{ns}}^*/\beta^*)^2} \tag{7.26}$$

and

$$L = \frac{L^*\beta^{*2}}{v_{\mathrm{ns}}^{*2}} \frac{v_{\mathrm{ns}}^2}{\beta^2} = c_L^2 \frac{v_{\mathrm{ns}}^2}{\beta^2} \tag{7.27}$$

As a different example, suppose the lowest velocity v_c for creating a tangle of quantized line in a channel of dimension d^* is found to depend only upon d^*. Then from (7.23) we can write

$$\frac{v_{c,0}}{v_{c,0}^*} = \frac{1}{\lambda} = \frac{d^*}{d} \tag{7.28}$$

Then

$$\frac{v_c d}{\beta} = \frac{v_c^* d^*}{\beta^*} = c_v \tag{7.29}$$

where again we use dimensionless velocities expressed as in Equation (7.11).

Consider now fluctuations in line density. Let us suppose that such fluctuations arise because of fluctuations in velocity described by a power spectrum density $S_v(f)$ where f is the frequency of observation. The power spectral density of line fluctuation $S_L(f)$ depends upon $(dL)^2$

which has dimensions cm^{-4}. If one observes that the high frequency fall-off depends only on frequency f, then

$$\frac{(dL)^2}{(dL^*)^2} = \lambda^4 = \left(\frac{t_0}{t_0^*}\right)^2 = \left(\frac{f_0^{*2}}{f_0^2}\right) \tag{7.30}$$

and

$$(dL)^2 = \frac{(dL^*)^2 f^{*2}}{\beta^{*2}} \frac{\beta^2}{f^2} = c_f^2 \frac{\beta^2}{f^2} \tag{7.31}$$

which suggests that the $S_L(f)$ of line fluctuations should vary inversely with the square of the frequency, but with logarithmic corrections.

Schwarz (1988) has observed that it is possible to consider the dynamical balance between vortex growth and decay. The instantaneous fractional rate of change of the line length at some particular point on the vortex is equal to $s' \cdot \partial s'/\partial t$. Thus an element of line of length $\Delta \xi$ is described by

$$\frac{1}{\Delta\xi} \frac{\partial \Delta\xi}{\partial t_0} = \alpha [\mathbf{v}_{ns,0} \cdot (\mathbf{s}' \times \mathbf{s}'') - |\mathbf{s}' \times \mathbf{s}''|^2] - \alpha' \mathbf{v}_{ns,0} \cdot \mathbf{s}'' \tag{7.32}$$

which leads to

$$\frac{\partial L}{\partial t_0} = \frac{\alpha}{\Omega} \mathbf{v}_{ns,0} \cdot \int \mathbf{s}' \times \mathbf{s}'' \, d\xi - \frac{\alpha}{\Omega} \int |s''|^2 \, d\xi \tag{7.33}$$

where the term in α' has been dropped for reasons of symmetry. In the steady state, the ensemble average of this equation must be zero. In addition, the assumption of homogeneity allows use of Equations (7.27) and (7.16) to obtain the equation

$$\frac{\partial L}{\partial t_0} = \alpha I_1 v_{ns,0} L^{3/2} - \alpha c_2^2 L^2 \tag{7.34}$$

where c_2 is defined by Equation (7.9). Strictly speaking, this is valid only when $\partial L/\partial t_0 = 0$, since we are assuming the scaling coefficients do not change much for deviations near the steady state. Setting $\partial L/\partial t_0$ to zero and comparing with Equation (7.27) implies that $c_L = I_1/c_2^2$, reflecting the fact that in the final analysis the equilibrium density of the tangle is achieved by balancing the mean anisotropy of the self-induced velocity $\mathbf{s}' \times \mathbf{s}''$ against its magnitude. Equation (7.34) can then be reexpressed in terms of scaling coefficients,

$$\partial L/\partial t_0 = \alpha I_1 [v_{ns,0} L^{3/2} - c_L^{-1} L^2] \tag{7.35}$$

A model for the development of the tangle of quantized vortex lines was introduced by Vinen in his pioneering papers (Vinen 1957*a*, *b*, *c*,

1958). He considered a spatially homogeneous distribution of vortex lines whose time rate of change is determined by competing growth and decay processes. Vinen derived the growth term by dimensional analysis and modelled the decay process after the decay of classical turbulence in the Kolmogorov cascade. He obtained

$$dL/dt = \chi_1 B\rho_n v_{ns} L^{3/2} - \chi_2 \kappa L^2 / 2\pi \qquad (7.36)$$

where χ_1 and χ_2 are undetermined parameters. The relationship of (7.36) to (7.35) is obvious. The derivation given by Schwarz reproduces the Vinen equation, but now relates the coefficients to quantities derivable from vortex dynamics. The Vinen parameters χ_1 are χ_2 are now seen to be given by

$$\chi_1 = I_1, \qquad \chi_2 = 2\pi\alpha\beta I_1/\kappa c_L \qquad (7.37)$$

The mutual friction force $\mathbf{F}_{ns}$ which is used in the equations of motion (2.2) and (2.3) and has the empirical (Gorter–Mellink) form (2.28) can be defined as an integral of the drag force $\mathbf{f}_D$ over the tangle. Using $\mathbf{f}_D$ from (3.8) we have

$$\mathbf{F}_{ns} = (-\rho_s \kappa \alpha / \Omega) \int \mathbf{f}_D \, d\xi$$

$$= (-\rho_s \kappa \alpha / \Omega) \int \mathbf{s}' \times [(\mathbf{s}' \times (\mathbf{v}_{ns} - \mathbf{v}_i)] \, d\xi \qquad (7.38)$$

the term in α' vanishing by symmetry (see Swanson and Donnelly (1985) p. 391). The resultant average force in the $\mathbf{v}_{ns}$-direction is

$$F_{ns} = \rho_s \kappa \alpha (c_L^2 I_{\parallel} - c_L^3 I_1) v_{ns}^3 / \beta^2 = \rho_s \kappa \alpha (I_{\parallel} - c_L I_1) L v_{ns} \qquad (7.39)$$

from (7.26).

The average drift velocity of the vortex tangle with respect to the superfluid rest frame is

$$\mathbf{v}_l = (1/\Omega L) \int \dot{\mathbf{s}} \, d\xi - \mathbf{v}_s \qquad (7.40)$$

where $\dot{\mathbf{s}}$ is given by (7.19), and averages to

$$v_l = [c_L(1 - \alpha') I_1 + \alpha' I_{\parallel}] v_{ns} \qquad (7.41)$$

in the direction of $\mathbf{v}_{ns}$. Here the term in α has vanished by symmetry.

When one wishes to go further than scaling allows, for example, to compute the temperature dependence of critical velocities, some new theoretical input is needed. So far, straight estimates from theory have not been developed. But computer simulations of a vortex tangle have been carried out in a lengthy investigation by Schwarz, which we describe in the next section.

7.3   Numerical simulation of a vortex tangle

The dynamics of vortices in helium II forms a particularly attractive problem to simulate numerically. The equations of motion, which we have discussed in Chapter 3 and used extensively in the previous two sections are useful in the thin-filament approximation where the Arms–Hama localized induction hypothesis is often employed as a simplification. We have referred briefly to the results of simulation studies in Section 5.6 on vortex pinning and in Chapter 6 on solitary and Kelvin waves.

The idea of applying vortex dynamics to the full problem of superfluid turbulence is a formidable one. Schwarz (1978, 1982, 1983, 1985, 1988) has nevertheless developed this approach in considerable detail, and has presented results which compare favorably with experiments, which we shall now discuss.

7.3.1   *Vortex reconnections*

Equation (3.8), which is used in (7.38) neglects all dynamical effects arising from other vortices or boundaries. Many years ago Feynman (1955) speculated that vortices approaching each other closely would reconnect as shown in Figure 7.3(a). On a small enough scale, this event is surely quantum mechanical. Schwarz has observed that vortex–vortex reconnection provides a mechanism by which vortex regularities can multiply (Figure 7.3(c)), and generalized this idea to include vortex–surface reconnections, as shown in Figure 7.3(b). He estimated how closely a line of radius of curvature R must come to a boundary before its image-induced motion becomes comparable to its self-induced motion. Calling the distance between a loop and its image Δ he found

$$\Delta \approx 2R/\ln\left(cR/a\right) \tag{7.42}$$

where c is a constant of order 1. When two vortices approach closer than a distance Δ, a local instability occurs in which the velocity field of each vortex acts to deform the other in such a way that the two vortices are driven together at a point where their vorticity vectors are oppositely directed. He assumed that the result of this is a sudden reconnection as shown in Figure 7.3 with immediate separation of the vortices to macroscopic distances.

On the quantum mechanical side, we have noted in Section 4.2 that Jones and Roberts (1982) have demonstrated that a vortex ring in the condensate can shrink only to a certain value of the ratio of ring radius

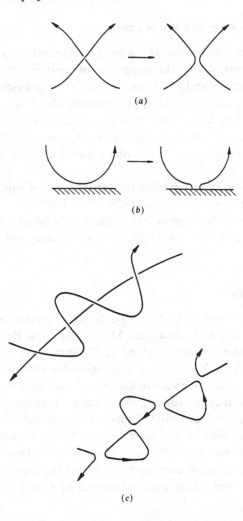

(a)

(b)

(c)

Figure 7.3 Illustration of (a) a possible reconnection sequence between vortex filaments in a tangle, (b) a vortex filament reconnection at a surface and (c) multiplication of singularities through the reconnection process. Here two vortex lines reconnect to form five (after Schwarz (1988)).

to core size. At that point the circulation disappears and the remnant of the vortex becomes solitary waves of compression. It is conceivable that it is this sudden cessation of circulation that allows reconnection to occur without violating Kelvin's circulation theorem (Equation (1.19)).

Schwarz noted that his studies demonstrate that the details of when and how the vortices are reconnected or of how they behave immediately thereafter have no significant influence on the behavior of the vortex tangle. He referred to a simulation consisting of Equation (3.17) and the reconnection assumptions of Figure 7.3 and Equation (7.42) as the *reconnecting vortex tangle model*.

From his experience Schwarz was able to give a qualitative picture of the self-sustaining vortex tangle state (Schwarz 1988):

The self-induced velocity causes a complicated three-dimensional internal motion of the vortex tangle, the whole thing being washed along by any applied superflow field v_s which may be present. Highly curved sections of line, and sections propagating opposite to v_{ns}, decay. Simultaneously, other parts of the vortex tangle where the self-induced motion is being overtaken by the v_{ns} field grow by ballooning outwards. The cross-stream nature of the vortex growth implies that in the steady state at least a certain fraction of the singularities is constantly being driven toward the walls. The line–line reconnections which occur as the vortex tangle undergoes its complicated dance play several important roles. First, they provide a mechanism by which new vortex singularities can be created [Figure 7.3], allowing the vortex tangle to be established and sustained against the loss of singularities at the walls. Secondly, and more subtly, since the vortex amplification process is essentially a two-dimensional outward motion in the plane perpendicular to v_{ns}, the reconnections and the subsequent motions along v_{ns} which result are necessary to maintain the three-dimensional random nature of the vortex tangle. Finally, the reconnections occur more often as the tangle becomes denser. The increasing frictional line loss associated with the creation of a more and more highly curved vortex tangle is the factor which eventually limits the tangle density. All of these complicated dynamical features interact self-consistently to produce the turbulent steady state.

7.3.2 Computational considerations

The evolution of a vortex tangle can be investigated beginning with an initial configuration such as is shown in Figure 7.4. The algorithms for stepping the vortex configuration forward in time are described by Schwarz (1985). Since the calculations are performed in a finite sample of fluid, it is necessary to specify how the boundaries are treated. The sample of fluid is always taken as a rectangular box with one set of faces perpendicular to $\mathbf{v}_{ns}$. This set of faces is subject to periodic boundary conditions, i.e., one line leaving the box appears to reenter it from the opposite face. This makes the fluid appear infinite in the direction of the flow. The other boundaries are treated in one of the three ways: as periodic, if the intent is to make the fluid unbounded in all directions;

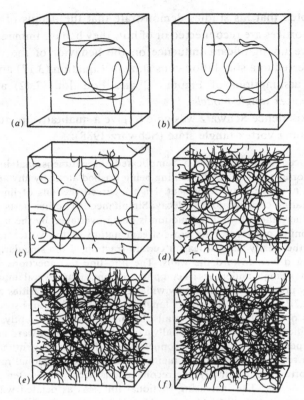

Figure 7.4   Case study of the development of a vortex tangle in a real channel. Here, $\alpha = 0.10$, corresponding to a temperature of about 1.6 K, and $v_{s,0} = 75$ cm^{-1} into the front face of the channel section shown. (a) $t_0 = 0$ cm^2, no reconnections; (b) $t_0 = 0.0028$ cm^2, three reconnections; (c) $t_0 = 0.05$ cm^2, 18 reconnections; (d) $t_0 = 0.20$ cm^2, 844 reconnections; (e) $t_0 = 0.55$ cm^2, 12 128 reconnections; (f) $t_0 = 2.75$ cm^2, 124 781 reconnections (after Schwarz (1988)).

as smooth, rigid boundaries, in which case a vortex line approaching the face will reconnect to the wall as shown in Figure 7.3(b), the end then gliding freely along the wall; or as rough rigid boundaries, in which case vortex lines terminating on the wall undergo a complicating pinning and depinning motion as they move along (see Section 5.5.3).

The evolution of a typical numerical experiment is shown in Figures 7.4 and 7.5. Here an initial configuration of six vortex rings is allowed to evolve in a rough channel under the influence of a pure superflow driving field. This situation is seen to evolve towards a self-sustaining

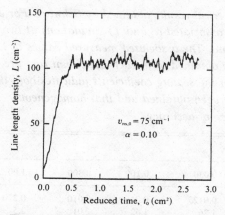

Figure 7.5 Line length per unit volume as a function of reduced time for the run shown in Figure 7.4. Note the occurrence of significant intrinsic fluctuations in $L(t)$ after the tangle has reached the steady state (after Schwarz (1988)).

chaotic steady state with well-defined average properties independent of the initial conditions.

The rough wall calculations are straightforward, but computationally expensive. Periodic boundary conditions in all directions are much more efficient, but run into difficulties which are removed by a special mixing step discussed in detail by Schwarz (1988). This procedure uses the assumption that lines which encounter each other closely will reconnect with essentially unit probability.

7.3.3   Results of numerical simulations

The results of calculations using periodic boundary conditions with a mixing step are given in Table 7.1. The results here, derived by numerical experiment, may be taken as the standard (starred) condition discussed above in connection with scaling. The calculations make a definite assumption of the variation of α with temperature (independent of velocity and frequency) which we reproduce in Figure 7.6.

The internal structure of the tangle can be seen from the numerical data of Table 7.1 to depend on temperature through α. For example, the tangle changes from being quite anisotropic at large α to a more isotropic structure as the friction constant is decreased. The variation of c_1 and c_2 shows that the vortex tangle becomes increasingly kinky as the friction constant is reduced. Figure 7.7 illustrates these trends.

Table 7.1. *Summary of numerical results obtained by Schwarz. For each friction constant (α and the associated α' and T) simulations at two velocities $v_{ns,0}$ were carried out. The associated measured values of L and of the scaling coefficients, as obtained from each individual run, are shown. The reproducibility of the scaling coefficients indicates both that reliable average values have been obtained and that homogeneous scaling holds to a high degree of accuracy*

	α				
	0.010	0.030	0.100	0.300	1.00
α'	0.005	0.0125	0.016	0.010	−0.270
T (K)	1.07	1.26	1.62	2.01	2.5
$v_{ns,0}$ (cm^{-1})	140	55	40	20	25
	190	80	135	30	35
L (cm^{-2})	24.5	16.0	31.7	19.4	45.8
	48.1	34.0	323.4	43.6	95.8
c_L	0.0353	0.0728	0.141	0.220	0.270
	0.0365	0.0729	0.133	0.220	0.280
c_1	2.91	2.00	1.41	1.02	0.718
	2.81	1.97	1.48	1.05	0.719
c_2	3.47	2.46	1.79	1.40	1.16
	3.33	2.42	1.84	1.43	1.14
$I_\parallel$	0.721	0.746	0.787	0.875	0.954
	0.719	0.749	0.770	0.870	0.952
$I_\perp$	0.437	0.454	0.461	0.440	0.355
	0.428	0.442	0.454	0.460	0.358
$J_\parallel$	0.825	0.841	0.870	0.927	0.975
	0.823	0.844	0.858	0.924	0.973
$J_\perp$	0.756	0.756	0.739	0.700	0.665
	0.766	0.755	0.745	0.705	0.667

7.4 Production, measurement and analysis of superfluid turbulence

7.4.1 Devices for producing turbulence in the superfluid

The simplest and by far the most common method of production of superfluid turbulence occurs when a channel, heated at a closed end, is operated at a heat flux exceeding a critical value, as shown in Figure 7.1. Counterflow channels break down into two broad categories, narrow and wide channels. Narrow channels can be studied only by means of temperature and/or pressure gradients, wide channels have the space to be equipped for second sound attenuation or ion measurements. Wide channels, then, tend to be several millimeters to a centimeter in diameter

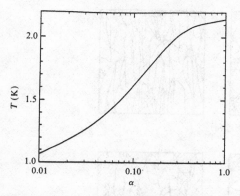

Figure 7.6 Assumed relation between the friction constant α and temperature (after Schwarz (1988)).

and narrow channels tend to be capillary tubes. The range of channel sizes in counterflow turbulence is dramatic: from microns to centimeters, or four orders of magnitude!

The principles of wide channel construction are simple. We show in Figure 7.1 a thermal counterflow apparatus for studying quantum turbulence equipped with second sound transducers to probe the vorticity transverse to the axis of the channel. Second sound attenuation is observed to begin only above a critical heat flux q_c (q is the ratio of power W supplied to the heater to the cross-sectional area of the channel A). The channel shown can be rotated to produce a regular array of vortices for calibration purposes or to conduct experiments with combined heat and rotation.

A fully instrumented wide channel used for both transverse and axial second sound attenuation measurements and axial temperature difference measurements is shown in Figure 7.8.

7.4.2 Second sound and temperature difference measurements in a wide channel

Pure counterflow measurements in a channel 1 cm × 1 cm, 40 cm long were pursued by the Oregon group for about ten years. The advantage of a wide channel is that second sound transducers can be installed to measure the attenuation of second sound owing to the vortices. The disadvantage is that the effective thermal conductivity of the fluid is so large because of the superfluidity that it is difficult to measure the temperature gradient.

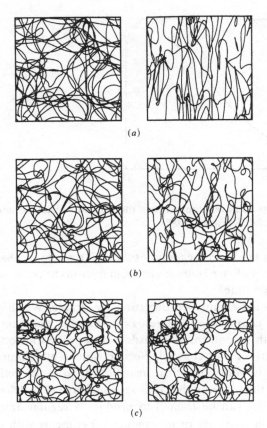

(a)

(b)

(c)

Figure 7.7 Illustration of the predicted changes in the internal structure of the vortex tangle as α is varied. Each figure on the left shows the computational volume at a particular instant, viewed along the flow direction. The corresponding figure on the right shows the same instant viewed across the flow direction. These tangles were generated using periodic boundary conditions with mixing, and the driving velocities were chosen to give L approximately equal to 47 in each distance. (a) $\alpha = 1.00$; (b) $\alpha = 0.10$; (c) $\alpha = 0.01$ (after Schwarz (1988)).

A counterflow channel was designed and built in the early 1980s which exhibits axially homogeneous line density. The channel is constructed of interchangeable units joined by indium 'o' ring seals, into some of which one can insert second sound transducers or thermometers (for details, see Barenghi (1982), Swanson (1985)). One possible arrangement of the channel is shown schematically in Figure 7.8.

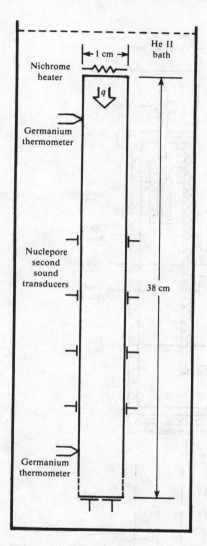

Figure 7.8 Schematic of one possible arrangement of wide channel interchangeable sections used at the University of Oregon. Both transverse and axial second sound transducers are shown, as well as thermometers to measure the axial temperature gradient (after Donnelly and Swanson (1986)).

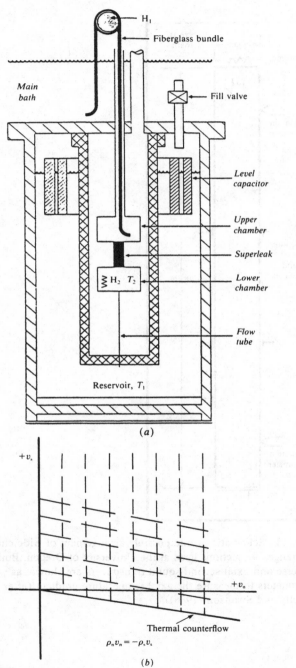

H_1

Fiberglass bundle

Main bath

Fill valve

Level capacitor

Upper chamber

Superleak

H_2 T_2

Lower chamber

Flow tube

Reservoir, T_1

(a)

$+v_s$

$+v_n$

Thermal counterflow

$\rho_n v_n = -\rho_s v_s$

(b)

Figure 7.9

The heater is a nichrome film deposited on a fused quartz substrate, joined by indium seals to the channel. Electrical contact to the nichrome is through superconducting indium along two opposite edges, providing a uniform current density. The heat capacity of the film is very low and hence can respond very rapidly to pulsed inputs.

The transverse second sound transducers are capacitor loudspeakers and microphones with Nuclepore membranes. The membranes have many 0.1 μm holes, allowing ready passage of superfluid but nearly no passage of normal fluid; membrane oscillations create counterflowing oscillations of the normal and superfluid, i.e., second sound. For details of their behavior, see Giordano (1984a, b), Giordano and Musikar (1984), and references therein.

The axial attenuation of second sound is measured by transducers at the bottom of the channel. The transmitter is made of electrical resistance board, and the receiver is a Nuclepore transducer as described above. The resistance board is able to produce a powerful low frequency burst of second sound needed to traverse the long axial path in the presence of high vortex line densities.

Second sound attenuation can be determined by the width of a resonance peak or the decay of echos of a second sound burst. The two methods agree to better than 2%. The best line density resolution one can currently achieve is about 20 cm^{-2} over a range of 0–200 000 cm^{-2}.

Axial temperature difference measurements are made by using two germanium resistance thermometers mounted 30 cm apart on the sidewalls, forming two legs of an impedance bridge. The thermometers are matched (i.e., their resistances as a function of temperature are nearly the same) so that correlated temperature fluctuations (e.g., bath temperature fluctuations) cause only a small second-order effect in temperature difference measurements. Temperature gradients smaller than 10^{-8} K/cm could be determined in this way.

Caption for Figure 7.9.

Figure 7.9 (a) Apparatus designed by Courts and Tough (1988) to produce flows with v_s and v_n in the same or opposite directions. (b) Illustration of the first and fourth quadrants of the (v_n, v_s)-plane showing the thermal counterflow trajectory (solid line) and two types of experimental trajectories (dashed lines) (after Courts and Tough (1988)).

7.4.3   Production of turbulence in narrow channels

An example of a modern apparatus to generate turbulence in narrow tubes is shown in Figure 7.9(a). This apparatus, designed by Courts and Tough (1988) allows a wide variety of flows of varying directions of v_n and v_s to be generated.

The flow tube has a 0.0134 cm inside diameter and connects a large reservoir at temperature T_1 to a small lower chamber at temperature T_2 containing a heater H_2. A further small upper chamber is connected by a superleak, allowing superfluid flow, but not normal flow, nor (of course) heat flow.

A bundle of fiberglass fibers wrapped on a heater H_1 is used to siphon superfluid from the main bath through the superleak and through the flow tube into the reservoir. The fiberglass bundle consists of 7200 fibers of approximately 10^{-3} cm diameter. These fibers become coated with a helium film of order 10^{-6} cm in thickness and provide a total perimeter for the film of order 22 cm. Heater H_1 acts as a control valve by increasing the local temperature of the helium film. The sensitivity of film flow to changes in temperature results in fine control of the superfluid transfer rate. The superfluid velocity through the film bundle alone is denoted V_{sm} (superfluid mass flow) and ranges from 0–25 cm/s at 1.5 K. The actual value of V_{sm} is determined by the rate of change of level in the reservoir as superfluid is transferred from the bath. A cylindrical level-sensing capacitor is monitored with a capacitance bridge. The heater H_2 produces a counterflow in the flow tube. Since no heat can pass through the superleak the average normal fluid velocity v_n is given by $q/\rho ST$ where q is the heat flux. The superfluid velocity V_{sCF} due to counterflow comes from the relationship $j = \rho_s v_s + \rho_n v_n = 0$. Appropriate combinations of counterflow and superflow can provide velocities with v_n parallel to v_s, and v_s opposite to v_n over a region limited by the counterflow relationship $j = 0$. Pure counterflow can be induced by leaving the heater H_1 off. The accessible regions of the (v_s, v_n)-plane are shown in Figure 7.9(b).

7.4.4   Analysis of counterflow experiments

Much of the study of quantum turbulence is done in steady state pipe flow. The time and cross-sectional averages of Equations (2.2) and (2.3) in steady state (the 'mutual friction approximation') are, to first order,

$$0 = -(\rho_s/\rho)\langle \nabla P \rangle + \rho_s S \langle \nabla T \rangle - \langle \mathbf{F}_{ns} \rangle \tag{7.43}$$

$$0 = -(\rho_n/\rho)\langle \nabla P \rangle - \rho_s S \langle \nabla T \rangle + \langle \mathbf{F}_{ns} \rangle + \eta \langle \nabla^2 \mathbf{v}_n \rangle \tag{7.44}$$

where the angle brackets denote time and cross-sectional averages. The quantity $\mathbf{F}_{ns}$ is the mutual friction force between normal and superfluid components and is zero in the absence of quantized vortices (Section 3.1). In the presence of superfluid turbulence, Gorter and Mellink (1949) proposed the form (2.28) where A is a function of T (and in principle v_{ns}) and is of order $50 \text{ cm s}^{-1}\text{ g}^{-1}$. Adding Equations (7.43) and (7.44), we find

$$\langle \nabla P \rangle = \langle \eta \nabla^2 \mathbf{v}_n \rangle \tag{7.45}$$

This suggests that the pressure gradient ∇P in turbulent superflow is not much different than in laminar (Poiseuille) flow, which is only very roughly true, and is called 'the Allen and Reekie rule'. Since it is known that turbulence in classical fluids has a drastic effect on the velocity profile, it is difficult to see why there should be similarity here to laminar flow.

With a laminar mean flow assumption, we have for the pressure gradient

$$\nabla P_L = -G\eta \mathbf{v}_n / d^2 \tag{7.46}$$

where G is a factor which one can obtain theoretically for Poiseuille flow for each channel shape and d is the channel size. Hereafter we omit the brackets and denote temporal and spatial averages by v_n and v_s. The gradient of the thermodynamic potential gradient per unit mass, often called the chemical potential in the helium II literature, vanishes in laminar flow:

$$\nabla \mu_L = -S \nabla T_L + \nabla P_L / \rho = 0 \tag{7.47}$$

consistent with dissipationless flow for the superfluid. Thus the laminar flow temperature gradient is

$$\nabla T_L = \nabla P_L / \rho S = -G\eta \mathbf{v}_n / \rho S d^2 \tag{7.48}$$

In turbulent flow, the total temperature gradient is

$$\nabla T = \nabla T_L + \nabla T_T \tag{7.49}$$

where ∇T_T is, neglecting the increase in ∇P above the laminar value,

$$\nabla T_T = \mathbf{F}_{ns} / S \rho_s \tag{7.50}$$

and since F_{ns} varies roughly as v_{ns}^3, the neglect of the extra pressure gradient becomes quantitatively reasonable for large flow rates. Equation (7.50) is the standard way to measure mutual friction in narrow tubes where the only observable may be the temperature gradient. The chemical potential gradient is given by

$$\nabla \mu = -\mathbf{F}_{ns} / \rho_s \tag{7.51}$$

This shows that dissipation owing to vortices produces a direct change in chemical potential.

The connection of the mutual friction coefficients to line length per unit volume may be seen by a simple analogy. In uniform rotation, from Equation (3.20), the coefficient of the dissipative term of $\mathbf{F}_{ns}$ parallel to $\mathbf{v}_{ns}$ has the magnitude $B\rho_s\rho_n\kappa L/2\rho$ since the vorticity $\omega = \kappa L$. For turbulent flow the assumption is that mutual friction acts in the same manner on each segment of vortex line in a tangle as it does on an array produced by rotation. It has been assumed traditionally that the tangle is isotropic, i.e., that an average of one-third of the vortex line segments in the tangle will be oriented parallel to the second sound propagation direction and not detected. Thus it has been customary to assume

$$\mathbf{F}_{ns} = -\rho_s\kappa\alpha\tfrac{2}{3}L\mathbf{v}_{ns} \tag{7.52}$$

We know, however, from experiments by Wang, Swanson and Donnelly (1987) and the discussion of Section 7.2 that the tangle is not isotropic and that an equation such as (7.39) should be used.

7.4.5   Shock wave apparatus

Second sound shock waves form an interesting, but perhaps underused, probe of quantum turbulence. Liepmann and Laguna (1984) describe shock wave experiments in helium II in some depth. Here we include only a brief description of the apparatus.

The gas dynamical cryogenic shock tube for liquid helium research was developed by Liepmann, Cummings and Rupert (1973) and Cummings (1974). A shock wave is produced in the helium vapor above the bath by modified standard techniques and allowed to strike the free surface of the bath, producing both a pressure and a temperature shock in the liquid helium II.

Temperature shock waves without significant accompanying pressure shock waves can be produced easily by the rapid heating of a thin conducting film. A series of such tubes has been constructed by Liepmann's group (Liepmann and Laguna, 1984). One of these tubes is shown in Figure 7.10. Part 1 is a brass housing for the apparatus, 2 is a spring loading for the heater, a 1000 Å thick nichrome film, vacuum deposited on a quartz substrate 3. The sensor, 4, is a thin film of superconducting material biased to be at its transition field by a superconducting magnet 5. Such a biased superconducting film has very fast and

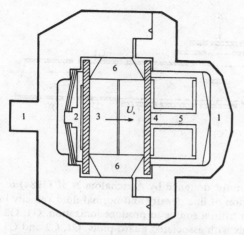

Figure 7.10 Thermal shock wave apparatus built at the California Institute of Technology. The operation is described in the text (after Liepmann and Laguna (1984)).

sensitive response to heat pulses. The shock wave tube, 6, is made of Teflon.

7.4.6 Ion measurements in a wide channel

Another tool, which has been profitably used, is the interaction of ions in liquid helium with vortices, which was discussed in Section 4.5.

Awschalom, Milliken and Schwarz (1984) have used pulsed ion techniques to study the vortex–line length density distribution and normal fluid velocity field in turbulent counterflow. Their apparatus is shown in Figure 7.11. An ion source and a variety of interchangeable grid and collector assemblies are built into the channel walls. In a typical measurement, a narrow pulse of negative ions is gated into the channel and allowed to propagate to a particular position under the action of a drift field E. The field is then switched off, allowing the pulse to remain at this position for as long as several seconds. Some of the ions in the pulse are trapped by quantized vortices at a rate determined by the local value of the line density L; the rest drift along with the local normal fluid velocity v_n. Later, E is turned back on and the received pulse is measured by means of one or more electrometers as it arrives at the collectors. From the amplitude of the observed pulse and where it falls on the collectors, local values of L and v_n can be readily determined as a function

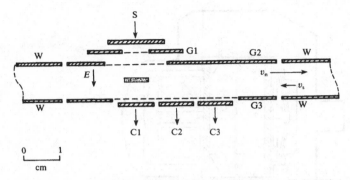

Figure 7.11 Apparatus designed by Awschalom *et al.* (1984) to study the distribution of line density and normal fluid velocity in a counterflow. S is a tritium source to produce ionization. G1, G2 and G3 are pulsed grids with associated guard plots, C1, C2 and C3 are collectors. W is the grounded channel well. The dotted region represents a typical pulse of charge and extends 0.8 cm into the plane of the figure. The heater producing the flow is on the left and is not shown.

of where the pulse was stopped. The spatial resolution achieved with this interrupted flight technique is better than 1 mm.

7.4.7   The normal jet

A flow which can be produced only in helium II is the jet emerging from a counterflow channel. This experiment attracted the interest of Kapitza in his pioneering work on superfluidity, then was ignored for over 30 years until taken up by Liepmann's group at California Institute of Technology. A review by Liepmann and Laguna (1984) traces much of this modern investigation. The conventional understanding of the jet is shown in Figure 7.12, where the normal fluid exits in a well-defined jet and the superfluid flows back into the orifice in potential flow. This picture leads to the conclusion that the counterflow velocity in the main part of the jet $v_{ns} \approx v_n$. High values of v_{ns} will lead to the formation of vortices in the submerged jet, and therefore vortices should be detectable by thermometry or by second sound.

Dimotakis and Broadwell (1973) carried out an experiment to measure the temperature profile of the jet by traversing a small carbon thermometer along the axis of the jet. The ($\sim\frac{1}{4}$ mm) tube shape was designed in such a way as to have the highest counterflow velocity near the orifice as

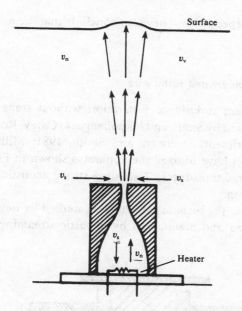

Figure 7.12 The counterflow jet shown in the standard picture in which the superfluid flows into the orifice toward the heater in potential flow and the normal fluid exits as a classical viscous jet which is shown impinging upon a free surface (after Liepmann and Laguna (1984)).

shown in Figure 7.12. The results showed that the temperature gradient extends only over the region confined by the channel walls near the exit.

In another experiment Laguna (1975) measured the attenuation of second sound in the jet. The jet passed through a plane parallel resonator for second sound, and the change in Q of the resonance was measured. The attenuation in the jet was found to be 15 times less than that expected in a thermal counterflow channel and varied as $q^{3/2}$ instead of q^2.

Liepmann and Laguna (1984) suggest that the measurements described above support the idea that the superfluid is entrained with the jet. This means that there must be a superfluid stagnation point at the orifice with superfluid flow into the channel and along with the normal fluid jet. Rather than attribute the small attenuation of second sound to vortices in the jet, the authors calculate the scattering of second sound by fluctuations in the jet on a model by Liepmann (1952) for the diffusion of a light ray passing through a turbulent boundary layer. They find nontrivial agreement with helium experiments and theory, and conclude that the superfluid must move with the normal fluid: using v_{ns} in place of v_n gives the wrong temperature dependence. They also conclude that the normal jet is turbulent.

7.4.8   *Ultrasonically genenerated turbulence*

A novel form of quantum turbulence generation without transverse boundaries has been devised by Smith and his colleagues (Carey, Rooney and Smith, 1978). Experiments (Schwarz and Smith, 1981; Milliken, Schwarz and Smith, 1982) have utilized the apparatus shown in Figure 7.13. The pair of ultrasonic transducers T set up a strong acoustic field in the cross-hatched region.

The authors believe that the turbulence, which is studied by negative ion trapping, is generated and maintained by acoustic streaming and

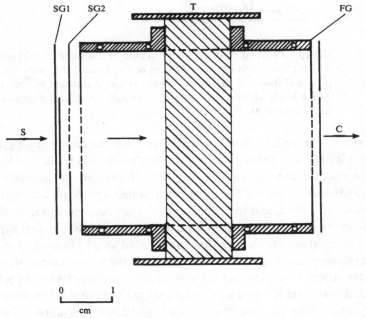

Figure 7.13 Apparatus used by Schwarz and Smith (1981) and Milliken *et al.* (1982) to study ultrasonically generated turbulence. Ions are produced by a radioactive tritium source S and manipulated by screen grids SG1 and SG2 to produce ion pulses. An ion pulse, drifting under the influence of the field determined by SG2 and the Frisch grid FG on the right can be stopped anywhere in the cell by making the drift field zero.

hence reflects some very complicated physics. The sound has a wavelength of about 10^{-2} cm, hence the experimental region shown cross-hatched in Figure 7.13 lies in the extreme Fresnel diffraction limit, and the level of acoustic excitation is expected to drop off sharply at the edges of the ultrasonic beam.

The grids SG1 and SG2 are manipulated to introduce a narrow ion pulse into the main drift space, which is detected when it reaches the collector C. The attenuation of the pulse caused by ion trapping is a measure of the vortex line density L. The authors, however, introduced a novel technique in which they stopped the pulse (by making the drift field zero) at some point x_0 on the path, and allowed the ions to interact with the local vortex field for an additional time τ_{off}. The drift field is then switched on again, and the remaining free ion pulses are detected

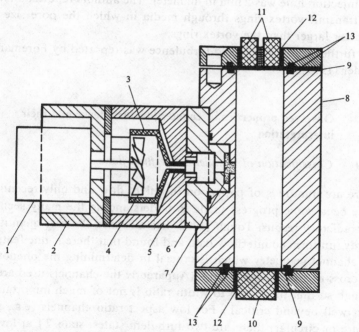

Figure 7.14 Experimental apparatus used by d'Humières and Libchaber (1978) to study injected turbulence: (1) motor; (2) motor support; (3) funnel; (4) propeller; (5) support of second sound transmitter and receiver; (6) support for porous medium; (7) porous medium; (8) cylindrical cavity; (9) aquadag layer on Nuclepore foil; (10) fixed electrode for emitter; (11) fixed electrodes for twin detectors; (12) stycast seal; (13) indium 'o'-rings.

at C. Measurement of the attenuation as a function of τ_{off} and at different locations x allowed a direct measure of the spatial structure of turbulent field. Careful experiments were performed to ensure that the capture process in the tangle was the same as for capture by uniform vortex lines in a rotating bucket. This information was used to extract the line density $L(x_0)$ from the data.

7.4.9   Injected turbulence

D'Humières and Libchaber (1978) have studied the self-diffusion of superfluid turbulence through a porous medium, arranged to allow no net mass transport of fluid. The motor driven propeller creates turbulence which is studied in a resonator equipped for second sound (Figure 7.14). The injection hole was 2 mm in diameter. The authors reported that they can transmit vortex rings through media in which the pore size is of diameter larger than the vortex rings.

A further study of injected turbulence was reported by Foreman and Snyder (1979).

7.5   Observed properties of counterflow turbulence and their interpretation

7.5.1   Classification of turbulent superfluid flows

There are hundreds of papers on turbulent flow and only recently has there been some progress made in understanding the many seemingly contradictory results. Tough (1982) has analyzed the vast quantities of steady, uniform counterflow data and found that there is one feature of the channel geometry which is crucial in determining the phenomena, the cross-sectional aspect ratio. (Apparently the channels used are long enough so that the length to width ratio is not of much importance, at least well beyond critical.) For low aspect ratio channels (e.g., nearly square or circular) there are two turbulent states, state TI at low heat fluxes and state TII at high heat fluxes, which are separated by a rather complicated transition and have different characteristic values of γ (see (7.55)). For high aspect ratio channels there is only one turbulent state, state $TIII$, which has values of γ similar to state TII. Tough also designated the turbulence in pure superflow as state TIV, a state which has mass flow and presumably a flat velocity profile. Recent measurements

(Opatowsky and Tough, 1981) show that γ in state TIV agrees quantitatively with γ in state TII.

There was some doubt for a long time as to the classification of wide channel flow, because a second critical heat flux was observed but not the first. Careful studies by Swanson (1985) using a technique invented by Vinen (1957a) established that there is indeed a lower critical velocity and that the Oregon 1 cm × 1 cm channel does indeed produce flows in the classification TII. The waiting time technique consists of establishing a low heat flux at the value q_c of interest, and measuring the relaxation time for the line density to achieve some final value which is considerably supercritical.

We have argued in Section 7.2 that dynamical similarity of tangles should be obtained by using the dimensionless velocity V given in Equation (7.11), or its equivalent in Equation (7.29). Let us apply these scaling ideas to the second critical velocity (the TI-TII transition) in small aspect ratio channels (Swanson and Donnelly, 1985). It has been customary to plot the quantity $v_{c_2}d$ where v_{c_2} is the second critical value of v_{ns}. We show the results of doing this for channels spanning two orders of magnitude in diameter in Figure 7.15. If the logarithmic parameter is included, then V_{c_2} is plotted, the data collapse in a much more consistent manner, as can be appreciated by comparing the two plots in Figure 7.15.

The solid line in Figure 7.15 comes from the following considerations. Tough (1982) reported that for values of v_{ns} just under v_{c_2} $L^{1/2}d \approx 12$. If we follow this hint from experiment and assume that $L_c d^2$ is temperature independent, then the temperature dependence of V_c is just the temperature dependence of c_1/u_{ic}. Swanson and Donnelly (1985) noted that

$$c_1/u_i = [4(1-\alpha'+\alpha)/\pi(\alpha+\alpha')]^{1/2}/h \qquad (7.53)$$

where h, a constant of order unity, provides a useful correlation for the data. There are at present no simulation results nor theories giving the magnitude and temperature dependence of critical velocities. From (7.11) and (7.29) we may summarize by stating that the second critical velocity for tubes of small aspect ratio is given by

$$V_{c_2} = v_{c_2}d/\beta = c_v = 12[4(1-\alpha'-\alpha)/\pi(\alpha+\alpha')]^{1/2}/h \qquad (7.54)$$

where the magnitude and temperature dependence of c_v is a problem for future research.

Now let us consider the situation beyond critical where the tangle can be considered to be well developed. Under these circumstances ion measurements by Awschalom, Milliken and Schwarz (1984) using the

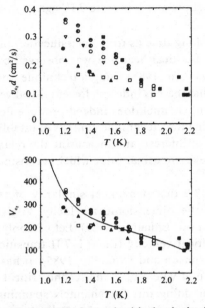

Figure 7.15 Second critical velocity in small aspect ratio channels at various temperatures. The upper plot is scaled with channel size, while the lower plot includes the logarithmic parameter as in Equation (7.11) (After Swanson and Donnelly (1985)).

apparatus shown in Figure 7.11 have produced the data shown in Figure 7.16 which shows that the line density and normal fluid velocity profile are uniform over at least 80% of a transverse section of the flow, which in this case is 1 cm. Second sound attenuation experiments at Oregon show that the line density is axially homogeneous (Barenghi, 1982; Swanson, 1985). This experimental information, coupled with the observation that mutual friction effects acting between the vortices and the normal fluid are much greater than shear forces acting within the normal fluid, tends to lead one to believe that homogeneous turbulence exists. Unlike classical wind-tunnel produced turbulence, the vortex tangle is produced locally by the counterflow at every axial position. This local production can be demonstrated experimentally. Barenghi (1982) was able to show that vortex lines can be created by shock waves and that the velocity at which the initiation of vorticity is propagated is the velocity of second sound. One can therefore be confident that the vortex line turbulence is homogeneous and that the scaling ideas of Section 7.2 are applicable.

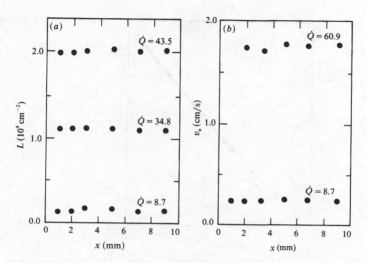

Figure 7.16 Data of Awschalom *et al.* (1984) on the transverse homogeneity of (*a*) the line density, (*b*) the normal fluid velocity in a turbulent counterflow experiment.

The dependence of the average line density L_0 upon v_{ns} near critical is complicated. But observers generally agree (see, e.g., Tough (1982), Vinen (1975a)) that beyond the critical region

$$L_0 \approx \gamma^2 (v_{ns} - v_0)^2 \qquad (7.55)$$

where γ and v_0 are functions of temperature.

We can interpret (7.55) in terms of the discussion of homogeneous turbulence in Section 7.2.4. According to (7.27)

$$L^{1/2} = C_L v_{ns}/\beta \qquad (7.56)$$

where β is given by (7.12) and (7.13). Then $L^{1/2}$ is not simply dependent upon v_{ns}, and if (7.56) is expanded about some mean value of v_{ns}, we immediately recover the form (7.55). Inclusion of the full logarithmic parameter gives the fit of Figure 7.17 which has also been corrected for the velocity dependence of mutual friction discussed in Section 3.3. The details are somewhat tedious (see Swanson (1985)) but the resulting agreement with experiment shown in Figure 7.17 is essentially perfect. The experimental quantity v_0, then, is seen to have no physical significance.

This observation allows us to make a significant improvement in the understanding of historical data. We show in Figure 7.18(*a*) data on state TII which has been interpreted according to (7.55). It appears from

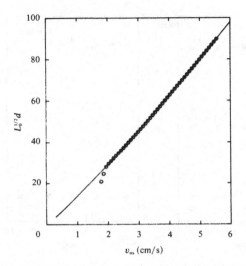

Figure 7.17 $L_0^{1/2}d$ as a function of v_{ns} at 1.7 K. The data are from Martin and Tough (1983). The line represents Equation (7.56) (after Swanson and Donnelly (1985)).

these data that γ_2 is a function of channel size. If we use the correct form (7.56) we obtain Figure 7.18(b). The details of this procedure are described by Swanson and Donnelly (1985). We see substantial improvement in the representation of the data.

The logarithmic parameter also has important implications for the Gorter–Mellink relationship between the temperature gradient and the applied heat flux (see (7.48)–(7.50)). Scaling shows that the correct relationship between line density, counterflow velocity and mutual friction is given by (7.39). Let us amend the usual Gorter–Mellink rule to write

$$\nabla T_{\mathrm{T}} = -\mathbf{F}_{ns}/ S\rho_s \propto q^m \qquad (7.57)$$

where $m = 3$ is the traditional value. Swanson and Donnelly (1985) show that it is possible to use a nonlinear power approximation to (7.56)

$$L^{1/2} \propto v_{ns}^n \qquad (7.58)$$

which is valid over a wider velocity range than the approximation (7.55). This leads to the result $m = 2n + 1$ in (7.57). If we include the velocity dependence of α in (7.39) as well (see Section 3.3), then we find that (7.57) is a good approximation about some heat flux q_0, with m a function of temperature and q_0 as shown in Figure 7.19. These values are consistent with a variety of experiments (Ahlers, 1969; Bon Mardion, Claudet and

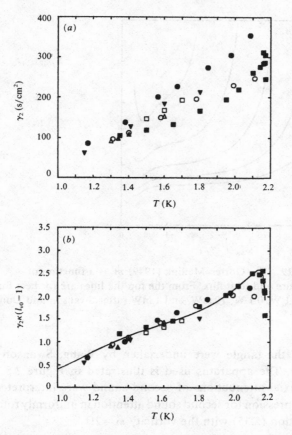

Figure 7.18 γ_2 as a function of temperature measured by various investigators. Diagram (*a*) shows γ_2 and diagram (*b*) shows γ_2 scaled by the implicit channel-size dependence (after Swanson and Donnelly (1985)).

Seyfert, 1978; Swanson, 1985) that have roughly determined m. The theory leading to this figure breaks down as m grows, essentially because the core becomes comparable to $L_0^{-1/2}$, but it is probably satisfactory while m is less than 4.

7.5.2 Anisotropy of the vortex tangle

We have cited considerable experimental evidence which suggests that in pure counterflow at least, the vortex line density is spatially homogeneous. Measurements designed to determine the isotropy and

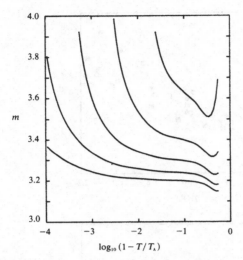

Figure 7.19 The Gorter-Mellink (1949) m as a function of temperature and heat flux. From the top the lines are for heat fluxes of 10 W, 1 W, 0.1 W, 10 mW, and 1 mW (after Swanson and Donnelly (1985)).

drift velocity of the tangle were undertaken by Wang, Swanson and Donnelly (1987). The apparatus used is illustrated in Figure 7.8. The transverse and axial attenuations of second sound were interpreted by analogy to the expression for second sound attention in uniformly rotating helium II, Equation (2.35) with the vorticity $\omega = 2\Omega$

$$\alpha = B\omega/4u_2 \tag{7.59}$$

Identifying the axial and transverse attenuation of second sound as α_A and α_T, we have the apparent line densities L_A and L_T given by

$$L_A = 4u_2\alpha_A/\kappa B_A = I_\parallel L \tag{7.60}$$

$$L_T = 4u_2\alpha_T/\kappa B_T = I_\perp L \tag{7.61}$$

where B_A and B_T are the values of B appropriate to the experimental frequencies of second sound used to probe each direction (see the discussion in Section 3.3). The details of the analysis are somewhat technical (Wang, Swanson and Donnelly, 1987), and the results for the apparent anisotropy of the tangle, which appears to be temperature dependent are shown in Figure 7.20. The ratio L_T/L_A is just the ratio $I_\perp/I_\parallel$ defined by Schwarz in Equations (7.14) and (7.15) and tabulated from the simulation in Table 7.1. Comparison with Schwarz's data from

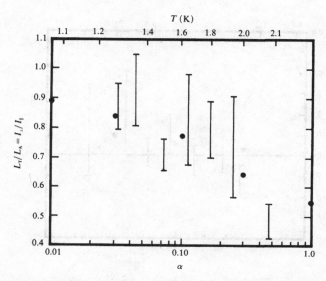

Figure 7.20 The ratio of $L_T/L_A = I_\perp/I_\parallel$ of line density as a function of temperature (Wang, Swanson and Donnelly, 1987). Even with the substantial uncertainties (which represent three standard deviations), it is evident that the tangle is not isotropic since L_T/L_A is not unity. Predictions of Schwarz (1988) are shown by solid circles.

his homogeneous simulation is shown by dots in Figure 7.20. The agreement is quite satisfactory. The authors show experimentally that the ratio L_T/L_A is independent of v_{ns} at $T = 1.5$ K.

The average drift velocity $\mathbf{v}_l$ of the tangle, which is defined in Equations (7.40) and (7.41) can be measured by supplementing the measurements of attenuation of second sound in the transverse and axial directions by measurement of the temperature gradient. Using (7.50) and (7.39) we have

$$\nabla T = \kappa \alpha (I_\parallel - C_L I_1) L v_{ns}/S$$
$$= \kappa \rho_n B_G (I_\parallel - C_L I_1) v_{ns}/2\rho S \qquad (7.62)$$

where B_G is the value of B corrected for the dc value of B as described in Section 3.3. The value of L can then be deduced using (7.17) and $C_L I_1$ can be derived from ∇T. We can approximate (7.41) by

$$\frac{v_l}{v_{ns}} = \frac{v_L - v_s}{v_{ns}} = c_L I_1 \qquad (7.63)$$

and this quantity is plotted in Figure 7.21. We see that the results show the tangle moves with the superfluid and that the simulation results of Schwarz give the same behavior.

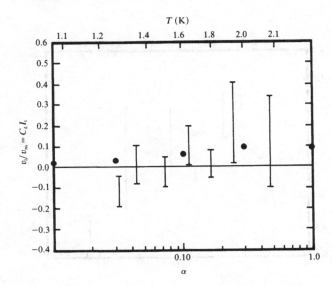

Figure 7.21 The ratio $v_l/v_{ns} = C_L I_1$ as a function of temperature. The bars represent the data of Wang, Swanson and Donnelly (1987) (the range shown is three standard deviations). The solid dots are the predicted vortex drift velocities by Schwarz as given in Table 7.1.

7.5.3   Intrinsic fluctuations

The first apparent observation of intrinsic fluctuations of the tangle was reported by Hoch, Busse and Moss (1975) and Mantese, Bischoff and Moss (1977). Further measurements were reported by Ostermeier, Cromar, Kittel and Donnelly (1980), and Smith and Tejwani (1984). Since then apparatus and detection capability have improved considerably and some doubt has arisen about the exact interpretation of these early measurements. For example, the drift of the tangle past a detector will itself produce fluctuations. The intrinsic fluctuations should reflect the dynamics of the tangle itself, and probably the size of the volume probed. On the other hand, the Vinen equations can be interpreted in a way which gives the response to induced fluctuations of v_{ns}. An experiment to test this prediction by Barenghi, Swanson and Donnelly (1982) fully confirms the predictions of Vinen's equations (7.35), (7.36).

A study of intrinsic fluctuations has been carried out by Griswold, Lorenson and Tough (1987). Their apparatus measured the chemical potential difference $\Delta\mu$ across a flow tube 132 µm in diameter and 1 cm long. The fluctuations in the fully developed TII state were observed to be less than 0.1% of the total potential difference.

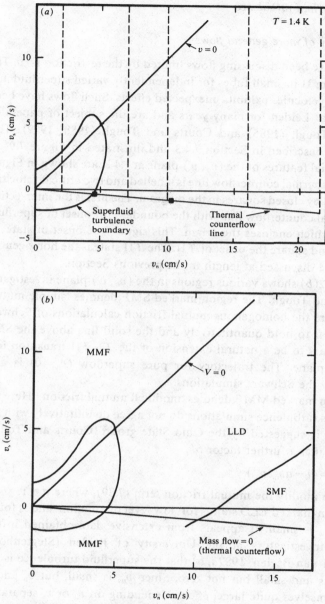

Figure 7.22 (a) Characteristics of the (v_n, v_s)-plane at 1.4 K. There is no superfluid turbulence within the boundary, and critical velocities for the onset of TI and TII as shown by solid squares along the counterflow line (after Courts and Tough (1988)). (b) The same plane as (a) showing the types of dissipation in each region. The line marked $v = 0$ corresponds to $v_s = v_n$ (after Courts and Tough (1988)).

7.5.4   Study of more general flows

So far, we have been discussing flows limited by the restriction $j = 0$. The flow of helium II in small tubes for independently varied superfluid and normal fluid velocities exhibits unexpected effects. Such flows have been investigated at Leiden for many years and are the subject of papers by Baehr and Tough (1985) and Courts and Tough (1988, 1989). This apparatus is described in Section 7.4.3 and illustrated in Figure 7.9(a).

The essential features of the (v_n, v_s)-plane at 1.4 K are shown in Figure 7.22(a). The thermal counterflow line is labelled and two critical velocities are indicated by closed squares in the diagram. The first is the intersection of the thermal counterflow line with the boundary for onset of superfluid turbulence which encloses the origin. This signals the onset of state TI and the second square the onset of TII. The TII state is the homogenous state we have discussed at length in the previous Section.

Figure 7.22(b) shows various regions in the (v_n, v_s)-plane investigated by Courts and Tough. The region marked SMF denotes 'simple mutual friction.' Here the homogenous mutual friction calculations of Schwarz (1988) appear to hold quantitatively and the solid line above the SMF region appears to be a natural extension of the TI–TII transition into the (v_n, v_s)-plane. The trajectory for pure superflow ($v_n = 0$) is also described by the Schwarz simulations.

The region marked MMF denotes 'modified mutual friction.' Here the homogenous turbulence simulations do not agree quantitatively with the data, and it is suggested by the Ohio State group (Courts and Tough, 1988, 1989) that a further factor α'',

$$\alpha'' = (1 - a v_n / v_{ns}) \tag{7.64}$$

is needed to modify the mutual friction term (7.39), where $a \approx 1$.

The region marked LLD stands for 'low level dissipation' and is found wherever v_{ns} is small. It appears from extensive data obtained in this region by investigators at the University of Leiden (Slegtenhorst, Marees and Van Beelen, 1982a, b) that the superfluid turbulence is not homogenous and small but not zero. Since v_{ns} is small, but v_n and v_s may be themselves quite large, effects depending on v_n or v_s separately may be quite large.

8   Thermal activation and nucleation of quantized vortices

In this chapter we consider the processes of the nucleation of quantized vortices. There have been a number of attempts to understand nucleation phenomena and the critical velocities associated with them, but it is probably fair to say that successes have been few and far between. Part of the difficulty, of course, is that we cannot visualize quantized vortices and one is led to guess just what the circumstances of vortex formation in a given flow really are.

This field is evolving steadily thanks to new experiments, new theories and computer simulations, and the account given here is intended to give the reader some idea of the present situation and directions for future research. Among the more important of these directions are the understanding of macroscopic tunnelling and the effects on ^{3}He on nucleation.

Thermal excitation of vortices plays a crucial role in the behavior of vortices in thin films. We conclude this volume with a discussion of vortex dynamics in thin films.

8.1   Extrinsic and intrinsic nucleation processes

Ideal superflow of helium II through a tube, for example, breaks down at quite low velocities of flow, much lower, for example, than given by the Landau critical velocity (Figure 2.2). It is generally believed that this breakdown is due to quantized vortices, although the way in which this breakdown occurs is, at the present time, obscure. It seems likely that there are often residual vortices such as were described in Chapter 5 (see Figure 5.22). When flow begins these remnant lines move and perhaps multiply, in a process called 'extrinsic nucleation'. The velocity of onset of dissipation by such processes is called an 'extrinsic critical velocity'. The creation of vortex line where none was present before appeals to one as perhaps the more fundamental problem, and is called 'intrinsic nucleation'. Intrinsic nucleation leads to 'intrinsic critical velocities' which may or may not be substantially different from extrinsic velocities. We shall show that intrinsic critical nucleation is opposed by an energy barrier and that at $T = 0$ K the only way to penetrate this barrier is by tunnelling. At finite temperatures, however, it is possible to surmount this barrier by thermal activation. We shall discuss these processes in Section 8.3, the motion of existing vortices in Section 8.4 the problem of nucleation by tunnelling in Section 8.5, and the influence of ^{3}He impurities in Section 8.6.

8.2   The nucleation energy barrier

Three examples of energy barriers will serve to illustrate some of the problems encountered in nucleation theories. These are shown in Figures 8.1–8.3.

A pair of hollow vortex filaments of opposite circulation κ has energy per unit length given by (1.44)

$$E = (\rho_s \kappa^2 / 2\pi) \ln (2z/\pi a) \tag{8.1}$$

and impulse per unit length

$$p = \rho_s \kappa (2z) \tag{8.2}$$

In the presence of a uniform superflow as shown in Figure 8.1(a) the energy is changed to

$$F = E - \mathbf{p} \cdot (\mathbf{v}_n - \mathbf{v}_s)$$
$$= (\rho_s \kappa^2 / 2\pi) \ln (2z/a) - 2\rho_s \kappa z v_s \tag{8.3}$$

where the last term, the '$\mathbf{p} \cdot \mathbf{v}_s$' interaction (we consider $\mathbf{v}_n = 0$ here) shifts the energy of an excitation of the superfluid in the presence of a superflow $\mathbf{v}_s$ (see, e.g., Wilks (1967) Section 6.5). Note that the negative sign comes from the angle π between $\mathbf{p}$ and $\mathbf{v}_s$. The situation just described is no different for a single vortex line a distance z above a plane, with the second vortex the image of the first. The second term reduces the energy, and the shape of the resulting free energy is shown in Figure 8.1(b). One can see that in a flow of 15 cm/s to draw a vortex line away from a boundary would require surmounting an energy barrier of $\sim 6 \times 10^8$ K/cm situated 400 Å from the boundary.

The motion of the vortex pair in this flow owing to the superflow above is given by $\mathbf{v} = \partial F / \partial \mathbf{p}$ (1.58) and is the slope of the free energy curve when plotted as a function of p instead of z. Thus for small z, the line

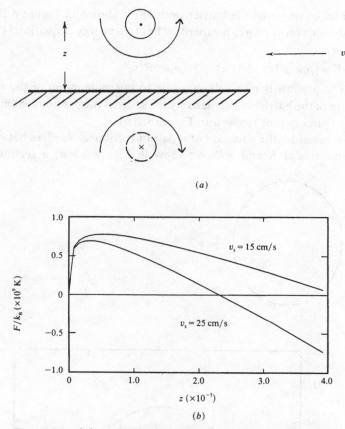

Figure 8.1 (*a*) A rectilinear vortex line situated a distance z above a boundary with its image behaves as a pair of vortices of opposite circulation a distance $2z$ apart. (*b*) Free energy expressed as degrees Kelvin per centimeter of vortex line at 1.2 K in the presence of flows of $v_s = 15$ cm/s and 25 cm/s. The top of the free energy curve is the barrier C.

and image move toward the superflow with z increasing until the maximum C is reached, beyond which the line moves with v_s (opposite to its momentum) and increases z by extracting energy from the flow. A special situation occurs at C and the configuration there is called 'critical' because any further increase in z will lead to indefinite expansion of the pair. The location C marks a saddle point in the free energy since in (8.3) any angle different from π will lead to a smaller $\mathbf{p} \cdot \mathbf{v}_s$ energy and hence a larger F.

A second example of the barrier problem is shown in Figure 8.2(a). Here a hollow vortex ring in a superflow has free energy (Equation (1.68) with $\Gamma = \kappa$)

$$F = \tfrac{1}{2}\rho_s\kappa^2 R[\ln\,(8R/a)-\tfrac{3}{2}]-\rho_s\kappa\pi R^2 v_s \qquad (8.4)$$

when the superflow is precisely opposite to the momentum of the ring. The nature of the barrier is qualitatively the same as the previous example, and a free energy plot is given in Figure 8.2(b).

Finally, consider the situation of a pair of rectilinear vortices between two boundaries at R and $-R$. We show in Figure 8.3(a) a section of

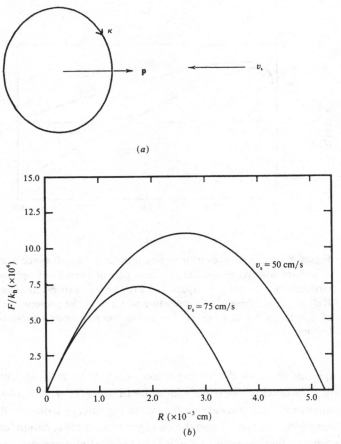

(a)

(b)

Figure 8.2   (a) Vortex ring of radius R with its impulse vector **p** oriented opposite to a superflow of velocity v_s. (b) Free energy expressed as Kelvin degrees for vortex rings in a superflow $v_s =$ 50 cm/s and 75 cm/s.

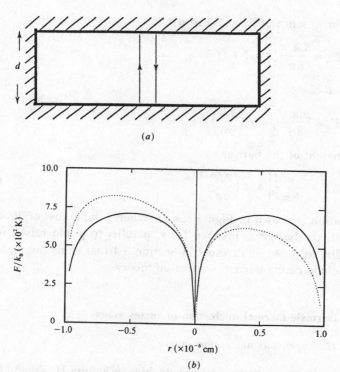

(a)

(b)

Figure 8.3 (a) A pair of vortices between boundaries at $\pm R$ in a flow v_s. (b) Free energy of the vortices in (a) in degrees Kelvin in flows of 0 (solid line) and 200 cm/s (dotted line) at $T = 1.2$ K. Here $R = d = 10^{-6}$ cm.

height d of these vortices which are assumed to remain straight in the presence of a superflow. This problem was first discussed by Gopal (1963) who showed that the energy of the pair at any symmetrical separation $2r$ is given by

$$E = \frac{\rho_s \kappa^2 d}{8\pi} \left\{ 4 \ln \left[\frac{\sin (\pi r/R)}{\pi a/2R} \right] + 1 \right\} \tag{8.5}$$

The resulting energy as function of r is shown in Figure 8.3(b). The impulse of the pair is $\rho_s \kappa 2r$ so that the free energy is

$$F = \frac{\rho_s \kappa^2 d}{8\pi} \left\{ 4 \ln \left[\frac{\sin (\pi r/R)}{\pi a/2R} \right] + 1 \right\} - \rho_s \kappa d 2 r v_s \tag{8.6}$$

An interesting feature of the free energy (8.6) is that there is a barrier even in the absence of flow. When $a \ll R$

$$E_A = \frac{\rho_s \kappa^2 d}{8\pi} [4 \ln 2 + 1]$$ (8.7)

when $r = R/2$

$$E_C = \frac{\rho_s \kappa^2 d}{8\pi} \left[4 \ln \left(\frac{2R}{\pi a} \right) + 1 \right]$$ (8.8)

and the height of the barrier

$$\Delta E = \frac{\rho_s \kappa^2 d}{8\pi} \left[4 \ln \left(\frac{2R}{\pi a} \right) \right]$$ (8.9)

The situation is shown in Figure 8.3(b). When a superflow of 200cm/s is present, the barrier is lowered for v_s parallel to p and raised for v_s antiparallel to p. We shall show in Section 8.4 that situations like this lead to a 'competing barrier' nucleation theory.

8.3   Intrinsic thermal nucleation of vortex rings

8.3.1   *Homogeneous nucleation theory*

In 1961 in a lecture on the critical velocities in helium II, Vinen (1963) noted that thermal fluctuation nucleation of vortex line might be a possibility, and produced the first estimates of the rate of such a process. He ended by noting, 'Therefore we may conclude that in practice thermal nucleation of vortex line can probably be ignored ... '. As we shall see, this conclusion has not been modified except perhaps very near the lambda transition where barriers are small since ρ_s is small. Vinen's estimates were considerably expanded in an elegant discussion by Iordanskii (1965), and in a qualitative way by Langer and Fisher (1967), producing considerable excitement as a logical explanation for critical velocities. This process, now called ILF after the above-mentioned authors, considers the growth of a microscopic vortex ring in the presence of a superflow. Since boundaries do not play a role in ILF theory, it is sometimes called the theory of homogeneous nucleation of quantized vortex rings. ILF theory has been discussed by Langer and Reppy (1970) and Donnelly and Roberts (1971).

The basic idea of vortex ring nucleation can be appreciated from the discussion of drag and Magnus forces on vortices given in Section 3.1. Referring to Figures 3.8 and 8.2 let us adopt a frame of reference at rest with respect to the normal fluid and assume that v_s is directed oppositely to v_i. Then, Equation (3.43) and Table 3.1 gives

$$\dot{R} = \frac{\gamma}{\rho_s \kappa} (v_s - v_i) \tag{8.10}$$

so that if $|v_s| > v_i$ the ring can grow $(R > 0)$ taking the energy to grow from the chemical potential driving the superflow. For small rings, however, the values of v_s needed to exceed v_i are very large leading to the barrier shown in Figure 8.2. The essence of the fluctuation nucleation theory is that lucky fluctuations expand a microscopic ring until its size reaches a critical value which (for a given v_s) can expand spontaneously in the flow according to (8.10). Thus, just as the Brownian particle in Figure 1.22 diffuses over a barrier in configuration space, to fall down the potential from C to B, so a microscopic vortex ring expands in momentum space (recall $p \propto R^2$), until it surpasses a free energy barrier at C and can spontaneously expand in the flow to nucleate a microscopic vortex ring. It is a matter of taste whether to work in configuration space or momentum space. One can see that this process occurs in a region of constant $(v_s - v_n)$ in the absence of boundaries and in the absence of any preexisting vortex line.

In order to find the appropriate friction coefficient, one can transform (8.10) into momentum space by taking $v_s = 0$, $v_i = \partial F / \partial p_i$, and multiplying both sides by $2\pi R \rho_s \kappa$:

$$dp_i / dt = -2\pi R \gamma \, \partial F / \partial p_i = -\Lambda(p) \, \partial F / \partial p_i \tag{8.11}$$

where

$$\Lambda = 2\pi R \gamma = 2\pi \gamma (p / \rho_s \kappa \pi)^{1/2} \tag{8.12}$$

is the drag coefficient on vortex rings.

Given the quantities derived above, Donnelly and Roberts (1971) show that the appropriate diffusion equation for the probability density $w(\mathbf{p})$ of vortex rings (see Section 1.9 for the definition of w) is the Fokker–Planck equation

$$\frac{\partial w}{\partial t} = \frac{\partial}{\partial p_i} \left[\Lambda \frac{\partial F}{\partial p} w + q(p) \frac{\partial w}{\partial p_i} \right] \tag{8.13}$$

where $q(\mathbf{p})$ is the diffusion coefficient which is evaluated by noting that in the steady state (8.13) must reduce to the Maxwell–Boltzmann solution

$$w = w_0 \exp (-F / k_B T) \tag{8.14}$$

where w_0 is a normalization constant. Hence

$$q = \Lambda k_B T = 2\pi \gamma R k_B T \tag{8.15a}$$

The diffusion coefficient q sets the time scale to diffuse a 'distance' Δp in momentum space as $(\Delta p)^2/q$ exactly as the diffusion constant for rings in bulk liquid helium D_b sets the time to diffuse a distance Δx as $(\Delta x)^2/D_b$. The value of D_b can be estimated by the same argument used for q: it is

$$D_b = \gamma k_B T/2\pi\rho_s^2\kappa^2 R \tag{8.15b}$$

and for vortex pairs (Figure 1.8(b)) in a bulk film of thickness d, it is

$$D_b = 2\gamma k_B T/\rho_s^2\kappa^2 d \tag{8.15c}$$

Equation (8.13) implies the steady state flux of vortex rings per unit 'area' in momentum space per unit time

$$j = -\left(\Lambda\frac{\partial F}{\partial p_i}w + q\frac{\partial w}{\partial p_i}\right) \tag{8.16}$$

and the calculation of the escape over the free energy barrier (nucleation) proceeds in much the same way as in Section 1.9 except that here we are considering escape over a three-dimensional barrier in momentum space, thus $F(p)$ is an ordinate in a four-dimensional vector space. Since (8.13) is a Fokker–Planck equation it is not governed by the restrictions of the Smoluchowski equation, and Donnelly and Roberts proceed to evaluate the probability of escape P for general values of velocity. Their result uses the expansion near C much as in (1.112):

$$F \approx F_C - \tfrac{1}{2}\omega_C^2(p_x - p_{Cx})^2 + \tfrac{1}{2}S_C^2(p_y - p_{Cy})^2$$
$$+ \tfrac{1}{2}t_C^2(p_z - p_{Cz})^2 \tag{8.17}$$

where x is along the direction of flow, $\theta = 0$. Since the microscopic vortex rings are quasi-particles, the number N_R in the well at A must come from a separate calculation much the same as is used to calculate the density of rotons in the Landau theory of superfluidity, and the final expression for the number of vortex rings nucleated per second per unit volume is the product of the number of candidates and the probability per candidate for escape:

$$\nu = NP_R = \frac{(2\pi)^{1/2}(k_B T)^{3/2}\Lambda_C\omega_C}{h^3 S_C^2}\exp\left(-F_C/k_B T\right)$$
$$= n_0\exp\left(-F_C/k_B T\right) \tag{8.18}$$

where the curvatures come from partial derivatives of F at C as indicated by (8.17):

$$\omega_C^2 \approx v_s/2p_C, \qquad S_C^2 \approx v_s/p_C \tag{8.19}$$

and Λ_C is computed for a ring of momentum p_c. The expression (8.18) coincides exactly with Equation (3.39) of Donnelly and Roberts (1971), and differs by a factor of $2^{1/2}\pi^2$ from that of Iordanskii (1965) who first derived it.

One should remark here that Iordanskii's derivation recognized one further feature of the nucleation calculation, the internal density of states. This comes about because the vortex rings can have degrees of freedom associated with their core such as the waves discussed in Section 1.7. The modification required is formally simple when the entropy S_i of internal degrees of freedom is known, one simply multiplies the result (8.18) by

$$g_i(R) = \exp\left[S_i(R)/k_B\right] \tag{8.20}$$

One can also amend the free energy (8.3) to read

$$F = E - \mathbf{p} \cdot (\mathbf{v}_n - \mathbf{v}_s) - TS_i \tag{8.21}$$

Iordanskii tried to estimate S_i by considering vortex waves on a ring. His calculation has been reexamined by Barenghi, Donnelly and Vinen (1985) who show that this idea encounters severe difficulties near T_λ (see Section 6.5).

8.3.2 Decay of a superflow in a toroidal geometry

Now suppose we consider the decay of a superflow in a thin toroidal gyroscope of mean radius R_g and volume V of helium II. Suppose that the superflow is prepared with n quanta of circulation, $n \gg 1$, then the average superfluid velocity will be given by

$$v_s = n\kappa/2\pi R_g$$

When a nucleation event occurs, a vortex ring will surmount the free energy barrier, and expand until it reaches the boundary of the toroid. In doing so, it cuts the phase as discussed in Section 2.8.5, and n is accordingly reduced to $(n-1)$. If ν such vortices are nucleated per second per unit volume, then

$$\frac{dv_s}{dt} = -\frac{\kappa}{2\pi R_g}\frac{dn}{dt} = -\frac{\nu\kappa V}{2\pi R_g} \tag{8.22}$$

In order to proceed to examine this relationship in detail we need to cast the result in dimensionless form. We do this by following a procedure devised by Donnelly and Roberts (1971). First we eliminate the radius of the critical ring R_C using (8.4) to obtain R_c from $\partial F/\partial R = 0$:

$$R_C = (\kappa/4\pi v_s) \ln(c\kappa/4\pi a v_s)$$
$$= (\kappa/4\pi v_s) \ln \lambda \tag{8.23}$$

where $c = 8/e^{1/2}$,

$$\lambda = c\kappa/4\pi v_s a \tag{8.24}$$

and where $\ln \lambda$ is taken to be constant in a superfluid decay experiment since v_s generally changes very slowly. In this approximation the barrier F_C becomes

$$F_C = (\rho_s \kappa^3/8\pi v_s) \ln \lambda [\ln (8R/a) - 2]$$
$$- (\rho_s \kappa^3/16\pi v_s)[\ln \lambda]^2 \tag{8.25}$$

Now

$$\ln (8R/a) - 2 = \ln [c\kappa/4\pi v_s a] \ln (c\kappa/4\pi v_s a)] - 1$$
$$= \ln (\lambda \ln \lambda) - 1$$
$$= \ln \lambda + \ln \ln \lambda - 1 \tag{8.26}$$

For a velocity v_s of 1 m/s, $\lambda = 234$, $\ln \lambda \sim 5.45$, $\ln \ln \lambda = 1.70$ and it is not a bad approximation to put

$$F_C/k_B T \sim [\rho_s \kappa^3/16\pi v_s k_B T](\ln \lambda)^2$$
$$= (v_{th}/v_s)(\ln \lambda)^2 \tag{8.27}$$

where the thermal characteristic velocity

$$v_{th} = \rho_s \kappa^3/16\pi k_B T \tag{8.28}$$

is a large number, and as Donnelly and Roberts (1971) point out, the origin of one difficulty with ILF theory. At 1 K, $v_{th} = 209$ m/s, close to the velocity of sound and much greater than the Landau critical velocity or indeed the velocity of second sound.

The quantity n_0 entering in (8.18) can be written in a form without R_c or v_s by substituting $R_c = (\lambda a/c) \ln \lambda$ and $v_s = c\kappa/4\pi\lambda$:

$$n_0 = 4\pi^{5/2} \rho_s^{1/2} (k_B T)^{3/2} (\lambda a/c)^{5/2} \gamma h^{-3} (\ln \lambda)^2 \tag{8.29}$$

Equation (8.22) can now be made dimensionless. We introduce the dimensionless vortex coupling constant

$$B = \rho_s \kappa^2 a/4 c k_B T \tag{8.30}$$

a friction parameter defined from (8.12), $\Lambda = \beta p^{1/2}$:

$$\beta = 2\pi\gamma(\rho_s \kappa \pi)^{-1/2} \tag{8.31}$$

and the constant

$$A = \frac{4\pi^2 \beta \rho_s V (k_B T)^{3/2} \kappa^{1/2} (a/c)^{7/2}}{R_g h^3} \tag{8.32}$$

Then with the dimensionless velocity

$$y = 1/B\lambda(\ln\lambda)^2$$
$$= v_s/v_{th}[\ln(\lambda)]^2 \qquad (8.33)$$

and dimensionless time

$$\tau = ft = AB^{-7/2}(\ln\lambda)^{-5}t \qquad (8.34)$$

Equation (8.22) becomes

$$dy/d\tau = -y^{-5/2}\exp(-1/y) \qquad (8.35)$$

Numerical integration of (8.35) is shown in Figure 8.4; it is seen that the flow does not reach zero but instead approaches it asymptotically. Indeed it can be shown that for τ large, the solution to (8.35) is

$$y = 1/\ln\tau \qquad \tau\to\infty \qquad (8.36)$$

and τ is indeed large in practice since $f\approx 10^{20}$.

Equation (8.36) has the property of being independent of the initial conditions of the flow, and provided the initial flow is large enough, the critical velocity seen in the laboratory depends on the instant of first observation: $\tau_0 = ft_0$, and (8.3) defines the 'saturated critical velocity' (i.e., the maximum flow velocity which can be set up by an applied

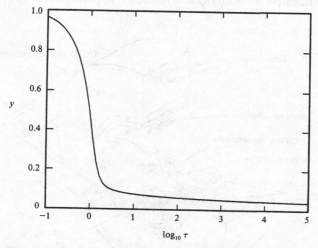

Figure 8.4 A universal decay curve for superflow in a closed channel. The curve is the solution of the differential Equation (8.35) where the variables y and τ are defined by (8.33) and (8.34) (after Donnelly and Roberts (1971)).

chemical potential gradient)

$$y_0 = 1/\ln \tau_0 \tag{8.37}$$

The fractional decay per decade of a persistent current is calculated from (8.36)

$$\frac{1}{y}\frac{dy}{d\ln \tau}\Big|_{\tau_0} = -\frac{1}{\ln \tau_0} \tag{8.38}$$

and the slope of the logarithmic decay curve is

$$\frac{dy}{d\ln \tau}\Big|_{\tau_0} = -y_0^2 \tag{8.39}$$

8.3.3 Comparison of the homogeneous nucleation theory with experiment

The superfluid gyroscope has been developed by Reppy's group (Langer and Reppy, 1970) to investigate flow in a toroidal geometry. An example is shown in Figure 8.5. The toroidal bucket has a rectangular cross-section typically $R_g = 2$ cm in radius with an area ~ 1 cm^2. Filter materials of sizes 2000 Å and 500 Å as well as porous Vycor glass with an estimated pore size of 38 Å were used in the gyroscope to raise persistent superfluid currents to velocities high enough to measure. The procedure was to orient the gyroscope horizontally and rotate the apparatus about a vertical

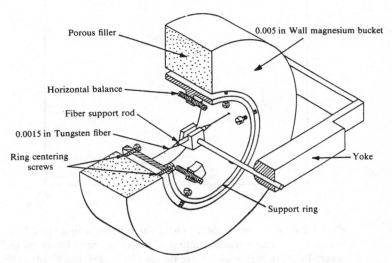

Figure 8.5 A typical superfluid gyroscope (after Kukich (1970)).

axis rapidly enough to obtain a saturated critical velocity. The gyroscope was then oriented to a vertical position and the apparatus rotated about its vertical axis at some angular speed ω_a. Since the angular momentum of the persistent current L_p is perpendicular to ω_a a torque $\omega_a L_p$ will deflect the gyroscope on its suspension fiber through an angle $\delta\theta = \omega_a L_p/k$, where k is the torsion constant of the tungsten fiber. Thus L_p and hence the superfluid velocity become known experimentally through the relationship $L_p = \rho_s V R_g v_s$.

Saturated critical velocities found by Kukich (1970) are shown in Figure 8.6. Figures 8.6(a) and (b) show the critical velocity at a nominal time, say $t_0 = 100$ s computed from (8.34) and (8.37); the velocity v_s in λ is obtained self-consistently by an iteration procedure. Velocities such as those below $T = 2$ K in Figure 8.6(a) are larger than are seen in laboratory experiments and have steeper temperature dependence.

The gyroscope also yielded important information on the decay of persistent currents. In order to observe the decay, a persistent current is set up as described above and allowed to come to equilibrium. On raising the temperature slightly, the flow is too large for the new temperature, and starts to decay. Examples of this decay are shown in Figure 8.7: they are linear in $\log_{10} t$, i.e.,

$$v_s = v_{s0} - r \log_{10} t \qquad (8.40)$$

It is clear that over several decades of time, the experimentally observed decay (8.40) which is linear in the logarithm of the elapsed time cannot be mistaken for the homogeneous nucleation decay given by (8.36) which varies as the reciprocal of the logarithm of the elapsed time.

The homogeneous theory gives superfluidity all the way to T_λ as measured (of course) in the bulk. All experiments with pore sizes <2000 Å show a clearly depressed onset temperature $T_0 < T_\lambda$, and this behavior cannot be described by an homogeneous theory.

The critical vortex ring of radius R_c is one in which the self-induced velocity v_i is just equal to the superfluid velocity v_s. Thus a knowledge of the saturated critical velocity v_c tells us the value of R_c. Data for v_i as a function of R_c are shown in Figure 8.8.

Kukich (1970) gives his numerical results for the Cornell gyroscope experiments described briefly in Langer and Reppy (1970). He quotes a critical velocity of 5.92 cm/s for 2000 Å powder at $T = 2.1681$ K and a comparable value for 500 Å powder. However, the data in Figure 8.8 show that at 2.169 K and 5 cm/s the ring radius is 10 000 Å, and hence larger than the pore sizes used by Kukich in his gyroscope.

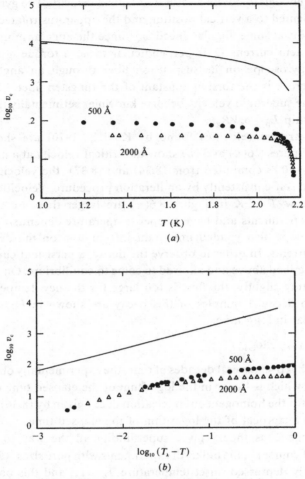

Figure 8.6 Critical velocities as a function of (a) temperature and (b) reduced temperature calculated from Equation 8.37. The data is from Kukich (1970) obtained with a superfluid gyroscope.

Comparison of the homogeneous decay calculations for fractional and absolute decay rates per decade are shown in Figures 8.9 and 8.10. We conclude that the homogeneous nucleation theory is not supported quantitatively by experimental results presently available. Nevertheless, the theory stimulated useful experimentation because it was the first theory to provide estimates of critical velocities that looked reasonable near T_λ. As shown in Figure 8.10(b).

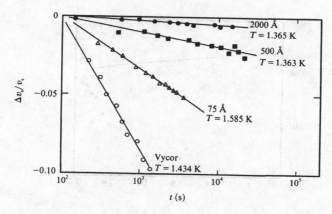

Figure 8.7 The logarithmic decay of some saturated persistent currents (after Kukich (1970)).

8.4 Vortex motion in porous media

8.4.1 The competing barrier model

Donnelly and Roberts (1971) pointed out the necessity of considering a permanent barrier ΔE for flow in a restricted geometry which is present, irrespective of any superflow. An example of such a barrier is shown in

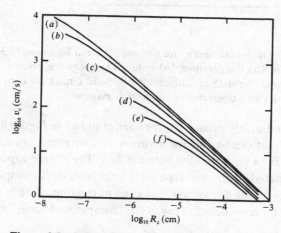

Figure 8.8 Relationship of the size of vortex rings R and their self-induced velocity v_i for several temperatures. (a) $T = 1$ K; (b) $T = 2$ K; (c) $T = 2.15$ K; (d) $T = 2.17$ K; (e) $T = 2.1715$ K; (f) $T = 2.1719$ K. The increase in a as $T \to T_\lambda$ (Figure 2.3 and Section 4.4) limits the size of the rings.

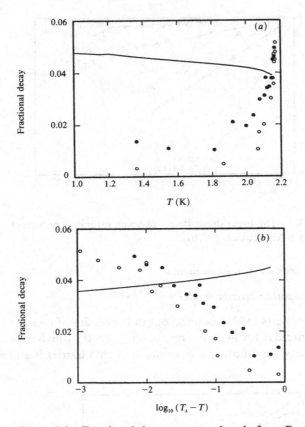

Figure 8.9 Fractional decay rate per decade from Equation (8.38) as a function of (*a*) temperature, (*b*) reduced temperature, shown compared with the data of Kukich (1970). Solid circles are data for 500 Å powder and open circles for 2000 Å powder.

Figure 8.3. With a relatively small flow, the barrier shown in Figure 8.3(*b*) will be shifted so that the barrier for a vortex pair oriented against the flow is lower than for a pair oriented with the flow. For a large superflow the process associated with one type of vortex pair will completely dominate the other, while for slower flows the nucleation of both types of vortex have to be considered. This is the 'competing-barrier' model of nucleation.

In this section we show that the competing-barrier model will qualitatively explain experiments on decay of persistent currents as reported, e.g., by Kukich (1970) and Langer and Reppy (1970). We confine ourselves to the broad issues using a simplified one-dimensional nucleation

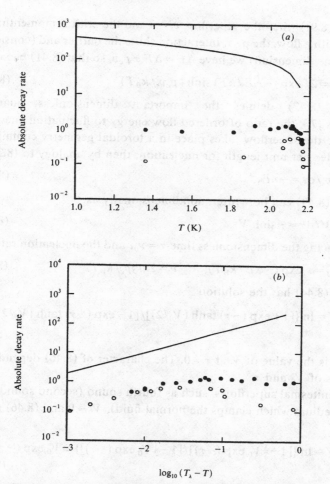

Figure 8.10 Absolute decay rate per decade from Equation (8.39) as a function of (*a*) temperature, (*b*) reduced temperature, compared with the data of Kukich (1970). Solid circles correspond to 500 Å powder, open circles to 2000 Å powder.

model, neglecting the geometrical differences among various sorts of porous materials (Donnelly, Hills and Roberts, 1979).

The nucleation probability, P, per unit time can be written

$$P = f \exp\left(-\Delta F / kT\right) \tag{8.41}$$

where f is the temperature-dependent attempt frequency discussed in the previous section and ΔF is the barrier height. For the homogeneous nucleation model, ΔF is due to the superflow alone. Here, $\Delta F = \Delta E$ (cf.

Equation (8.9)) when the superflow $v_s = 0$ and the critical momentum is then p_c. With a flow, the $\mathbf{p} \cdot \mathbf{v}_s$ interaction shifts the barrier and (considering only one dimension) we have $\Delta F = \Delta E \pm p_c v_s$ so that (8.41) becomes

$$P = 2f \exp\left(-\Delta E / kT\right) \sinh\left(p_c v_s / k_B T\right) \tag{8.42}$$

Equation (8.42) defines the important dimensionless quantity $V (\equiv p_c v_s / k_B T)$, the ratio of ordered flow energy to fluctuation energy.

Suppose the superflow takes place in a toroidal geometry containing n candidates per unit length for nucleation; then by analogy to (8.22)

$$dv_s / dt = -nP\kappa, \tag{8.43}$$

Equation (8.43) can be written nondimensionally as

$$dV / d\tau = -\sinh V \tag{8.44}$$

by introducing the dimensionless time $\tau = \nu_0 t$ and the nucleation rate ν_0:

$$\nu_0 = \nu \exp\left(-\Delta E / k_B T\right), \qquad \nu = 2n\kappa f p_c / k_B T \tag{8.45}$$

Equation (8.44) has the solution

$$V = \ln\left\{[1 + \exp\left(-\tau\right) \tanh\left(V_0/2\right)]/[1 - \exp\left(-\tau\right) \tanh\left(V_0/2\right)]\right\}, \tag{8.46}$$

where V_0 is the value of V at $\tau = 0$. The character of (8.46) depends on the ranges of V_0 and τ.

For infinitesmal superflows, such as fourth sound (second sound in a porous medium which clamps the normal fluid), $V_0 \to 0$ and (8.46) gives for all τ

$$V \cong \ln\left\{[1 + \tfrac{1}{2} V_0 \exp\left(-\tau\right)]/[1 - \tfrac{1}{2} V_0 \exp\left(-\tau\right)]\right\} \cong V_0 \exp\left(-\tau\right) \tag{8.47}$$

Thus all small superflows decay exponentially.

For $V_0 \to \infty$ and $V \to \infty$ simultaneously, $\exp\left(-V\right) \cong \exp\left(-V_0 + \tau/2\right)$ so

$$V \cong V_0 \qquad \text{(small } \tau\text{)} \tag{8.48a}$$

$$V \cong \ln\left(2/\tau\right) \qquad \text{(large } \tau\text{)} \tag{8.48b}$$

and the dividing case occurs at τ_L obtained by equating the two estimates in (8.48):

$$\tau_L = 2 \exp\left(-V_0\right) \tag{8.49}$$

For fixed τ (>0), V is independent of V_0 in the limit $V_0 \to \infty$:

$$V \cong \ln\left\{[1 + \exp\left(-\tau\right)]/[1 - \exp\left(-\tau\right)]\right\} = \ln \coth\left(\tfrac{1}{2}\tau\right)$$

so that

$$V \cong \ln (2/\tau) \quad \text{(small } \tau) \tag{8.50a}$$
$$V \cong 2 \exp (-\tau) \quad \text{(large } \tau) \tag{8.50b}$$

and the dividing case occurs at τ_E obtained by equating the two estimates in (8.50):

$$\tau_E \cong 1 \tag{8.51}$$

The dimensionless times τ_L and τ_E are fundamental to our discussion: τ_E is universal and sets the lifetime of all superflows, while τ_L depends on V_0. The flow observed depends on the magnitude of τ. For $\tau \ll \tau_L$ the flow is almost steady; for $\tau_L \ll \tau \ll \tau_E$ the flow shows logarithmic decay behavior; for $\tau \gg \tau_E$ the flow decays exponentially. The number of decades of logarithmic behavior is given by $\log_{10}(\tau_E/\tau_L) \cong 0.43 V_0 - 0.37$.

Figure 8.11 shows several examples of (8.46), the time evolution of finite superflows. In particular, the flow for $V_0 = 15$ has $\log_{10} \tau_L = -6.2$. Larger initial flows are independent of V_0 at $\log_{10} \tau = -6.2$, and an experimenter observing at this τ would term $V_0 = 15$ the 'saturated critical velocity.' Hence, the condition at time τ for a saturated critical velocity is given by $\tau = \tau_L$ and thus the notion of a saturated critical velocity depends crucially on the time of observation.

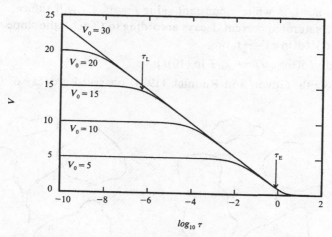

Figure 8.11 Plot of Equation (8.46) assuming $k_B T/p_c = 4$ and various values of the initial dimensionless velocity V_0. The dimensionless time showing the beginning of exponential decay is common to all curves. τ_L is shown for $V_0 = 15$. Initial flows greater than 15 have the same velocity near $\log_{10}\tau = -6$, initial flows less than 15 are 'steady' for increasing periods as V_0 decreases (after Donnelly, Hills and Roberts, (1979)).

An experimenter using fourth sound as a probe for evidence of super-fluidity would observe nothing for times larger than τ_E. Thus the condition for 'onset of superfluidity' is given by $\tau = \tau_E$.

8.4.2   Vortex nucleation in porous media

In order to make numerical calculations, we need estimates of ΔE, n, and f. Imagine a channel containing packed powder with a mean open dimension d and a vortex stretched between two grains as shown in Figure 8.12. A vortex stretched between two grains will have an energy $E_A = (\rho_s \kappa^2 d/4\pi) \ln (d/a)$, where a is the vortex core parameter. When the line is moved into a semicircle, it will just touch the next grain, and then its energy is $E_c = (\rho_s \kappa^s d/4) \ln (d/a)$. The 'barrier' is given by $\Delta E = E_c - E_A$ and $p_c = \frac{1}{2}\rho_s \kappa \pi d^2$. After each nucleation event, two vortices are left which are candidates for the next event. The vortices 'creep' through the porous medium. To order 1, we adopt for simplicity

$$\Delta E = (\rho_s \kappa^2 d/4\pi) \ln (d/a) \qquad (8.52a)$$

$$p_c = \rho_s \kappa \pi d^2 \qquad (8.52b)$$

In the same spirit, since the preexponential factors are not important, we assume that n corresponds to one trapped vortex between each pair of grains, $n = 1/d$, while a constant value $f = 10^8 \text{ s}^{-1}$ will suffice.

When a saturated current decays according to (8.48b), the slope of the decay is $dV/d(\ln \tau) = -1$, or

$$dv_s/d(\log_{10} t) = -k_B T \ln (10)/p_c \qquad (8.53)$$

Kojima, Veith, Guyon and Rudnick (1974) observed a decay of 0.63%

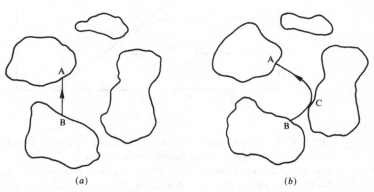

$$(a) \qquad\qquad\qquad\qquad (b)$$

Figure 8.12 Geometry for discussing the competing barrier model for vortices in a porous medium (after Donnelly, Hills and Roberts (1979)).

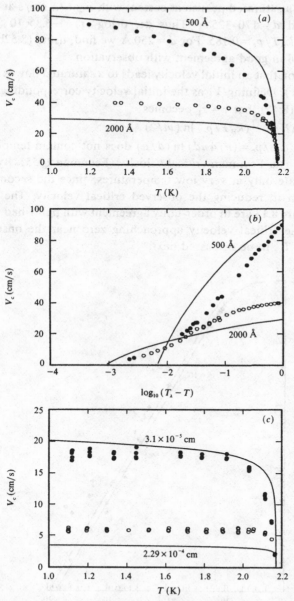

Figure 8.13 Saturated velocities as a function (*a*) temperature (*b*) reduced temperature obtained from Equation (8.54) at $t = 100$ s for two channel sizes. Data from Kukich (1970). (*c*) Flow data for two precision channels compared to (8.54). Data from Keller and Hammel (1968).

per decade of a saturated persistent current with $\nu = 67.7$ cm/s at $t = 1$ s, $T = 1.3$ K, and $d = 170\text{–}325$ Å. Thus $\mathrm{d}v_s/\mathrm{d}(\log_{10} t) = -6.3 \times 10^{-3} \times 67.7$ or, by (8.53), $k_B T/p_c = 0.185$. For $d = 250$ Å we find, using (8.52b), that $k_B T/p_c = 0.208$ in good agreement with observation.

The condition that an initial velocity leads to a saturated flow is $\tau = \tau_L$, or $V_0 = \ln(2/\tau)$. Defining V_s^0 as the initial velocity corresponding to V_0, we find from (8.45) that $\tau = \tau_L$ becomes

$$v_s^0 = \Delta E/p_c - (k_B T/p_c) \ln(\nu t/2) \tag{8.54}$$

The first term $\Delta E/p_c = (\kappa/4\pi d) \ln(d/a)$ does not contain temperature explicitly and is a Feynman critical velocity (Feynman 1955). By itself, it is appropriate only at very low temperatures, since the second term subtracts from it, reducing the observed critical velocity. The results shown in Figure 8.13 are in order-unity agreement with published results. They show the critical velocity approaching zero near the onset temperature $T_0 < T_\lambda$, to be discussed next.

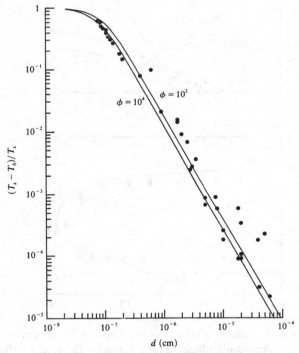

Figure 8.14 Onset temperatures T_0 obtained from various channel sizes. Data from many sources (after Donnelly, Hills and Roberts (1979)).

The condition for the onset of superfluidity is $\tau = \tau_E$. If we express our observation time in terms of frequency $\varphi = t^{-1}$, this condition by (8.45) is $\nu_0 = \varphi \tau_E$ or

$$\ln (d/a)/\ln (\nu/\varphi\tau_E)](\rho_s d/T_0) = 4\pi k_B/\kappa^2 \qquad (8.55)$$

which must be solved by iteration since ν and a are themselves temperature dependent. We show in Figure 8.14 the results of T_0 as a function of d compared with the results of many experiments, taking typical values of $\varphi = 10^2 - 10^4$ Hz. The general agreement is quite satisfactory – below about 0.6 K our results are questionable since the corresponding films are very thin. T_0 increases weakly with φ. We shall discuss thermal activation theory in thin films in detail in Section 8.7.

8.5   Nucleation of vorticity by tunnelling

8.5.1   *Motion of ions in helium II*

The experimental study of the motion of ions in superfluid ^{4}He has led to many interesting, instructive, and important experimental results (see, e.g., Section 2.5). Both positive and negative ions can be formed in the liquid by means, for example, of a suitable radioactive source. Another technique is field emission, where a large negative potential applied to a sharp metal tip in liquid helium causes electrons to tunnel from the tip and proceed towards a collecting electrode. The negative ion exists as a bare electron localized within an otherwise empty bubble, the bubble radius ranging from about 17 Å at the vapour pressure to about 11.5 Å near the melting pressure. The positive ion, He$^+$, exists at the centre of a sphere of solid helium, formed by electrostriction, with a radius of about 0.6 nm and with a large pressure gradient round the sphere. Both ions can be thought of as spherical structures, with sizes that are relatively large compared with an interatomic spacing, and both can move through the liquid, under the influence, for example, of an electric field.

Suppose that an ion moves through superfluid ^{4}He at a low velocity. In its interaction with the superfluid component the ion will behave like a sphere moving through an ideal, inviscid, incompressible fluid; there will be no drag, but the effective mass of the sphere is increased by an amount equal to half the mass of the fluid displaced (Section 1.1). Interaction with the normal fluid will arise from the scattering by the ion of the thermally-excited quasi-particles (phonons and rotons, see Section 2.2) and any ^{3}He atoms naturally present or introduced into the liquid. These scattering processes will give rise to a drag on the ion and will therefore determine the low field mobility of the ion. When the ion moves with high velocity, ideal superflow round the ion may break down. Recall the development of a wake behind a sphere in a viscous fluid illustrated in Figure 1.3. For ions at low pressures the breakdown involves the creation of quantized vortices (as discussed in Section 2.5.1); for negative ions at higher pressure the breakdown is believed to involve the creation of rotons as well. Breakdown by the creation of rotons has been studied in great detail in the experiments of McClintock and his colleagues (Allum, McClintock, Phillips and Bowley, 1977; Ellis, McClintock, Bowley and Allum, 1980); this process is interesting because it is the only known example of the breakdown of superflow by the mechanism originally discussed by Landau and shown in Figure 2.2. The creation of a quantized vortex by an ion seems usually to result in the formation of a vortex ring which ultimately traps the ion. The trapping mechanism is discussed in Section 4.5. Vortex creation by a moving ion is an example of nucleation processes in superfluid helium. However, as we have noticed in Sections 8.3 and 8.4 the details of this type of process are not well understood: some processes come from extrinsic rather than intrinsic nucleation. These difficulties might be expected to be absent in vortex nucleation by a moving ion: there seems no reason why an ion should have attached to it any nucleating remnant of vortex; and flow round an ion, at least at not too large a velocity, ought to be of the simple dipolar form for an ideal incompressible fluid.

Experiments (see Table 8.2) suggest that the critical velocity for the nucleation of vorticity by a moving ion is of order 40–60 m/s, that the nucleation process appears to be stochastic in nature and that even then the nucleation process will tend to be inhibited by a potential barrier (Section 8.2). This problem has been addressed by Muirhead, Vinen and Donnelly (1984). Their discussion includes calculations of the critical velocity, of the shape and size of the potential barrier, and of the manner and rate at which this barrier can be overcome. We reproduce their arguments in some detail, since they are illustrative of current thinking about quantized vortices in ^{4}He.

The authors assume initially that both positive and negative ions can be regarded as ideal smooth rigid spheres and that flow of the superfluid in the neighbourhood of these spheres is, prior to nucleation, the same as that of an ideal continuous incompressible fluid.

They consider first the situation at zero temperature, where the moving ion can create vorticity only if the ion–liquid system does not thereby need to gain energy, and when the momentum (or, more accurately, the impulse) of the ion–liquid system must be conserved. The nucleation process appears to involve the formation of a loop of vortex line out of the side of the ion and a small potential barrier impedes formation of the loop. The novel contribution of Muirhead *et al.* is to suggest that quantum tunnelling through this barrier might occur at a significant rate.

Strictly speaking, the energy and impulse of any configuration of vortex line depends on the structure of the core of the vortex, which is not known (see Chapter 4). It turns out that this is not a serious matter as long as the ratio R_0/a is reasonably large compared with unity, where a is an effective vortex core radius, and R_0 is the radius of curvature of the vortex loop. When these inequalities are not satisfied there is considerable uncertainty in the energy and the impulse. Using the results of calculations of Jones and Roberts discussed in Section 4.2, Muirhead *et al.* attempted to estimate the energy and impulse of the relevant vortex line configurations when R_0/a is small.

8.5.2 Calculations of critical velocities and energy barriers at $T = 0$

Muirhead *et al.* (1984) consider two configurations of quantized vortex which might be formed near the ion. The first is that considered by Schwarz and Jang (1973); i.e., a vortex ring encircling the ion in a plane parallel to the equatorial plane (Figure 8.15(*a*)). The second is a circular vortex loop, attached to the sphere and meeting it normally, as shown in Figure 8.15(*b*). This geometry was discussed by Donelly and Roberts (1971).

As explained by Schwarz and Jang, the energy and impulse of ion-vortex complexes of this kind can be calculated from formulae that involve surface integrals of the velocity field and the associated vector

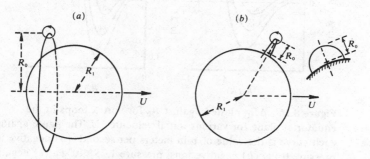

Figure 8.15 Geometry of the nucleating vortices: (*a*) the encircling ring; (*b*) the attached loop (after Muirhead, *et al* (1984)).

potential over the surfaces of the sphere and of the vortex core, and Muirhead *et al.* used the same basic formulae in their own calculations. They ensured that the boundary conditions at the surface of the sphere are satisfied by using the method of images.

The detailed results of the calculations reported by the authors show that the vortex loop configuration of Figure 8.15(*b*) is the lowest energy configuration and leads to the lowest critical velocities. Results are shown in Figure 8.16 for negative and positive ions for several initial velocities and two pressures.

An ion-vortex complex can be formed at $T = 0$ K only if $\Delta E = 0$ i.e., there must be a value of $R_0 > a$ at which the energy curves of Figure 8.16 reach zero. U_c is defined as the minimum ion velocity at which an ion–vortex complex can be formed in this way. The authors show that

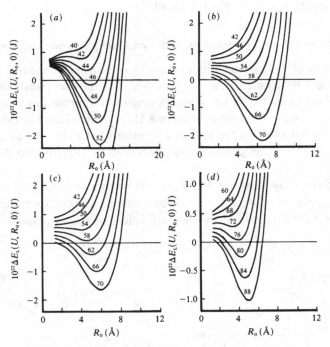

Figure 8.16 ΔE_L plotted against R_0 for vortex loops in the equatorial plane for various initial velocities U. The figure against each curve is the value of v in meters per second: (*a*) negative ions, pressure 0 Pa, (*b*) negative ions, pressure 1.75×10^6 Pa (*c*) negative ions, pressure 2.5×10^6 Pa (*d*) positive ions, pressure 0 Pa (after Muirhead *et al.* (1984)).

the most favorable (minimum energy) position for the loop is always in the equatorial plane. Corresponding values of U_c are shown in Figure 8.17. At velocities higher than U_c vortex loops can be formed outside the equatorial plane. Detailed calculations confirm that the formation of an ion–vortex loop complex at velocities greater than U_c requires that a potential barrier be overcome. The barrier for negative ions has a typical width of 4 Å, and a typical height of probably 5×10^{-23} J. This estimate of the height is based on the assumption that the curves of $\Delta E_L(U, R_0)$ do not rise in the region $R_0 < 1$ Å; i.e., where R_0 has become significantly less than the vortex core diameter. This assumption seems reasonable in view of the fact that $\Delta E_L(U, R_0)$ must fall toward zero as the vortex loop disappears altogether.

The calculations quoted above show that nucleation of a vortex loop becomes energetically possible when the ion velocity exceeds a critical value, but that nucleation is impeded by a potential barrier. A realistic calculation of the production of vortex loops by quantum tunnelling is difficult, and Muirhead *et al.* (1984) therefore attempt to solve a simpler problem that involves the same essential physics, and from which they could make a reasonable guess about the tunnelling rate for the vortex loops. This simpler problem relates to the nucleation of a vortex line at the edge of a thin flowing superfluid film at absolute zero.

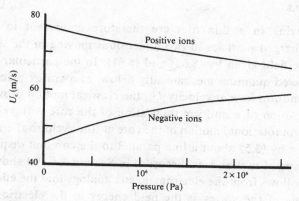

Figure 8.17 Critical velocities, U_c at which it becomes energetically possible to create a vortex, plotted against pressure for positive and negative ions. Note, however, that the positive ion velocities are higher than the Landau critical velocity (after Muirhead *et al.* (1984)).

8.5.3   Equations of motion and quantum mechanics for vortices in a thin film

Let d be the thickness and ρ_s the average superfluid density of the film. The vortex energy per unit length ε given by (2.16). Each vortex has its axis normal to the plane of the film. The core of each vortex has an effective mass $m_c = md$; for a hollow core md is simply the mass of liquid displaced by the core (i.e., $m = \pi\rho_s a^2$). The equation of motion of such a vortex (circulation κ) in a local velocity field v_s is

$$m\ddot{\mathbf{r}} = \rho_s(\mathbf{v}_s - \dot{\mathbf{r}}) \times \boldsymbol{\kappa} \tag{8.56}$$

assuming that the vortex is subject only to a Magnus force and not to any frictional forces. The equation of motion may be rewritten in the form

$$m\ddot{\mathbf{r}} = \rho_s \kappa \mathbf{v}_s \times \mathbf{s}' - \rho_s \kappa \dot{\mathbf{r}} \times \mathbf{s}' \tag{8.57}$$

and compared with that of an electrostatic line charge (q per unit length) attached to a line mass (m per unit length) moving in a direction normal to its length in an electromagnetic field ($\mathbf{E}, \mathbf{B}$). The equation is

$$m\ddot{\mathbf{r}} = q\mathbf{E} + q\dot{\mathbf{r}} \times \mathbf{B} \tag{8.58}$$

and we see that there is an analogy if we identify

$$q \to \kappa \tag{8.59}$$

$$\mathbf{E} \to \rho_s v_s \times \mathbf{s}' \tag{8.60}$$

$$\mathbf{B} \to -\rho_s \mathbf{s}' \tag{8.61}$$

Vortices moving in a thin film are therefore equivalent in their behaviour to charged particles in two dimensions moving in the steady electromagnetic field given by (8.60) and (8.61). In the particular case, which is discussed quantum mechanically below, of a vortex placed in an otherwise uniform flow at velocity U_s, the classical motion proves to be the superposition of a uniform translation of the core with velocity U_s and a uniform rotational motion of the core at (the cyclotron) angular frequency given by (3.5) about a line parallel to the core but displaced from it. This type of motion was discussed in Section 3.1 and shown in Figure 3.1. It follows from the electromagnetic analogy that the effective potential energy of the vortex is the field energy in the electric field, which is equivalent to the field energy in the velocity field v_s.

The quantum mechanics of a vortex (regarded as a particle moving in two dimensions) in the presence of an imposed uniform superfluid velocity, U_s, in the film is analogous to that of an electrostatic line charge (given by (8.59)) in a magnetic field given by (8.61) and an electric field given by

$$\mathbf{E} = \rho_s \mathbf{U}_s \times \mathbf{s}' \tag{8.62}$$

Choosing Cartesian coordinates in which the z-axis is normal to the film (and therefore parallel to the **B**-field) and the x-axis is parallel to E (i.e., perpendicular to both $\boldsymbol{\kappa}$ and $\mathbf{U}_s$) as shown in Figure 8.18 and selecting a gauge in which the magnetic vector potential is $(0, Bx, 0)$, the wave equation for the vortex will take the form

$$\frac{\partial^2 \psi}{\partial x^2} + \left(\frac{\partial}{\partial y} + \frac{i \rho_s \kappa \, dx}{\hbar} \right)^2 \psi + \frac{2md}{\hbar^2} [\varepsilon + \rho_s \kappa \, dU_s(x-b)] \psi = 0 \tag{8.63}$$

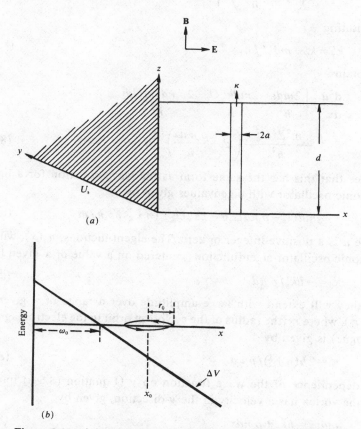

Figure 8.18 (a) Geometry of discussion of vortex nucleation in a thin film showing the electromagnetic analogy; (b) diagram illustrating the nucleation of a vortex at the edge of a film by quantum tunnelling (after Muirhead *et al.* (1984)).

in exact analogy to that for a charged particle moving in crossed $\mathbf{E}$ and $\mathbf{B}$ fields, a standard problem in quantum mechanics. (The scalar potential from which $\mathbf{E}$ is derived is chosen so that it is zero at $x = b$.)

In order to solve (8.63) we first put

$$\psi = \exp(ik_y y) u(x) \tag{8.64}$$

and find that $u(x)$ is given by

$$\frac{d^2 u}{dx^2} + \left[\frac{2m\, d\varepsilon}{\hbar^2} + \frac{2\rho_s \kappa\, d^2 m U_s}{\hbar^2} (x - b) \right.$$

$$\left. - \left(k_y + \frac{\rho_s \kappa\, dx}{\hbar} \right)^2 \right] u = 0 \tag{8.65}$$

Substituting

$$k_y' = k_y - m d U_s / \hbar \tag{8.66}$$

we obtain

$$\frac{d^2 u}{dx^2} + \left[\frac{2md\varepsilon}{\hbar^2} - \frac{2mdk_y' U_s}{\hbar} - \frac{2\rho_s \kappa d^2 m U_s b}{\hbar^2} \right.$$

$$\left. - \frac{m^2 d^2 U_s^2}{\hbar^2} - \left(k_y' + \frac{\rho_s \kappa\, dx}{\hbar} \right)^2 \right] u = 0 \tag{8.67}$$

We see that this has the same form as the wave equation for a linear harmonic oscillator with eigenvalues given by

$$\varepsilon - \tfrac{1}{2} m d U_s^2 - \rho_s \kappa d U_s b - \hbar k_y' U_s = (n + \tfrac{1}{2}) \hbar \rho_s \kappa / m \tag{8.68}$$

where n is a positive integer or zero. The eigenfunctions, $u_n(x)$, will be harmonic oscillator eigenfunctions centered on a value of x given by

$$x_0 = -\hbar k_y' / \rho_s \kappa d \tag{8.69}$$

and they will extend with large amplitude over a range of x given by $(x_0 \pm r_0)$, where r_0 (the radius of the cyclotron orbit in the electromagnetic analogue) is given by

$$r_0^2 \approx 2\hbar (n + \tfrac{1}{2}) / \rho_s \kappa d \tag{8.70}$$

The dependence of the wave function on y (Equation (8.64)) implies that the vortex has a velocity in the y-direction, given by

$$m dv_y = \hbar k_y + \rho_s \kappa dx \tag{8.71}$$

If x is made equal to its mean value, x_0, for the vortex, we find that the mean velocity of the vortex in the y-direction is simply U_s. These quantum mechanical results are consistent with the known classical behavior.

In order to deal with the tunnelling problem, we shall need to know how the function $u_n(x)$ behaves for values of x outside the range of the classical orbit, i.e., outside the range $(x_0 \pm r_0)$. Using the WKB approximation we find that the required form of $u_n(x)$ is

$$u_n(x) \sim \exp\left(\frac{-1}{\hbar} \int_{x_0 \pm r_0}^{x} \left\{ 2md \left[\frac{\rho_s^2 \kappa^2 d}{2m} (x_0 - x)^2 \right. \right. \right.$$
$$\left. \left. \left. - (n + \tfrac{1}{2}) \frac{\hbar \rho_s \kappa}{m} \right] \right\}^{1/2} dx \right) \qquad (8.72)$$

i.e.,

$$u_n(x) \sim \exp\left\{ \frac{-\rho_s \kappa d}{\hbar} \int_{x_0 \pm r_0}^{x} [(x_0 - x)^2 - r_0^2]^{1/2} \, dx \right\} \qquad (8.73)$$

where we have made use of Equation (8.70).

8.5.4 The nucleation of a vortex at the edge of a thin film by quantum tunnelling

Muirhead *et al* (1984) now consider the production of a vortex line at the edge of a film in which there is a uniform superfluid velocity field U_s as shown in Figure 8.18(a). We take the edge of the film to be the y-axis, and the velocity U_s is directed along the positive y-axis. If the vortex is formed at a distance x from the edge, the change in effective potential energy of the system is given by (8.3)

$$\Delta V = \frac{\rho_s \kappa^2 d}{4\pi} \ln\left(\frac{2x}{a}\right) + \rho_s \kappa dx U_s \qquad (8.74)$$

provided that $x/a \gg 1$. The form of ΔV as a function of x for $U_s = 40$ m/s is shown in Figure 8.1. We see that a potential barrier opposes the creation of the vortex. Vortex creation at the edge of the film can take place at $T = 0$ K only as a result of quantum tunnelling through this barrier.

In order to simplify the calculations the authors replace the real potential energy of the vortex (Figure 8.1) by the form shown by the solid line in Figure 8.18(b), which is given by

$$\Delta V = \rho_s \kappa d U_s'(w_0 - x) \qquad (8.75)$$

where U_s' is an effective superfluid velocity (equal to about half U_s) and w_0 is the width of the potential barrier. They demonstrate that the change from (8.74) to (8.75) should have little effect on the results.

Strictly speaking, a superfluid film with an edge is not realistic, and one must be careful to define exactly the conditions that are imagined to exist at this edge. Assume initially that the film is formed on a perfectly smooth substrate, and that it is bounded at its edge by a kind of two-dimensional solid wall as suggested in Figure 8.18(*a*). The vortex must then be formed as the result of an interaction that leads to an exchange of momentum between the helium and this wall. The authors presume that such an interaction can take place only if the wall is rough; i.e., has protuberances that destroy the translational invariance of the system in the direction parallel to the edge of the film.

The production of the vortex can now be considered by reference to Figure 8.18(*b*). In order to conserve energy the total energy of the vortex must be zero; and it must be created in one of the states described in the preceding section. It will therefore be in a quantum state corresponding to an orbit with radius r_0 and centre x_0, the whole orbit moving in the y-direction with an average velocity U'_s. Suppose first that the wall contains a single localized protuberance, leading to a perturbation V acting on the helium. The resulting rate of production of vortices will be given by the 'golden rule'

$$\nu_0 = (2\pi/\hbar) \left| \int \psi_f^* V \psi_i \, d\tau \right|^2 g(\varepsilon) \tag{8.76}$$

where ψ_i and ψ_f are the initial (vortex-free) and final (with vortex) states of the helium, and $g(\varepsilon)$ is the final density of states associated with the vortex. The rate of vortex production depends therefore on the magnitude of the perturbation V, which is thought to be a strong perturbation, and that furthermore in the absence of the potential barrier such a perturbation will lead to vortex creation on a timescale equal to the reciprocal of the 'cyclotron frequency', $\rho_s \kappa / m$, since this is the minimum time between the collisions of an orbiting vortex and a small fixed scattering centre. In the presence of the tunnelling barrier this rate will presumably be reduced by a factor of order the square of the extent to which the wave function of the vortex in its final state overlaps the perturbing protuberance on the wall; i.e., by the factor

$$\exp\left\{ \frac{-2\rho_s \kappa d}{\hbar} \int_0^{x_0 - r_0} [(x_0 - x)^2 - r_0^2]^{1/2} \, dx \right\} \tag{8.77}$$

where Equation (8.73) has been used. It follows that the rate of vortex production by a single protuberance will be given by

$$\nu_0 \sim \frac{\rho_s \kappa}{m} \exp\left\{ \frac{-2\rho_s \kappa d}{\hbar} \int_0^{x_0 - r_0} [(x - x_0)^2 - r_0^2]^{1/2} \, dx \right\} \tag{8.78}$$

Recalling that the vortex must be formed with zero energy, so that x_0

and n must be related by the equation

$$x_0 = w_0 + (n + \tfrac{1}{2})\hbar/mdU_s + mU_s'/2\rho_s\kappa \tag{8.79}$$

where we have used Equations (8.68) and (8.69) with $\varepsilon = 0$ and $b = w_0$. For values of w_0, d and U_s' that are comparable with analogous quantities for the ion problem, we find that

$$\hbar/mdU_s' \sim w_0 \tag{8.80}$$

$$mU_s'/2\rho_s\kappa \ll w_0 \tag{8.81}$$

In making an approximate evaluation of (8.78) we can therefore first neglect tunnelling into states with $n > 0$, since such states would yield much smaller overlap integrals than those with $n = 0$, and secondly we can put

$$x_0 = w_0 + \hbar/2mdU_s' \tag{8.82}$$

Furthermore, if $n = 0$, then $r_0^2 = \hbar/\rho_s\kappa d \ll w_0^2$, it follows that

$$\nu_0 \sim \frac{\rho_s\kappa}{m} \exp\left[\frac{-\rho_s\kappa d}{\hbar}\left(w_0 + \frac{\hbar}{2mdU_s'}\right)^2\right] \tag{8.83}$$

This production rate refers to the effect of a single localized proturberance at the wall. If we assume that the wall is rough on an atomic scale, we can take the number of such protuberances per unit length of wall to be of order $1/s$, where s is an interatomic spacing, and we can assume further that the different proturberances act incoherently. It follows that the total rate of vortex production per unit length of wall will be given by

$$\nu = \frac{\nu_0}{s} \sim \frac{\rho_s\kappa}{ms} \exp\left[\frac{-\rho_s\kappa d}{\hbar}\left(w_0 + \frac{\hbar}{2mdU_s'}\right)^2\right] \tag{8.84}$$

It turns out that, in situations comparable with those relevant to the ion problem, the rate ν does not depend strongly on the velocity U_s'; this is why the authors believe that the replacement of (8.74) by (8.75) will not have a serious effect.

So far we have assumed that the substrate on which the film is formed is smooth. If it is not smooth, there could be two effects. First, there could be an added perturbation on the helium, leading to an increased nucleation rate. (This perturbation could also give rise to a different nucleation process - the creation of a vortex-antivortex pair - for which the wall is unnecessary.) Secondly, it could give rise to a frictional force opposing the motion of a vortex in the film. Such a force could decrease the nucleation rate (Caldeira and Leggett, 1981). Suppose that the frictional force takes the form of a scattering process with relaxation time

Table 8.1. *Predicted vortex nucleation rates for negative ions at* $T = 0\,K$ *(Muirhead et al. 1984).*

Pressure (bar)	Initial ion velocity (m/s)	Nucleation rate (s^{-1})
0	46	7.84×10^{-64}
	52	2.92×10^{-11}
	56	7.15×10^{-3}
17.5	58	1.01×10^{-7}
	62	1.06×10^{4}
	66	9.22×10^{7}

τ. If the results of Caldeira and Leggett still apply in the present situation, we would then expect the tunnelling rate to be reduced by a factor

$$\exp\left(-Amd\omega_0^2/\hbar\tau\right) \tag{8.85}$$

where A is a factor of order unity.

8.5.5   Application to nucleation of vorticity by a moving ion

The two-dimensional nucleation process at a film edge is not dissimilar in principle from that required for the ion, and therefore it seems reasonable to base a guess for the ion–vortex nucleation rate on the following modification of Equation (8.83): (i) We replace w_0 by the computed barrier width for the ion; (ii) we take d equal to the length (πw_0) of the tunnelling loop; (iii) we take the prefactor to be of order the 'cyclotron' frequency, $\rho_s \kappa / m$, multiplied by the number of atoms on the surface of the ion. The results of this procedure are shown in Table 8.1.

When we come to compare the theory of Muirhead *et al.* (1984) with experiment, we shall find that to date all observed nucleation rates relate to an ion which prior to nucleation is not travelling with constant velocity, as assumed in their theory, but is instead subject either to collisions with existing thermal excitations or to roton emission processes, the average velocity of the ion being maintained constant by means of an electric field. The most satisfactory experimental results relate to an ion for which vortex nucleation is accompanied by roton emission, this latter process occurring at a rate that is much larger than the vortex nucleation rate. It is important to know whether these two processes can be regarded as independent.

The authors suggest that the emission of a roton is equivalent to a scattering of the nucleating vortex, and therefore that the roton emission will reduce the vortex nucleation rate by the factor (8.85), where τ is now the average time interval between the emission of rotons.

Note that in a quantum description of a vortex line, the core of the line is no longer necessarily localized in position. In the film problem the eigenstates correspond to vortices that are completely delocalized in the direction parallel to the edge of the film. In the case of vortex loops on an ion there will be a similar delocalization, with a loop in the equatorial plane delocalized in the direction of changing azimuth.

Note that for a vortex loop the height of the potential barrier is typically 5×10^{-23} J, which corresponds to a temperature of about 3.6 K. It follows that thermal excitation over the barrier requires only the intervention of a single, high energy phonon or a single roton. Owing to the shape of the phonon–roton dispersion curve in helium, the density of rotons in thermal equilibrium always greatly exceeds the density of high energy phonons, and therefore at a finite temperature we can expect a contribution to the nucleation rate that is proportional to the density of rotons. Furthermore, if nucleation involves the absorption of a roton, the condition that the process conserve energy and momentum requires modification, and there is a resulting reduction in critical velocity. The details of this modification have been described by Bowley *et al.* (1982).

8.5.6 *The experimental situation*

The experimental investigation of the nucleation of vortex rings from ions has proved to be a long and difficult task, full of subtle problems with a number of clever solutions. As we shall see there are few measurements in existence which directly address the theory described here. Discussions in the literature over the years, include articles by Careri (1961), Donnelly and Roberts (1971), Zoll (1976), Bowley *et al.* (1982). A brief guide to some of the relevant experimental work on ion motion follows.

The first indication that ion motion could be connected to vortex rings was reported by Meyer and Reif (1961). They discovered that ions at temperatures below 1 K failed to behave as they did at higher temperatures, and could not be stopped by simply reversing the potential gradient at some local position in the ion beam path. They named this phenomenon 'runaway behaviour'. Rayfield and Reif (1964) were able unambiguously to identify the runaway ions as charged quantized vortex rings as described in Section 2.5. Meyer and Reif also observed that application of pressures above 15 bar tended to inhibit ring production.

This early work stimulated a whole series of investigations by Careri, Cunsolo, Mazzoldi and Santini (1965), Rayfield (1966, 1967, 1968a, b), Neeper (1968) and Neeper and Meyer (1969). These experiments showed that as the electric field E is increased, the average ion velocity U rises linearly at first, then approaches a maximum value V_c at $E = E_c$. For fields beyond E_c, U decreased very rapidly (Careri named it the 'giant fall') and it was evident that the ions were dragging along vortex rings. It was observed that V_c lay in the region 30–50 m/s for either positive or negative ions for all pressures, and was relatively unaffected by temperature. A good discussion of this behavior, and its modification by adding ^{3}He is contained in the paper by Zoll (1976). In particular, Zoll discusses the effects on the nucleation rates which result when pressure is applied to the negative ion and roton emission occurs simultaneously with ring creation.

Strayer and Donnelly (1971) reported a series of experiments in which U and signal currents were measured as a function of electric field E. They demonstrated that the production of vortex rings is stochastic and the fractions of ions moving without vortex rings could be written

$$N/N_0 = \exp(-\nu t) \tag{8.86}$$

where N_0 is the number of ions in the bunch at $t = 0$.

Following Strayer and Donnelly were Titus and Rosenshein (1973), Zoll and Schwarz (1973) and Zoll (1976) who successively refined the techniques to measure ν and U. It became customary to define the critical velocity V_c as that value of U for which ν takes a small but finite value (say 10^3 s^{-1}). Some values of V_c for positive and negative ions at zero pressure are given in Table 8.2(a).

The more fundamental quantity is, of course, the transition rate ν itself. Zoll and Schwarz (1973) were able to use an 'interrupted flight technique' which allowed them to measure ν over four orders of magnitude: 10^2–10^6 s^{-1}. They also showed that the transition rate was nearly a universal function when plotted against E/E_c, where E_c is the field at which ν becomes large. Figure 1 of their paper shows that such a plot tends to make results independent of temperature, pressure, ion species, and even ^{3}He concentration for the so-called 'typical' transition. A typical transition is one in which a large discontinuity in ion velocity occurs indicating that when a vortex ring is formed it grows to large size before its drag comes into equilibrium with the applied field.

Table 8.2 (*a*) *Critical velocities for ions at P = 0.*

$T(K)$	V_c^+ (m/s)	V_c^- (m/s)	Source
0.395	40.3	33.3	Zoll (1976)
0.60	37.9	32.2	Strayer (1971)
0.65	37.5	31.6	Strayer (1971)
0.70	37.0	31.0	Strayer (1971)
0.75	36.0	—	Strayer (1971)

Table 8.2 (*b*). *Pressure dependence of critical velocities at 0.65 K after Strayer (1971).*

	P(bar)				
	0	5	10	12.5	15
V_c^+ (m/s)	37.5	36.1	35.2	34.5	34.1
V_c^- (m/s)	31.6	37.8	43.4	45.9	47.7

Table 8.2 (*c*). *Pressure dependence of critical velocities at 0.395 K after Zoll (1976).*

	P(bar)				
	0	5	10	15	20
V_c^+ (m/s)	40.1	37.1	—	35.9	—
V_c^- (m/s)	34.2	41.4	45.7	47.5	47.5

Zoll and Schwarz (1973) and Zoll (1976) pointed out that the ion velocities U reported are, of course, averages over a distribution of velocities, because the ions are continually colliding with excitations, the average velocity being kept constant by the field E. The nucleation rate, ν, calculated in the theory under discussion refers to ions moving prior to nucleation with a constant velocity U. We might hope that a calculation of $\nu(U)$ would give the instantaneous nucleation rate as a function of instantaneous ionic velocity, even when U is changing with time, although this may not be the case, in the presence of roton emission or nucleation rate. However, to calculate $\nu(U)$ from a knowledge of $\nu(U)$ (or vice versa) requires at least a knowledge of the ionic velocity distribution function, which is not generally available. That this is a major problem can be seen when it is realized that the spread in ionic velocities is typically a few meters per second, within which ν could change by several orders of magnitude. The most satisfactory remedy would be to carry out experiments at temperatures sufficiently low that collisions with excitations can be ignored, but there are formidable problems in controlling ionic motion under these conditions because the ion mean path becomes macroscopically long.

The most extensive set of measurements on ion–vortex nucleation have addressed this problem in a different way. They come from McClintock and his colleagues at Lancaster University and span nearly a decade of systematic research. The approach of this group was to go back to the early observation of Meyer and Reif (1961) that application of pressure above 15 bar quenched ring production for negative ions. Application of pressure changes the radius of the negative ion (Springett and Donnelly, 1966), reduces the roton gap Δ (Brooks and Donnelly 1977) and hence reduces the Landau critical velocity for roton emission below the vortex nucleation threshold. The Lancaster nucleation studies, then, are entirely devoted to negative ions at elevated pressures, where roton emission and nucleation are competing processes. A potential advantage is that the distribution function of velocities in the presence of roton emission is known theoretically (Bowley and Sheard, 1977) so that some progress can be made in extracting $\nu(U)$ from the observed $\nu(U)$. There remains, however, the problem of whether roton creation and vortex nucleation are independent processes. Another development was a new method for measuring ν based on electrostatic induction (described below) which allows electric fields to be employed two orders of magnitude greater than used previously. Thus U could be extended beyond V_c.

Perhaps the greatest surprise from the experimental point of view was the discovery by the Lancaster group that even the concentration of ^{3}He present in helium gas from wells is enough to alter ν profoundly (Bowley, McClintock, Moss and Stamp, 1980). This was accomplished by devising a way to sweep out almost all saturated ^{3}He atoms in ^{4}He. This discovery casts doubt on the significance of all earlier nucleation studies, including those quoted in Table 8.2. Indeed a comparison of the results of Bowley *et al.* with those of earlier workers suggests that the critical velocities quoted in Table 8.2 may have been reduced by as much as 10 m/s by the effect of ^{3}He.

8.5.7   *Direct experiments on nucleation by tunnelling*

McClintock's group in Lancaster have performed an experiment which

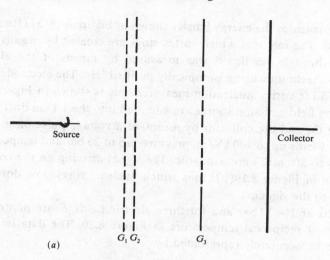

(a)

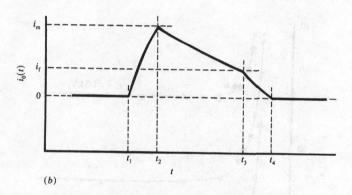

(b)

Figure 8.19 (*a*) Electrode structures used for vortex nucleation measurements. Ions from the field emission source are brought in by a voltage pulse applied to grids G_1 and G_2 to form a disk of charge. The signal is induced in the collector while the charges travel from G_3 to the collector in a string uniform electric field (after Hendry *et al.*, (1988)). (*b*) Idealized collector signal to be expected when a charge disk of length $(t_2 - t_1)$ in time travels from G_3 to the collector, plotted as collector current I, as a function of time. The charge disk is entering the induction space between t_1 and t_2; between t_2 and t_3 the disk moves freely, decaying exponentially owing to vortex ring production; between t_3 and t_4 the charge passes into the collector (after Bowley *et al.*, (1982)).

directly measures the energy barrier shown in Figure 8.16(*b*) (Hendry *et al.*, 1988). The rate ν at which vortex rings are created by negative ions moving through the liquid was measured by means of the electric-induction technique using isotopically purified ^{4}He. The electrode structure used for vortex nucleation measurements is shown in Figure 8.19. Ions for a field emission source are admitted into the 1.1 cm drift region between G_3 and the collector by means of a voltage pulse on grids G_1 and G_2. Fields up to 100 kV/m, pressures up to 25 bar and temperatures from 50 to 500 mK were available. The signal arriving on the collector is shown in Figure 8.19(*b*). Ions which nucleate rings slow down and are lost to the signal.

Results at $P = 12$ bar and for three electric fields E are plotted as a function of reciprocal temperature in Figure 8.20. The data in Figure 8.20 can be accurately represented by

$$\nu = \nu(0) + A \exp\left(-\varepsilon / k_B T\right) \tag{8.87}$$

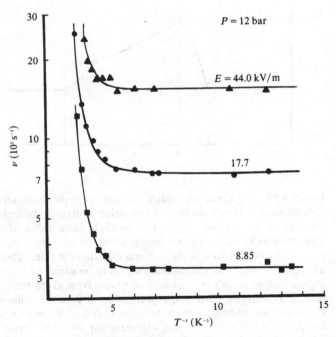

Figure 8.20 Vortex nucleation rate ν for negative ions in isotopically pure ^{4}He plotted as a function of reciprocal temperature for three electric fields E. The curves are fits of Equation (8.87) to the data (after Hendry *et al.*, (1988)).

where $\nu(0)$ represents the temperature-independent value of ν for $T <$ 200 mK and A and ε are constants. The mean obtained from $\nu(T)$ data for several values of E was $\varepsilon/k_B = 3.1 \pm 0.1$ K, which is considerably smaller than the roton gap shown in Figure 2.2. The authors interpret Equation (8.87) as representing quantum mechanical tunnelling through the barrier $(\nu(0))$ and thermal activation over it $(A \exp(-\varepsilon/k_B t))$; and suggest that the barrier is of the form shown in Figure 8.16 and magnitude as predicted in Section 8.5.5, i.e., 3.6 K. The critical velocity V_c must take account of the distribution of ionic velocities in the moving charge disk. Using a theory by Bowley (1976) the authors found $V_c = 59.5 \pm 0.3$ m/s, within 10% of the value predicted by Muirhead *et al* (1984), Figure 8.16(*b*).

8.6 The effect of dissolved ³He on vortex nucleation from ions

8.6.1 Ions and ³He

It has been known for many years that vortex nucleation by a moving negative ion in superfluid ⁴He is affected significantly by the presence of dissolved ³He (Rayfield, 1967, 1968*a, b*; Kuchnir, Ketterson and Roach, 1972) and it was suggested by Dahm (1969) that this effect might be associated with the adsorption of ³He atoms on the surface of the electron bubble that constitutes the negative ion. It has been discovered that the nucleation process is sensitive to even very small concentrations of ³He (Bowley *et al.*, 1980; McClintock, Moss, Nancolas and Stamp, 1981); a detailed study at a pressure of 23 bar has yielded results that are consistent with the idea that the effect is indeed due in some way to the adsorption of ³He atoms onto the surface of the bubble, and that the adsorption of a single ³He atom is sufficient to reduce the critical velocity for vortex nucleation by some 4 m/s and to enhance the nucleation rate by a factor of about 10^3 (Bowley *et al.*, 1984; Nancolas, Bowley and McClintock, 1985; we shall refer to these two papers as BNM). Muirhead, Vinen and Donnelly (1985) have advanced a relatively simple extension of the basic theory developed in their 1984 paper (and described in Section 8.5) in order to account for this effect.

According to the nucleation theory outlined above the vortex nucleation rate is given by an equation of the form

$$\nu = A \exp(-B) \tag{8.88}$$

where A depends on the perturbation that is applied to the helium at the surface of the ion, and B depends on the dimensions of the potential barrier opposing nucleation. The presence of a ^{3}He atom adsorbed on the surface of the ion could lead to changes in both A and B (the authors show that there is a negligible change in the bubble radius).

It has been suggested by both Dahm (1969) and BNM that the increased nucleation rate might be associated with a change in the boundary conditions at the surface of the bubble caused by the presence of the adsorbed ^{3}He atom. Such a change in boundary condition could alter A, but it is difficult to account quantitatively for changes in A. However, the ^{3}He atom can also alter the value of B, in a predictable way and, because of the form of Equation (8.88), this could have a much greater effect on the nucleation rate than changes in A.

The explanation of the change of B advanced by Muirhead *et al.* (1985) is based on two ideas: that a ^{3}He atom can be bound either to the surface of the negative-ion bubble or to the core of a vortex line, with maximum binding energies E_B and E_C respectively; and that, when bound to the surface of the ion, the ^{3}He atom can exist in a number of excited states (Shikin levels) that are thermally populated at the temperatures used in the experiments. The binding processes and the excited states are discussed below; the binding energies E_B and E_C are estimated, and evidence is presented that E_C exceeds E_B by about 0.5 K. When a negative ion with a trapped ^{3}He atom nucleates a vortex, the ^{3}He atom could be transferred either to the vortex core or simply to a lower Shikin level; in either case there would be a release of energy. The availability of this extra energy could lead both to a reduction in the minimum velocity at which vortex nucleation becomes energetically possible (the critical velocity) and, *via* a modification to the form of the potential barrier, to an increase in the nucleation rate at supercritical velocities.

8.6.2   *The binding of a ^{3}He atom to the negative ion bubble*

It is known from experiment that small concentrations of ^{3}He produce a striking reduction in the surface tension of liquid ^{4}He at low temperatures (see, e.g., Edwards and Saam (1978)), and it was suggested by Andreev (1966) that this effect is due to the adsorption of ^{3}He atoms into bound states on the liquid surface. Such adsorbed atoms will exist as quasi-particles of effective mass m_s^* with the energy–momentum relation

$$\varepsilon(p) = -\varepsilon_{30} - \varepsilon_B + p^2/2m_s^* \tag{8.89}$$

where ε_{30} is the normal bulk binding energy of ^{3}He in ^{4}He, ε_B is the maximum extra binding energy associated with the surface state, and p

is the momentum associated with the motion of the quasi-particle parallel to the surface. When a ^{3}He atom moves from the bulk ^{4}He into the surface it gains potential energy because it interacts with fewer ^{4}He atoms, but it also loses zero point kinetic energy because it is less closely confined by the neighbouring ^{4}He atoms. Various theories (reviewed by Edwards and Saam (1978)) show that the net effect must be a loss of energy, resulting in a positive binding energy ε_B of about 2 K. Values of ε_B and the effective mass m_s^* can be obtained from experiment by fitting an appropriate theory to the observed values of both the surface tension and the velocity of surface second sound, with the results

$$\varepsilon_B / k_B = (2.22 \pm 0.03)\ \text{K}$$
$$m_s^* / m_3 = 1.45 \pm 0.1 \tag{8.90}$$

If a ^{3}He atom can be bound to the free surface of liquid ^{4}He then it can also be bound to the surface of the negative ion bubble, as was suggested by Dahm (1969). Evidence for this from the observed ionic mobility in ^{3}He-^{4}He solutions was provided by Kuchnir et al. (1972). The maximum binding energy E_B of a single ^{3}He atom to the bubble surface would be expected to be close to the value (8.90), although not exactly the same because the bubble surface is sharply curved and because the pressure just below this surface differs from that at the horizontal free surface. In the interpretation of their results on vortex nucleation in solutions containing only small traces of ^{3}He, BNM use models in which E_B enters as a parameter, and they are able to determine its value from their experimental results. The precise value depends a little on the particular model, but of those that they do use, the model giving the best fit to the experimental data yields

$$E_B / k_B = (2.52 \pm 0.09)\ \text{K} \tag{8.91}$$

For an external pressure of 23 bar, and we shall use this value in our later discussion.

The ^{3}He atom can also exist in bound excited states as we have already mentioned; they are characterized by an angular momentum quantum number l, with energies

$$-E_B + \frac{l(l+1)\hbar^2}{2m_s^* R_I^2} \tag{8.92}$$

where R_I is the ionic radius (Shikin 1973). Each of these states has degeneracy $(2l+1)$, but, as shown by BNM, this degeneracy is slightly lifted by the surrounding superfluid velocity field when the ion is moving.

8.6.3   Binding of a ^{3}He atom to a vortex line

A dissolved ^{3}He atom is attracted towards the core of a vortex line for two reasons. The atom displaces some of the ^{4}He atoms, so reducing the kinetic energy of superfluid flow associated with the line; and the Bernoulli pressure drop in the vortex gives rise to a decrease in the minimum energy of the ^{3}He atom in the liquid ^{4}He. The total attraction is sufficiently large to cause binding, and we define the binding energy, E_c, as the difference between the minimum energy of a ^{3}He atom bound to the vortex and the minimum energy of a ^{3}He atom in the bulk of the liquid.

Correcting and extending an earlier analysis of Senbetu (1978) the authors find that bound states of the ^{3}He quasi-particle have eigenfunctions of the form

$$\psi_l(r, \theta, z) = R_l(r) \exp (il\theta) \exp (ikz) \tag{8.93}$$

with eigenvalues of the form

$$E_l = -E_l^0 - \varepsilon_0(p_0) + \hbar^2 k^2 / 2m_3^*. \tag{8.94}$$

The maximum binding energy, E_c, introduced above, will then be given by the largest eigenvalue E_l^0.

The authors determined the eigenvalues E_l^0 by numerical methods for pressures, P_0, of 0 and 15 bar they found that the low pressure bound states exist for values of l equal to 0, -1, and -2; the lowest level corresponds to $l = -1$ and has a binding energy given by

$$E_{-1}^0 / k_B = 3.51 \text{ K} \qquad (P_0 = 0) \tag{8.95}$$

At high pressure bound states exist for values of l equal to 0, -1, -2, and -3; the lowest level corresponds again to $l = -1$ and has a binding energy given by

$$E_{-1}^0 / k_B = 3.00 \text{ K} \qquad (P_0 = 15 \text{ bar}) \tag{8.96}$$

It is interesting that the energies (8.95) and (8.96) correspond to ground states with nonzero angular momenta. The ^{3}He atom is therefore in an orbital state round the vortex core, with a function that peaks at a finite radial distance, r_m, which turns out to be equal to about $1.3a$.

The authors show that the classical analogue of the state with $l = -1$ is a state in which the ^{3}He atom is in orbital motion round the line at a radius r_m with angular velocity, Ω, given by

$$m_3^* r_m^2 \Omega = -\hbar + \hbar \delta m_3 / m_4 \tag{8.97}$$

At a pressure of 15 bar, $\delta m_3 / m_4$ is approximately 1.3, so that Ω is positive.

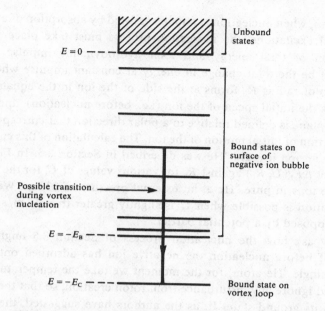

Figure 8.21 States of a ³He atom in association with a negative ion bubble (after Muirhead *et al* (1985)).

This means that the orbital motion of the atom is in the same direction as the circulation in the vortex.

There seems to be no direct experimental evidence relating to the binding energy of a single ³He atom to a vortex. However, there is much evidence that ³He atoms are attracted towards the core of a vortex, see Section 4.7. It appears that at a sufficiently high concentration a condensation, or phase separation, of the ³He atoms can occur onto the core.

It is clear that neither theory nor experiment has yet given us an entirely reliable estimate of the maximum binding energy of a ³He atom to a vortex line. However, such evidence as there is points consistently to a value close to 3 K at all pressures. The authors note that this value exceeds the binding energy to the surface of the negative ion bubble by about 0.5 K.

The energy level diagram of Figure 8.21 summarizes the results.

8.6.4 *Vortex nucleation by a negative ion containing one or two adsorbed ³He atoms*

According to the theory described above the most favorable vortex nucleus is in the form of a loop out of the side of the ion. At low temperatures, when nucleation cannot be assisted by absorption of existing thermal excitations, the nucleation process must take place with conservation of total energy and total hydrodynamic impulse. Let $\Delta E_L(U, R_0)$ be the total change in energy at constant impulse when a vortex loop of radius R_0 forms at the side of the ion in the equatorial plane; U is the initial speed of the ion (i.e., before nucleation), and the equatorial plane is defined relative to a polar direction that corresponds to the direction of initial motion of the ion. The calculation of this energy change for an ion in pure ^{4}He was described in Section 8.5. In Figure 8.16 we plot $\Delta E_L(U, R_0)$ against R_0 for various values of U, for the case of negative ions in pure ^{4}He at an external pressure of 23 bar. We see that nucleation is possible when U is slightly greater than 58 m/s, but that it is opposed by a potential barrier.

We now ask how the nucleation process of Section 8.5 might be modified if before nucleation the negative ion has adsorbed onto its surface a single ^{3}He atom; for the moment we take the temperature to be zero and ignore any disturbance from roton creation, so that the ^{3}He atom is in its ground state. If, as the authors have suggested, the ^{3}He atom is more strongly bound to a vortex core than to the negative ion bubble, then on nucleation the ^{3}He atom can move from the ion to the core of the vortex nucleus with the release of an amount of energy equal to the difference, $(E_c - E_B)$, between the two binding energies. This release of energy must be taken into account in calculating $\Delta E_L(U, R_0)$, and we see that its effect on Figure 8.16 must be simply to shift all the curves downwards by an amount equal to $(E_c - E_B)$. This will lead to a reduction in critical velocity, to a reduction in the height and width of the potential barrier, and hence to an increase in nucleation rate at supercritical velocities. Owing to the exponential dependence of the nucleation rate on barrier dimensions the effect could be quite large. These qualitative conclusions are consistent with the experiments of BNM.

8.6.5   *The predicted nucleation rates and a comparison with experiment*

The details of the calculations are discussed by Muirhead *et al.* (1985). They use calculations done for 23 bar yielding results much like Figure 8.16(c), but shifted by $(E_c - E_B)$ to calculate barrier dimensions for various initial ion velocities and initial Shikin states. They then use the theory of Section 8.5 to estimate the corresponding nucleation rate. Finally, for each value of the initial velocity they take an average over the possible initial Shikin states with weightings equal to the populations that are present in thermal equilibrium. The results are shown in Figure 8.22, compared with the results of the analysis by BNM of their own experimental results.

The agreement appears to be quite good. The predicted variation of nucleation rate with velocity under supercritical conditions is not a step function; other, smoother, forms of variation would probably fit the experimental results equally well. The change in critical velocity caused by one ³He atom is more or less correctly predicted; and the very large increase in nucleation rate at supercritical velocities is at least qualitatively in accord with the theory. There is evidence from the experiments of BNM that a second ³He atom does indeed have a large effect, giving rise, in fact, to an increase in nucleation rate by a second factor of about 10^3.

The theories of nucleation outlined in this chapter span a wide range of interesting phenomena and have, as one can see, some limited success. The main problem in helium II, however, the unexpectedly low critical velocities for pure superflow encountered in many experiments, is not yet understood. Problems such as spin-up, critical velocities in channels and near oscillating boundaries are not understood in any detail. One

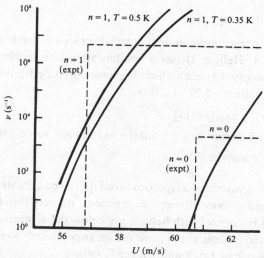

Figure 8.22 Predicted nucleation rate ν plotted against U for different values of the number, n, of ³He atoms adsorbed on the negative ion at a pressure of 23 bar. The broken lines are taken from the analysis of BNM of their experimental results (after Muirhead *et al* (1985)).

of the main frontiers of research in superfluidity is still to be successfully explored.

8.7   Vortex arrays and vortex dynamics in thin films

8.7.1   *Two-dimensional superfluidity*

Throughout this book we have been dealing with quantized vortex lines in either an unbounded fluid, or one with boundaries which are macroscopically distant compared to, say, the vortex core parameter. The question naturally arises, what would happen if the boundaries on an otherwise unbounded system such as we discussed in Section 5.4 should become closer and closer together? The limit of such a process would be a two-dimensional rather than three-dimensional problem in superfluidity.

The first question to be asked, naturally, would be whether there is such a thing as two-dimensional superfluidity. Superfluid behavior may be studied by considering the correlation function

$$G(r) = \langle \psi(r)\psi^*(0) \rangle \tag{8.98}$$

In three dimensions if there is no long range order then $G(r)$ decays exponentially

$$G(r) \sim \exp\left[-r/\zeta(T)\right] \tag{8.99}$$

where $\zeta(T)$ is a temperature-dependent correlation length: such must be the case in helium I. Helium II has a condensate which exhibits long range order in momentum space (which is phase correlated throughout the liquid – see Equations (2.72)–(2.76))

$$\langle \psi(r) \rangle \sim (\rho)^{1/2} \exp\left[i\varphi(r)\right] \tag{8.100}$$

and the correlation function has a nonzero asymptotic value for $G(r)$:

$$\lim_{r \to \infty} G(r) = \text{constant} \neq 0 \tag{8.101}$$

In 1967, Hohenberg provided a rigorous proof that a condensate cannot exist in two dimensions. Nevertheless, experiments on very thin films of helium II prepared in contact with helium gas below the saturated vapor pressure (unsaturated helium films) have shown superfluidity as far back as the pioneering work of Long and Meyer (1950).

Theoretical work (see Nelson (1983)) has shown that thermal fluctuations in two dimensions can lead to a decay law for $G(r)$ much like three-dimensional systems near a critical point $G(r) \sim 1/r^{\eta(T)}$ where $\eta(T)$ is a nonuniversal temperature-dependent exponent. This type of system is said to have 'algebraic long range order' and exhibits super-fluidity.

The developments which have come in the past few years have occurred so rapidly that superfluidity in thin films has become an entire subfield of low temperature physics by itself. Since the thermal fluctuations appear to be vortex-antivortex pairs, some substantial discussion of their behavior is appropriate in a book such as this. The language of thermal activation is used in important ways in this discussion. The reader should consult Sections 1.9, and 8.1–8.4 for ideas pertinent to this topic. In the subsections below we shall give brief accounts of some current ideas and directions for research. But the reader should be aware that some of the details may soon be superseded by newer research.

8.7.2 *Kosterlitz–Thouless theory*

Kosterlitz and Thouless (1973), recognizing that phase fluctuations are energetically favored over amplitude fluctuations in two-dimensional systems, predicted that vortex-antivortex pair excitations would dominate the fluctuations at sufficiently high temperatures, and could initiate a phase transition to a nonsuperfluid state. They used a simple free energy argument based on the form of the energy of a straight vortex line, Equation (2.16). In what follows, we reproduce, in part, a brief summary of the problem by Adams and Glaberson (1987), and Adams (1986).

Ignoring contributions from three-dimensional excitations of the line such as waves, the entropy is just proportional to the logarithm of the total number of different independent positions that can be occupied by the line:

$$S = k_B \ln (R/a)^2 = 2k_B \ln (R/a) + \text{constant} \qquad (8.102)$$

where k_B is Boltzman's constant and R^2 is the area of the system. The free energy in a film is then (by (2.16))

$$F = E - TS = (d\rho_s \kappa^2/4\pi) \ln (R/a) - 2k_B T \ln (R/a) \qquad (8.103)$$

where d is the thickness of the film. It follows that it is energetically favorable for a line to be present when the temperature is above some critical temperature T_{KT} given by (Nelson and Kosterlitz, 1977; Kosterlitz and Thouless, 1978)

$$T_{KT} = \sigma_s^0 \kappa^2/8\pi k_B \qquad (8.104)$$

where $\sigma_s^0 (= \rho_s d$ for thick, or bulk, films) is the areal superfluid density in the film. At temperatures below T_{KT} there is some thermal distribution of oppositely oriented bound vortex pairs. Above T_{KT}, vortex pairs will dissociate completely, leaving a finite number of free vortices (see Figure 8.23). The spontaneous appearance of free vortex lines implies loss of long range order, dissipation by mutual friction for any superflow and hence a breakdown of superfluidity. This 'normal' state, however, is not 'normal' in the sense of helium I. It is still superfluid in the sense that uniformly rotating helium II is superfluid. The transition from the normal state above T_{KT} to the superfluid state below T_{KT} is accompanied by a discontinuous jump in the effective superfluid density given by Equation (8.104). Of course, even below T_{KT}, the presence of superflow will induce some vortex pair dissociation. Furthermore the presence of a large number of vortex pairs will modify the energy of any particular pair so that the vortex number and the superfluid density σ_s must be 'renormalized' from its simplest estimate $\rho_s d$.

The energy required to create an isolated vortex, Equation (1.38), is much greater than that required to create a vortex–antivortex pair, Equation (1.44), thus a thermally activated pair is much more likely to appear than a single isolated vortex. The energy of a rectilinear pair varies logarithmically in pair separation r, (see Equation (1.44))

$$E_p = 2\pi\sigma_s^0\left(\frac{\hbar}{m}\right)^2 \ln\left(r/r_0\right) + E_c \tag{8.105}$$

where r_0 is the minimum separation and E_c the potential energy of the

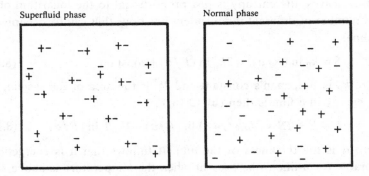

Figure 8.23 Schematic representation of the superfluid and normal phases of a two-dimensional film of ^{4}He. Below T_{KT} there are closely bound vortex–antivortex pairs designated by + and − to denote the direction of circulation. Above T_{KT} there is a plasma of dissociated vortex lines which destroy phase coherence.

pair in the sense of Section 1.3. E_p represents a long range interaction which can lead to dramatic cooperative effects in a many-body system.

Kosterlitz and Thouless used renormalization techniques to solve the statistical problem of a two-dimensional gas of logarithmically interacting vortex–antivortex pairs. They accounted for the attenuating effect of smaller pairs on the interaction between the respective members of larger pairs by introducing a scale-dependent dielectric constant $\tilde{\varepsilon}$ into Equation (8.105). As we noted in Section 1.3, two-dimensional vortices are characterized by points in a plane, and these points with their signs resemble a charged plasma. A large test pair polarizes the intervening pair plasma which, in turn, screens some of the bare interaction. The resulting effective interaction can be written as

$$E_p = 2\pi\sigma_s^0 \left(\frac{\hbar}{m}\right)^2 \int_{r_0}^{r} \frac{dr'}{r'\tilde{\varepsilon}(r')} + E_c \qquad (8.106)$$

Note that Equation (8.106) is logarithmic in r/r_0 when $\tilde{\varepsilon} = 1$. To obtain a self-consistent determination of $\tilde{\varepsilon}(r)$, one starts with the general relation between $\tilde{\varepsilon}$ and the susceptibility $\chi(r)$, familiar from electromagnetic theory:

$$\tilde{\varepsilon}(r) = 1 + 4\pi\chi(r) \qquad (8.107)$$

The susceptibility is related to the polarizability $\alpha(r)$ by

$$\chi(r) = \int_{r_0}^{r} \alpha(r')\, dn(r') \qquad (8.108)$$

where $n(r)$ is the number density of pairs. Equations (8.107) and (8.108) can, in principle, be used to determine $\tilde{\varepsilon}(r)$ once $\alpha(r)$ and $n(r)$ are known.

The polarizability per pair of a charged plasma is given by

$$\alpha(r) = q \frac{\partial}{\partial E} \langle r \cos\theta \rangle |_{E=0} \qquad (8.109)$$

where q is the charge of a pair member, E is the applied electric field, and θ is the angle between the dipole moment and E. This system is analogous to the vortex pair system with the following transcription:

$$2q^2 \leftrightarrow 2\pi\sigma_s^0(\hbar/m)^2$$

$$qE \leftrightarrow 2\pi\sigma_s^0 \frac{\hbar}{m} \mathbf{1}_z \times (\mathbf{v}_n - \mathbf{v}_s) \qquad (8.110)$$

where $\mathbf{v}_n$ and $\mathbf{v}_s$ are the normal fluid and superfluid velocities, respectively. Equation (8.109) can easily be evaluated by assuming that the field giving rise to the polarization (i.e., the field of the largest test pair) is relatively constant over the dipole pairs. Under this assumption, the differentiation with respect to E and the limit $E \to 0$ can be performed before integrating out the thermal average. After carrying out the differentiation and the limit, we can easily evaluate the average over the Boltzmann factor,

$$\exp\left[-E_p(r)/k_B T\right] = (r/r_0)(2q^2/k_B T\tilde{\varepsilon}) \exp\left(Eqr\cos\theta/k_B T\right)$$

$$\times \exp\left(-E_c/kT\right) \tag{8.111}$$

in which case

$$\alpha(r) = q^2 r^2 / 2k_B T \tag{8.112}$$

which translates to

$$\alpha(r) = \pi K_0 r^2 / 2 \tag{8.113}$$

in the two-dimensional helium system where

$$K_0 = \sigma_s^0 (\hbar/m)^2 / k_B T \tag{8.114}$$

The probability P of a pair existing in an area R^2 is equal to the sum of

$$\exp\left[-E_p(|\mathbf{r}_i - \mathbf{r}_j|/r_0)/k_B T\right] \tag{8.115}$$

over all possible $\mathbf{r}_i$ and $\mathbf{r}_j$. We can transform the sum into a double integral by assuming that $\exp(-E_c/k_B T) \ll 1$ and simply counting the possible configurations. There are $(R/r_0)^2$ possible locations for the first pair member and $2\pi r \, dr / r_0^2$ possible locations for the second in an annulus with radius r and width dr. Therefore,

$$dP(r) = \frac{R^2}{r_0^2} \left\{ \frac{2\pi r \, dr}{r_0^2} \exp\left[\frac{-E_p(r)}{k_B T}\right] \right\} \tag{8.116}$$

and we identify

$$dn(r) = dP(r)/R^2 \tag{8.117}$$

in which case Equation (8.108) becomes

$$\chi(r) = \int_{r_0}^{r} \alpha(r) \exp\left[\frac{-E_p(r)}{k_B T}\right] \frac{2\pi r \, dr}{r_0^4} \tag{8.118}$$

Equations (8.107) and (8.108) can now be combined for a self-consistent determination of the effective interaction. The result is the famous Kosterlitz–Thouless recursion relations from which a scale-dependent dielectric constant $\tilde{\varepsilon}$ and vortex pair excitation probability y^2

can be extracted

$$
\left.
\begin{aligned}
K^{-1}(l) &= K_0^{-1} + 4\pi^3 \int_0^l y^2(l')\, dl' \\[2ex]
y^2(l) &= y_0^2 \exp\left[4l - 2\pi \int_0^l K(l')\, dl' \right]
\end{aligned}
\right\}
\tag{8.119}
$$

where

$$
\begin{aligned}
l &= \ln (r/r_0) \\
K^{-1}(l) &= \tilde{\varepsilon}(l)/K_0 \\
y_0^2 &= \exp (-E_c/k_B T)
\end{aligned}
\tag{8.120}
$$

and y_0 is called the fugacity of the vortices.

The measured superfluid density σ_s in this theory is

$$
\sigma_s = \sigma_s^0(T)/\tilde{\varepsilon}
\tag{8.121}
$$

where $\sigma_s^0(T)$ is the temperature-dependent background areal superfluid density – of order $\rho_s(T)d$, where d is the film thickness. The vortex excitation probability y^2 is related to the density of pairs per unit area of separation $\delta n(r)$ by

$$
\delta n(r) = y^2(r)/r^4
\tag{8.122}
$$

In principle, the recursion relations of Equation (8.119) can be used to predict the static Kosterlitz–Thouless transition in an unbounded system by iterating them to $l = \infty$ from a locus of initial conditions

$$
\begin{aligned}
y^2(l = 0, T) &= y_0^2(T) \\
K(l = 0, T) &= K_0(T)
\end{aligned}
\tag{8.123}
$$

The static transition temperature T_{KT} is defined as the smallest temperature for which

$$
\lim_{l \to \infty} y^2(l, T_{KT}) \neq 0
\tag{8.124}
$$

or equivalently by

$$
\lim_{l \to \infty} \tilde{\varepsilon}(l, T_{KT}) = \pi K_0/2
\tag{8.125}
$$

and is the temperature at which pairs of infinite separation unbind. Above T_{KT} both $\tilde{\varepsilon}$ and y^2 become large, indicating that σ_s^0 is being renormalized to zero and that very large pairs are becoming significantly probable. Note that Equation (8.125) implies that

$$
\sigma_s(T_{KT}) = \frac{2k_B m^2}{\pi \hbar^2} T_{KT}
\tag{8.126}
$$

This is the central result of the static theory. Equation (8.126) predicts a universal jump in superfluid density at the static transition temperature, and is well verified by experiment.

8.7.3   Linear dynamic theory

Since many experiments are performed at finite frequencies, in which there is an oscillating superfluid flow, it is important to incorporate the static theory outlined above into a more comprehensive theory that accounts for the ac response of the vortex plasma. The motion of vortices is assumed to be diffusive in these systems so that it is intuitively clear that the vortex diffusion length r_D should play an important role in the dynamics. This length, in fact, determines the cross-over between smaller pairs which can equilibrate to the oscillating field and larger pairs which cannot. The theory outlined below, which was developed by Ambegaokar, Halperin, Nelson, and Siggia (1978, 1980) (AHNS), relates r_D to the response of the superfluid in terms of a complex dynamical dielectric constant ε defined by

$$\mathbf{v}_s(\omega) = [1 - \varepsilon^{-1}(\omega)]\mathbf{v}_n(\omega) \tag{8.127}$$

where ω is the frequency of oscillation. This generalized dielectric constant is analogous to $\tilde{\varepsilon}$ in the static theory and has contributions from both bound pairs and free vorticity.

Ambegaokar and Teitel (1979) derived the response function of a test pair in an oscillating field and extracted the corresponding contribution to the dielectric constant. They considered the equivalent problem of the diffusive motion of charged rods in an oscillating electric field. The motion of the rods is governed by the Langevin equation,

$$\frac{d\mathbf{r}}{dt} = \frac{-2D}{k_B T}\frac{\partial U}{\partial r} + \eta(t) \tag{8.128}$$

where $\mathbf{r}$ is the vector along the length of the rod, D is the rod diffusivity, U is the potential energy of the rod, and η is a Gaussian noise source satisfying

$$\langle \eta^\alpha(t)\eta^\beta(t') \rangle = 4D\delta_{\alpha\beta}\delta(t - t') \tag{8.129}$$

This noise term has contributions from phonons, rotons, ripplons (quantized surface waves) and even from surrounding vorticity. The potential U is

$$U(r) = 2q^2 \int_{r_0}^{r} \frac{dr'}{r'\tilde{\varepsilon}(r')} - q\delta\mathbf{E}\cdot\mathbf{r} - 2u_0 \tag{8.130}$$

where $2u_0$ is the chemical potential for the creation of a rod and δE is a small external (oscillating) electric field. Again, the vortex language is recovered using Equation (8.110).

The response function is a solution to the Fokker–Planck equation, which is obtained from Equation (8.128)

$$\frac{d\delta n(r, t)}{dt} = \frac{2D}{k_B T} \frac{\partial}{\partial \mathbf{r}} \cdot \left[\frac{\partial U}{\partial \mathbf{r}} \delta n(r, t) \right] + 2D \frac{\partial^2 \delta n(r, t)}{\partial r^2} \tag{8.131}$$

where $\delta n(r, t)$ is the density of pairs per unit area of separation. Ambegaokar and Teitel solved this equation to first order in δE. They then used the resulting expression for $\delta n(r, t)$ to calculate ε_b – the bound pair contribution to ε. In general,

$$\varepsilon_b(\omega) = 1 + 2\pi q \int d^2 r \cdot \partial \delta m / \partial E \tag{8.132}$$

which when combined with the solution of Equation (8.131) becomes

$$\varepsilon_b(\omega) = 1 + \int_{r_0}^{\infty} dr \left(\frac{d\tilde{\varepsilon}}{dr} \right) g(r, \omega) \tag{8.133}$$

where $g(r, \omega)$ is a complex pair response function approximated by

$$g(r, \omega) \approx \frac{14D}{14D - i r^2 \omega} \tag{8.134}$$

and $d\tilde{\varepsilon}/dr$ is determined from the recursion relations above. The response function varies rapidly compared to $d\tilde{\varepsilon}/dr$ so that the former may be approximated by

$$\begin{aligned} \mathrm{Re}\,[g(r, \omega)] &= \theta(14D - r^2\omega) \\ \mathrm{Im}\,[g(r, \omega)] &= \tfrac{1}{4}\pi r \delta(r - r_D) \end{aligned} \tag{8.135}$$

where θ is a step function defined by

$$\theta(x) = \begin{cases} 1 & \text{for } x > 0 \\ 0 & \text{for } x \leq 0 \end{cases} \tag{8.136}$$

and r_D is the diffusion length

$$r_D = (14D/\omega)^{1/2} \tag{8.137}$$

Inserting Equations (8.135) into (8.133), the resulting expression for ε_b in the vortex language is

$$\mathrm{Re}\,\varepsilon_b = \tilde{\varepsilon}(l_\omega) \tag{8.138}$$

$$\mathrm{Im}\,\varepsilon_b = K_0 \pi^4 y^2(l_\omega) \tag{8.139}$$

where

$$l_\omega = \tfrac{1}{2} \ln (14D/\omega r_0^2) \tag{8.140}$$

Thus the physical interpretation of Equation (8.138) is that pairs larger than r_D do not contribute to the renormalization of σ_s^0 so that the iteration must be truncated at $l = l_\omega$. Equation (8.139) accounts for pairs with separation of order r_D. These pairs are maximally out of phase with the oscillating field and give rise to dissipation.

Free charge (vorticity) is assumed to diffuse in the macroscopic electric (superfluid) field. This gives rise to an additional imaginary contribution to the dielectric constant

$$\varepsilon_f = i 4\pi n_f q^2 D / (k_B T \omega) \tag{8.141}$$

where n_f is the areal density of free charges. In the vortex language

$$\varepsilon_f = i 4 n_f \pi^2 K_0 D / \omega \tag{8.142}$$

There are two possible sources of free vorticity: that arising from pair dissociation and that induced by rotation. Pair dissociation occurs when pairs are thermally activated over the separation at which they are still bound. Below T_{KT}, all pairs are bound (in an infinite system), at least for infinitesimal external fields. Above T_{KT}, a correlation length ξ_+ can be defined which represents the length scale at which pairs begin to unbind. The thermally activated free vortex density is believed by AHNS to be related to ξ_+ by

$$n_f = F / \xi_+^2 \tag{8.143}$$

where F is a parameter of $O(1)$. In terms of Kosterlitz–Thouless theory, ξ_+ is defined by the value of l at which $y^2(l)$ iterates back to order y_0^2 (see Figure 8.24). AHNS have shown that in the temperature domain t,

$$\xi_+ \approx r_0 \exp\left(2\pi / b t^{1/2}\right) \tag{8.144}$$

where t is the reduced temperature

$$t = T / T_{KT} - 1 \tag{8.145}$$

and b is a nonuniversal parameter. There is no simple physical interpretation of b – it presumably reflects the temperature dependence of the bare superfluid density and vortex fugacity.

8.7.4   *Finite amplitude effects*

The theory outlined above assumes that infinitesimal external fields are used to probe the vortex plasma. In practice this assumption is easily violated – requiring an analysis of nonlinear finite amplitude effects. In the charged plasma formalism, the energy of a pair in an electric field $\mathbf{E}$ is

$$U(r) = 2q^2 \int_{r_0}^{r} \frac{dr'}{r' \tilde{\varepsilon}(r')} - q\mathbf{E} \cdot \mathbf{r} - 2u_0 \tag{8.146}$$

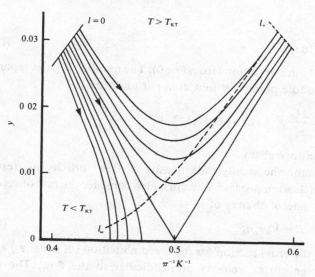

Figure 8.24 A schematic diagram of solutions to the Kosterlitz-Thouless recursion relations. The $l = 0$ line represents the initial conditions and the $l = l_\omega$ line represents the diffusion cut off. The l_+ line is determined by $y(l_+) = y(l = 0)$ and defines $\xi^+ = r_0 \exp(l_+)$ (after Adams and Glaberson (1987)).

The dipole term in Equation (8.146) leads to two major modifications of the linear theory. The first is related to the fact that the potential energy $U(r)$ has a saddle point at $r = (r_c, 0)$ where,

$$r_c = 2q/E\tilde{\varepsilon}(r_c) \qquad (8.147)$$

AHNS have used the two-dimensional Coulomb plasma analyses of McCauley (1977) and Huberman, Myerson and Doniach (1978) together with the nucleation theory of Langer and Reppy (1967) to calculate the rate R at which pairs nucleate over r_c and dissociate. The calculation, not unlike that in Section 8.3.1, begins by obtaining the Fokker–Planck equation given by Equation (8.131). This is then solved by expanding $U(r)$ in the neighborhood of $(r_c, 0)$

$$\exp\left(\frac{-U}{k_B T}\right) \approx \frac{r_0^4}{r_c^4} y^2(l_c) \exp\left[2\pi K(l_c)\right]$$

$$\times \exp\left[\frac{\pi K(l_c)(\delta x)^2}{r_c^2}\right]$$

$$\times \exp\left[\frac{-\pi K(l_c)(\delta y)^2}{r_c^2}\right] \qquad (8.148)$$

where

$$l_c = \ln(r_c/r_0) \tag{8.149}$$

and δx and δy are deviations from $(r_c, 0)$. The number of pairs separating across the saddle point, per unit area per unit time, is

$$R = \frac{2D}{r_c^2} y^2(l_c) \exp[2\pi K(l_c)] \tag{8.150}$$

in the vortex formalism.

To determine the steady-state density of free vorticity n_c, resulting from this nucleation process, one must also consider the rate of recombination. The rate of change of n_c is

$$\dot{n}_c = R - V_p \sigma_c n_c^2 \tag{8.151}$$

where σ_c is the cross-section for pair recombination (of order r_c) and V_p is the average vortex velocity perpendicular to the flow. The vortex velocity is obtained from Equation (8.127)

$$V_p \propto 2\pi K(l_c) D/r_c \tag{8.152}$$

and the steady-state solution to Equation (8.151) is

$$n_c = [R/2\pi K(l_c)]^{1/2} \tag{8.153}$$

The second modification of the linear theory comes in the derivation of the recursion relations outlined above. Clearly, a finite flow field will change the pair distribution – causing pairs to 'stretch' and polarize. Therefore, the dipole term in Equation (8.146) must be included in the Boltzman factors which arose in the calculation of the plasma polarizability $\alpha(r)$ and separation density $\delta n(r)$. Gillis, Volz and Mochel (1985) derived a set of modified recursion relations for a vortex pair in a finite (oscillating) superfluid flow field. The energy of such a pair, Equation (8.146), can readily be used in the Boltzmann factors if it is assumed that a typical pair lifetime is much longer than the oscillation period of v_s so that the dipole term can be averaged over all possible orientations of the pair. The susceptibility is then

$$\chi(r) = \pi^2 K_0 \int_{r_0}^{r} r^2 \langle \delta n(r) \rangle_\theta \, dr^2 \tag{8.154}$$

where

$$\langle \delta n(r) \rangle_\theta = \frac{y_0^2}{r_0^2} \exp\left[-2\pi K_0 \int_{r_0}^{r} \frac{dr'}{r'\tilde{\varepsilon}(r')}\right] I_0\left(\Gamma \sigma_s^0 V_s \frac{r}{k_B T}\right) \tag{8.155}$$

and I_0 is a zeroth-order modified Bessel function. The modified recursion relations resulting from Equations (8.154) and (8.155) are

$$K^{-1}(l) = K_0^{-1} + 4\pi^3 \int_0^l y^2(l') \, dl' \tag{8.156}$$

$$y^2(l) = y_0^2 \exp\left[4l - 2\pi \int_0^l K(l') \, dl' \right]$$
$$\times I_0(2\pi K_0 \exp(l) v_s / v_0) \tag{8.157}$$

where

$$V_0 = \hbar / m r_0 \approx \Gamma / 2\pi r_0 \tag{8.158}$$

Note that the finite-amplitude recursion relations differ from the linear relations only in the I_0 factor in Equation (8.157) and that setting V_s equal to zero recovers Kosterlitz–Thouless relations.

8.7.5 *Vortex diffusivity in two dimensions*

In a three-dimensional system, the diffusion of ring vortices was initially discussed in Section 3 of Donnelly and Roberts (1971) and is used in section 8.3.1. The diffusion coefficient for vortex rings is given in (8.15a) in momentum space, and in (8.15b) in configuration space. The diffusion coefficient for a vortex–antivortex pair in a relatively thick two-dimensional layer is (8.15c)

$$D = 2\gamma k_B T / \rho_s^2 \kappa^2 d \tag{8.159}$$

The behavior of (8.159) as a function of temperature is very rapid near T_λ, being dominated by the factor ρ_s^2 in the denominator. When the layer becomes thinner the meaning of quantities in (8.159) will change in important ways. For example, the friction parameter γ may be influenced by factors such as ripplons, pinning and the fact that the superfluid density is renormalized and will reach zero at temperatures often much less than T_λ. One would nevertheless expect to find that D increases as d decreases, and that D diverges rapidly as T_{KT} is approached. When one thinks about how the vortex pairs in thin films will behave for thick films, the vortex pair is likely to be subject to the Crow instability of Section 1.7.3.

The importance of understanding the origin and temperature dependence of diffusivity in two-dimensional helium systems is apparent when one realizes that almost all finite frequency experiments performed on the Kosterlitz–Thouless transition have been interpreted via the AHNS theory outlined above. Though D is the primary transport parameter of this theory, its value was at first assumed to be of order $\hbar/m$; a value arrived at purely on dimensional grounds by AHNS.

8.7.6 *Experiments on thin films*

Diffusivity appears in two ways in the dynamic theory: it comes in logarithmically in the cut-off l_ω to the bound pair contribution to ε and it is a prefactor in the free vortex contribution to ε. Early efforts to fit the former to experimental data were not very successful because of l_ω's logarithmic dependence on D. The first such effort was that of Bishop and Reppy (1980) who used an oscillating substrate to measure the dissipation and superfluid density of thin films of helium as a function of temperature through the transition. Their experiment essentially measured the real and imaginary parts of ε, as discussed above, to which they compared the predictions of the AHNS theory. They fit their data to an approximate solution of the recursion relations which is valid for small reduced temperatures

$$\left.\begin{array}{l} \mathrm{Re}[\varepsilon_b] \propto 1 - 0.5X(l_\omega) \\ \mathrm{Im}[\varepsilon_b] \propto Y^2(l_\omega) \end{array}\right\} \tag{8.160}$$

where

$$\left.\begin{array}{l} X(l_\omega) \approx 0.5t^{1/2}\coth(0.5bt^{1/2}l_\omega) \\ 4\pi Y(l_\omega) \approx 0.5bt^{1/2}\mathrm{cosch}(0.5bt^{1/2}l_\omega) \text{ for } t<0 \end{array}\right\} \tag{8.161}$$

and b is the nonuniversal parameter of Section 8.7.3 which absorbs information about $K_0(T)$ and $y_0^2(T)$. Agnolet, McQueeney and Reppy, (1989) give an analytical expression for b. (The hyperbolic functions in Equation (8.161) become sinusoidal for $t>0$.) A typical measurement of the reduced period, which is roughly proportional to the superfluid density, and the excess dissipation of Bishop and Reppy's oscillator is shown in Figure 8.25. The solid lines are fits to the data using Equations (8.160) and (8.161) for ε_b and Equation (8.143) for the thermal free vortex contribution. The fit uses six parameters: $\sigma_s(T_{KT})$, b, l_ω, F, T_{KT}, and ε' (ε' is to account for the anomalous dissipation tail on the cold side of the peak). The fit shown in Figure 8.25 is for $l_\omega = 12$ which

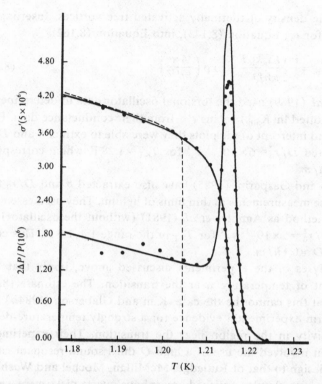

Figure 8.25 Data from Bishop and Reppy (1980). The reduced period and excess dissipation for a superfluid transition at the dynamical critical temperature $T_c = 1.215$ K is shown. The solid lines are fits using the dynamic theory of AHNS. The dashed line is the static theory prediction. Note how rapidly the superfluid density drops at T_c.

corresponds to $D \approx 20\hbar/m$. This is roughly an order of magnitude larger than conjectured by AHNS.

Diffusivity has also been extracted from thermal conductivity data in a manner similar to Bishop and Reppy's approach. Agnolet, Teitel and Reppy (1981) have measured the thermal conductance of helium films in conjunction with torsional oscillator measurements of the superfluid density. The film conductance as given by the AHNS theory is

$$K_{\text{film}} = mL\chi S_g K_B T / 2\pi\hbar D n_f \qquad (8.162)$$

where S_g is the entropy per unit mass of the vapor above the film, L is the latent heat of evaporation per unit mass, χ is a cell geometrical factor, and n_f is the density of thermally activated free vortices. Inserting the expression for n_f, Equation (8.143), into Equation (8.162)

$$K_{film} = \frac{m\chi L s_g k_B T}{2\pi\hbar D} r_0^2 \exp\left(\frac{4\pi}{bt^{1/2}}\right)$$ (8.163)

Agnolet *et al.* (1989) used the torsional oscillator data to determine T_{KT} and then plotted $\ln(K_{film})$ versus $t^{1/2}$ from their conductance data. From the slope and intercept of the plots they were able to extract b and D/r_0^2. They reported $D/r_0^2 \approx 6 \times 10^{-11}\,\text{s}^{-1}$ for $T_{KT} \approx 1.28$ K which corresponds to $D \approx 0.8\hbar/m$.

Finotello and Gasparini (1985) have also extracted b and D/r_0^2 from conductance measurements in thin films of helium. They used essentially the same method as Agnolet *et al.* (1981) (without the oscillator) and reported $D/r_0^2 \approx 5 \times 10^{-12}\,\text{s}^{-1}$ for T_{KT} in the range 1.4–1.9 K. This corresponds to $D \approx 0.1\hbar/m$.

The analyses of the experiments discussed above, imply that D is independent of temperature near the transition. The estimate (8.159) suggests that this cannot be the case. Kim and Glaberson (1984a) have provided firm experimental evidence for a strongly temperature-dependent diffusivity in the region near the transition. Their experimental arrangement involved the use of a high Q third sound resonant cavity, similar in design to that of Rutledge, McMillan, Mochel and Washburn (1978). Two thin circular polished amorphous quartz disks were welded together at their edges. Helium was allowed to diffuse through the disks at room temperature. At low temperatures, the helium condenses into films coating the inner quartz surfaces and constitutes a third sound resonant cavity. The third sound cavity was rotated at angular velocity Ω about an axis perpendicular to the film surfaces. The quality factor Q and resonance frequency of various modes were then monitored as a function of rotation speed and temperature.

At temperatures not too far below T_{KT}, it was possible to observe the damping of the third sound resonance associated with motion of the rotation-induced vortices. The vortex diffusivity could then be directly extracted from the contribution of the rotation-induced vortices to the third sound damping and is given by the expression

$$D = \frac{k_B T \, \Delta\omega}{4\pi^2(\hbar/m)^2 n_\Omega \sigma_s}$$ (8.164)

where $\Delta\omega$ is the excess resonance (full) width associated with rotation and $n_\Omega = m\Omega/\pi\hbar$.

It turned out that a convenient way of representing the data is as a function of T/σ_s. The diffusivity appears to collapse to a universal curve and could be reasonably well represented, for T not too low, by the relation (see Figure 8.26)

$$D \approx 0.17(\hbar/m)(k_B T/\sigma_s)^2 \tag{8.165}$$

and has the value $0.4\ \hbar/m$ at T_{KT}.

At low temperatures ($T/\sigma_s < 0.3(T/\sigma_s)_{KT}$), the diffusivity falls off more rapidly than at higher temperatures. The fall off occurs at the upper limit of T/σ_s where the authors observe vortex creep into and out of the film with the characteristic time behavior observed in persistent current decay experiments (Ekholm and Hallock, 1979, 1980; Browne and Doniach, 1982). At still lower values of T/σ_s the vortices are strongly pinned and substantial persistent currents could be achieved. Where necessary, to make sure that the rotation-induced vortex density was indeed its equilibrium value, $\Omega m/\pi\hbar$, and that there was no persistent current present – which could, in principle, result in vortex pair breaking and therefore extra sound damping – the cell was cooled to the measuring temperature from some high temperature, above T_{KT}, while rotating. In practice, this procedure made no perceptible difference. Petschek and Zippelius (1981) have carried out a calculation of the variation in the vortex diffusivity

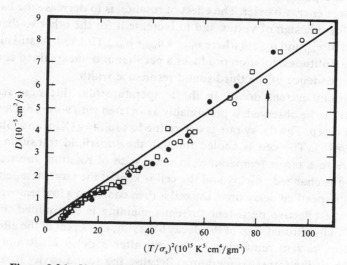

Figure 8.26 Vortex diffusivity in thin film plotted as a function of $(T/\sigma_s)^2$. The solid line is the relation $D = 0.17(\nu/m)^{-3}(k_B T/\sigma_s)^2$. The arrow indicates the location of the Kosterlitz–Thouless transition (Kim and Glaberson, 1984a).

near T_{KT} due to the existence of bound vortex pairs. They predict that the interaction of free vortices with the small vortex pairs should lead to a small decrease in the diffusivity comparable in magnitude to the increase in the dielectric constant. It is clear from the data of Kim and Glaberson, which show a much more rapid variation with the opposite sign, that either this prediction is wrong or else the data reflect a rapid variation of the bare – unrenormalized – diffusivity. Assuming the latter, experimental verification of the relatively small universal change predicted by Petschek and Zippelius would require a much more extensive investigation than has been carried out.

The contribution of rotation-induced vorticity to third sound damping is linear in the rotation speed only for relatively low sound amplitudes. For higher sound amplitudes, free vortices arising from the breaking of otherwise bound vortex pairs by the superflow also contribute to the damping. Treating the resonance as though it were Lorentzian and taking the resonance width as a measure of the effective free vortex density, that density is found to be reasonably well represented by the relation

$$n_{\text{free}} \sim (n_{\text{rotation}}^2 + n_{\text{pair breaking}}^2)^{1/2} \qquad (8.166)$$

This is interpreted in terms of a 'law of mass action', in which vortices in a bound pair can dissociate in the presence of superflow by diffusing over a free energy barrier. The effect of rotation is to decrease the barrier height for one sign of vortex and to increase it for the other so that the product of n_{up} and n_{down} (where $n_{\text{free}} = n_{\text{up}} + n_{\text{down}}$) is fixed. At still higher sound amplitudes, rotation produces a peculiar and unexplained chaotic time dependence of the third sound resonance width.

Persistent current decay, in the temperature/film thickness regime where it can be observed, is presumably associated with vortex nucleation and/or creep. The decay rate is determined as follows (Kim and Glaberson, 1984b). The cell is cooled through the superfluid transition temperature to a target temperature in some state of rotation; the state of rotation is changed quickly and the cell is kept at the target temperature for some specified delay time; the cell is then cooled to a low temperature where the effective persistent current remaining is determined. Figure 8.27 is an isochronal map of the decay behavior, i.e., a plot of the effective persistent current remaining in the cell after a delay of 30 min as a function of the target temperature. Because the sound cavity is sealed, the film thickness decreases as the temperature increases. The thickness in helium atomic layers is indicated at the top of the figure. The circles correspond to vortices entering the film (the cell is accelerated from rest

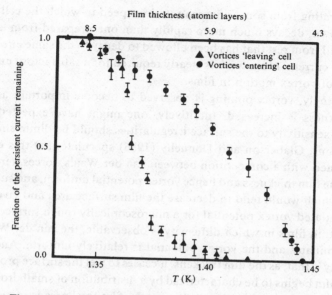

Figure 8.27 Persistent current decay behavior in films (after Kim and Glaberson (1984*b*)).

to 8.4 rad/s at the target temperature) and the triangles correspond to vortices leaving the film (the cell is cooled to the target temperature while rotating and then brought to rest).

At relatively low temperatures, the decay is logarithmic in time as is characteristic of vortex hopping from pinning site to pinning site (Browne and Doniach, 1982). For thinner films, particularly for decays associated with vortices entering the film, nonlogarithmic time behavior is observed. The very clear difference between the vortex inflowing and outflowing decay behavior in Figure 8.27 suggest that in the latter the decay rate is limited by vortex hopping whereas in the former the decay rate is dominated by a much slower nucleation rate. The decay behavior is, in many respects, qualitatively similar to that observed by Ekholm and Hallock (1979, 1980). By deliberately introducing vortices into the system by rotation, it may have been possible to disentangle vortex nucleation from vortex creep.

One serious problem with the suggestion that the vortex inflow decay is limited only by nucleation is the implication that the vortex distribution is then always reasonably uniform in this situation. It follows that at a given temperature the decay rate should only be a function of the persistent current. It is found, however, that a persistent current prepared by accelerating from some nonzero rotation speed in which the cell was cooled down, decays much more rapidly than one prepared from a cell accelerated from rest that has been allowed to decay to the same effective persistent current. More work is clearly required for a satisfactory understanding of vortex motion in films.

Surprisingly, vortex pinning is observed to become important as the film thickness is increased. Intuitively, one might have expected the contrary: sensitivity to the surface irregularities should be diminished in thicker films. Glaberson and Donnelly (1986) speculate that this effect is associated with a competition between van der Waals' forces, tending to keep the film thickness and hence vortex potential uniform, and surface tension which would tend to decrease the film surface area and give rise to a modulated vortex potential for a microscopically rough surface. For the very thin films in which diffusivity is observable, the van der Waals' force dominates and the vortex potential is relatively uniform. Another possibility is that, as the film thickens, it ceases to wet the surface properly and the film begins to be characterized by a distribution of small droplets. A more intriguing speculation is that, as the film thickness is increased,

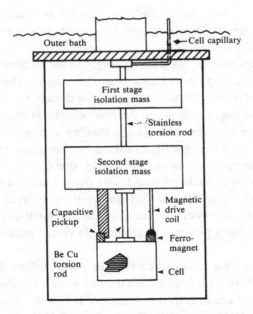

Figure 8.28 Apparatus used by Adams and Glaberson (1987) to study vortex dynamics in superfluid films.

the vortices gradually freeze into a regular array and therefore become more and more sensitive to the presence of isolated pinning sites.

Adams and Glaberson (1987) have utilized the oscillating substrate method of Bishop and Reppy (1980) in a geometry especially designed to study the vortex dynamics in superfluid helium films with rotation as an added experimental variable. Their apparatus is shown in Figure 8.28. The cell was made of magnesium and was approximately 2.5 cm high and 2.5 cm in diameter. It contained a stack of some 9100 0.1 mil Mylar disks which formed the substrate ($\sim 8 \times 10^4$ cm^2) for the films. The cell was set into oscillation by means of a superconducting coil, attached to the lower of two isolation masses. The isolation masses reduced coupling of vibrations of the cryostat to the oscillator and, in turn, loss of vibrational energy of the cell to the cryostat. The Q of the oscillator was $\sim 10^5$ which allowed period changes of the order of part in 10^9 to be made. This corresponds to a sensitivity of $\sim 1/100$ layer of helium atoms or the substrate. Since the superfluid does not couple to the substrate, changes in the period of the oscillator allowed a direct measurement of ρ_s. Observation of the amplitude and period of oscillation were made by two pickup capacitors (only one is shown in Figure 8.28). These results,

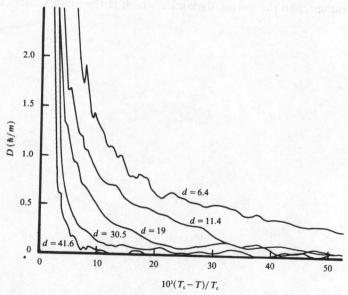

Figure 8.29 Diffusivity measurements by Adams and Glaberson (1987) as a function of reduced temperatures for various film thicknesses in angstroms.

in turn, allowed the quantities $\text{Re}[\varepsilon]$ and $\text{Im}[\varepsilon]$ of the complex dielectric constant of Equation (8.127) to be measured and hence ε to be measured as a function of temperature. The vortex diffusivity D is determined by measuring the contribution of a known free vortex density $n_\Omega = 2\Omega/\kappa$ (see (2.17)). According to (8.142) the difference between these rotating and nonrotating measurements of $\text{Im}[\varepsilon]$ is proportional to D:

$$D = (\text{Im}[\varepsilon]_\Omega - \text{Im}[\varepsilon]_0)\,\hbar\omega/4\pi m K_0\Omega \qquad (8.167)$$

Using this apparatus, Adams and Glaberson (1987) made a wide range of tests of the theory described above. Their diffusivity measurements do indeed show a very rapid increase near the dynamical critical point T_c, and an increase with decreasing thickness as suggested by (8.159). Figure 8.29 shows the relevant data in units of $\hbar/m$. Adams and Glaberson's data, however, do not show the same collapse of data $D \propto (T/\sigma_s)^2$ independent of thickness as reported by Kim and Glaberson (1984a, b).

All of the experiments reported above have been done above 1K. Agnolet, McQueeney and Reppy (1989) have recently completed a study of the superfluid response of helium films with transition temperatures ranging from 70 mK to 500 mK. The details of this study involve very thin film coverages and concern some of the problems of superfluidity itself rather than the vortex dynamics which is the main subject of this book.

References

Preface

Atkins, K. R. 1959. *Liquid Helium* Cambridge University Press, Cambridge.

Benneman, K. H. and Ketterson, J. B. 1976. *The Physics of Liquid and Solid Helium* Part 1 Wiley-Interscience, New York.

Donnelly, R. J., Glaberson, W. I. and Parks, P. E. 1967. *Experimental Superfluidity* University of Chicago Press, Chicago.

Keesom, W. H. 1942. *Helium* Elsevier, Amsterdam.

Keller, W. E. 1969. *Helium-3 and Helium-4* Plenum, New York.

London, F. 1954. *Superfluids* John Wiley, New York.

Putterman, S. J. 1974. *Superfluid Hydrodynamics* North Holland, Amsterdam.

Chapter 1

Andereck, C. D., Liu, S. S. and Swinney, H. L. 1986. Flow regimes in a circular Couette system with independently rotating cylinders *J. Fluid Mech.* **164**, 155–83.

Arms, R. J. and Hama, F. R. 1965. Localized-induction concept on a curved vortex and motion of an elliptic vortex ring *Phys. Fluids* **8**, 553–9.

Behringer, R. P. 1985. Rayleigh-Bénard convection and turbulence in liquid helium *Revs Mod. Phys.* **57**, 657–87.

Behringer, R. P. and Ahlers, G. 1982. Heat transport and temporal evolution of fluid flow near the Rayleigh-Bénard instability in cylindrical containers *J. Fluid Mech.* **125**, 218–58.

Chandrasekhar, S. 1943. Stochastic problems in physics and astronomy *Revs. Mod. Phys.* **15**, 1–89.

Chandrasekhar, S. 1961. *Hydrodynamic and Hydromagnetic Stability* Clarendon Press, Oxford.

Crow, S. C. 1970. Stability theory for a pair of trailing vortices *AIAA Journal* **8**, 2172–9.

Dennis, S. C. R. and Walker, J. D. A. 1971. Calculation of the steady flow past a sphere at low and moderate Reynolds numbers *J. Fluid Mech.* **48**, 771–89.

Donnelly, R. J. 1959. Experiments on the hydrodynamic stability of helium II between rotating cylinders *Phys. Rev. Lett.* **3**, 507–8.

Donnelly, R. J. and Fetter, A. L. 1966. Stability of superfluid flow in an annulus *Phys. Rev. Lett.* **17**, 747-50.

Donnelly, R. J. and LaMar, M. M. 1988. Flow and stabilty of helium II between concentratic cylinders *J. Fluid Mech.* **186**, 163-98.

Donnelly, R. J. and Roberts, P. H. 1969. Stochastic theory of the interaction of ions & vortices in helium II *Proc. Roy. Soc.* **A312**, 519-51.

Dyson, F. W. 1893. The potential of an anchor ring, I and II *Phil. Trans. Roy. Soc.* **A184**, 43-95, 1041-106.

Fetter, A. L. 1976. Vortices and ions in helium in *The Physics of Liquid and Solid Helium* Part I (K. H. Bennemann and J. B. Ketterson, eds.) John Wiley and Sons, New York, Chapter 3, pp. 207-305.

Fetter, A. L. 1982. Onset of convection in dilute superfluid ^{3}He-^{4}He mixtures I. Unbounded slab. II. Closed cylindrical container *Phys. Rev.* **B26**, 1164-81.

Fraenkel, L. E. 1970. On steady vortex rings of small cross-section in an ideal fluid *Proc. Roy. Soc. London* **A316**, 29-62.

Fraenkel, L. E. 1972. Examples of steady vortex rings of small cross-section in an ideal fluid *J. Fluid Mech.* **51**, 137-53.

Glaberson, W. I. and Donnelly, R. J. 1986. Structure, distributions & dynamics of vortices in helium II. In *Progress in Low Temperature Physics* IX (D. F. Brewer, ed.) North-Holland, Amsterdam Chapter 1, pp. 1-142.

Hasimoto, H. 1972. A soliton on a vortex filament *J. Fluid Mech.* **51**, 477-85.

Hopfinger, E. J., Browand, F. K. and Gagne, Y. 1982. Turbulence and waves in a rotating tank *J. Fluid Mech.* **125**, 505-34.

Jenson, V. G. 1959. Viscous flow round a sphere at low Reynolds numbers (<40) *Proc. Roy. Soc.* **A249**, 346-66.

Kramers, H. A. 1940. Brownian motion in a field of force and the diffusion model of chemical reactions *Physica* **7**, 284-304.

Lamb, H. 1945. *Hydrodynamics*, sixth edition, Dover Publications, New York.

Landau, L. D. and Lifshitz, E. M. 1959. *Fluid Mechanics* Pergamon Press, London. (second edition, 1987.)

Lugt, H. J. 1983. *Vortex Flow in Nature and Technology* John Wiley and Sons, New York.

Maxworthy, T. 1972. The structure and stability of vortex rings *J. Fluid Mech.* **51**, 15-32.

Maxworthy, T., Hopfinger, E. J. and Redekopp, L. C. 1985. Wave motions on vortex cores *J. Fluid Mech.* **151**, 141-65.

McCormack, P. D. and Crane, L. 1973. *Physical Fluid Dynamics* Academic Press, New York.

Milne-Thompson, L. M. 1962. *Theoretical Hydrodynamics* fourth edition, Macmillan, London.

Muirhead, C., Vinen, W. F. and Donnelly, R. J. 1984. The nucleation of vorticity by ions in superfluid ^{4}He I. Basic theory *Phil. Trans. Roy. Soc.* **A311**, 433-67.

Norbury, J. 1973. A family of steady vortex rings *J. Fluid Mech.* **57**, 417-31.

Pocklington, H. C. 1895. The complete system of the periods of a hollow vortex ring *Phil. Trans. Roy. Soc. London* **A186**, 603-19.

Rimon, Y. and Cheng, S. I. 1969. Numerical solution of a uniform flow over a sphere at intermediate Reynolds numbers *Phys. Fluids* **12**, 949-59.

Roberts, P. H. 1972. A Hamiltonian theory for weakly interacting vortices *Mathematika* 19, 169-79.

Roberts, P. H. and Donnelly, R. J. 1970. Dynamics of vortex rings *Phys. Lett.* 31A, 137-8.

Saffman, P. G. and Baker, G. R. 1979. Vortex Interaction in *Annual Review of Fluid Mechanics* 11 (M. van Dyke, J. V. Wehausen and J. L. Lumley, eds.). Annual Reviews Inc., Palo Alto, pp. 95-122.

Serrin, J. 1959. Mathematical principles of classical fluid mechanics in *Handbuch der Physik* Vol. VIII/1 (S. Flugge and C. Truesdell, eds.) Springer-Verlag, Berlin, 125-263.

Schwarz, K. W., 1981. Vortex pinning in superfluid helium *Phys. Rev. Lett.* 47, 251-4.

Stauffer, D. and Fetter, A. L. 1968. Distribution of vortices in rotating helium II *Phys. Rev.* 168, 156-9.

Taylor, G. I. 1921. Experiments with rotating fluids *Proc. Roy. Soc.* A100, 114-21.

Taylor, G. I. 1923. Stability of a viscous liquid contained between two rotating cylinders *Phil. Trans. Roy. Soc. Lond.* A223, 289-343.

Thomson, J. J. 1883. *A Treatise on the Motion of Vortex Rings* Macmillan, London.

Thomson, W. 1880. Vibrations of a columnar vortex *Phil Mag.* 10, 155-68.

Tritton, D. J. 1982. Discussion of 'vorticity and rotation' *Am. J. Phys.* 50, 421-4.

Widnall, S. E. 1975. The structure and dynamics of vortex filaments in *Annual Review of Fluid Mechanics* 7 (M. Van Dyke, W. G. Vincenti and J. V. Wehausen, eds.). Annual Reviews Inc., Palo Alto, pp. 141-65.

Widnall, S. E. and Tsai, C. Y. 1977. The instability of the thin vortex ring of constant vorticity *Phil. Trans. Roy. Soc.* A287, 273-305.

Chapter 2

Amit D. and Gross, E. P. 1966. Vortex rings in a Bose fluid *Phys. Rev.* 145, 130-6.

Anderson, P. W. 1966. Considerations on the flow of superfluid helium. *Rev. Mod. Phys.* 38, 298-310.

Andronikashvili, E. L. and Mamaladze, Y. G. 1967. Rotation of helium II in *Progress in Low Temperature Physics* V (C. J. Gorter, ed.) North-Holland, Amsterdam, Chapter 3, pp. 79-160.

Andronikashvili, E. L. and Tsakadze, D. S. 1960. The propagation of oscillations along vortex lines in rotating helium II *Sov. Phys. JETP* 10, 227-8

Ashton, R. A. and Glaberson, W. I. 1979. Vortex waves in superfluid ⁴He *Phys. Rev. Lett.* 42, 1062-64.

Atkins, K. R. 1959. *Liquid Helium* Cambridge University Press, Cambridge.

Bendt, P. J., 1967a. Attenuation of second sound in helium II between rotating cylinders *Phys. Rev.* 153, 280-4.

Bendt, P. J., 1967b. Superfluid flow transitions in rotating narrow annuli *Phys. Rev.* 164, 262-7.

Bendt, P. J. and Donnelly, R. J. 1967. Threshold for superfluid vortex lines in a rotating annulus *Phys. Rev. Lett.* 19, 214-16.

Benneman, K. H. and Ketterson, J. B. 1976. *The Physics of Liquid & Solid Helium* Part I, John Wiley and Sons, New York.

Benneman, K. H. and Ketterson, J. B. 1978. *The Physics of Liquid & Solid Helium* Part II, John Wiley and Sons, New York.

Bogoliubov, N. N. 1947. On the theory of superfluidity *J. Phys. USSR* 11, 23–43.

Brooks, J. S. and Donnelly, R. J. 1977. The calculated thermodynamic properties of superfluid helium-4 *J. Phys. Chem. Ref. Data* 6, 51–104.

Careri, G., McCormick, W. D. and Scaramuzzi, R. 1962. Ions in rotating liquid helium II *Phys. Lett.* 1, 61–3.

Donnelly, R. J., and Fetter, A. L. 1966. Stability of superfluid flow in an annulus *Phys. Rev. Letters* 17, 747–50.

Donnelly, R. J., Donnelly, J. A. and Hills, R. S. 1981. Specific heat and dispersion curve for helium II *J. Low Temp. Phys.* 44, 471–89.

Donnelly, R. J., Glaberson, W. I. and Parks, P. E. 1967. *Experimental Superfluidity*, University of Chicago Press, Chicago.

Ellis, T. and McClintock, P. V. E. 1985. The breakdown of superfluidity in liquid ^{4}He. V. Measurement of the Landau critical velocity for roton creation *Phil. Trans. Roy. Soc.* A315, 259–300.

Fetter, A. L. and Donnelly, R. J. 1966. On the equivalence of vortices and current filaments *Phys. Fluids* 9, 619–20.

Fetter, A. L. and Walecka, J. D. 1971. *Quantum Theory of Many-Particle Systems* McGraw-Hill, New York.

Feynman, R. P. 1955. Application of quantum mechanics to liquid helium, in *Progress in Low Temperature Physics* I (C. J. Gorter, ed.) North-Holland Publishing Co., Amsterdam, Chapter 2.

Ginzburg, V. L. and Landau, L. D. 1950. On the theory of superconductivity *Zh. Eksp. Teor. Fiz.* 20, 1064–82.

Ginzburg, V. L. and Pitaevskii, L. P. 1958. On the theory of superfluidity *Sov. Phys. JETP* 7, 858–61.

Glaberson, W. I. and Donnelly, R. J. 1966. Growth of pinned quantized vortex lines in helium II *Phys. Rev.* 141, 208–10.

Gross, E. P. 1961. Structure of a quantized vortex in Boson systems *Il Nuovo Cimento* 20, 454–77.

Gross, E. P. 1963. Hydrodynamics of a Superfluid Condensate *J. Math. Phys.* 4, 195–207.

Hall, H. E. 1958. An experimental and theoretical study of torsional oscillations in uniformly rotating liquid helium II *Proc. Roy. Soc. London* A245, 546–61.

Hall, H. E. 1960. The rotation of helium II *Phil. Mag. Suppl.* 9, 89–146.

Hall, H. E. and Vinen, W. F. 1956a. The rotation of liquid helium II, I. Experiments on the propagation of second sound in uniformly rotating helium II *Proc. Roy. Soc.* A238, 204–14.

Hall, H. E. and Vinen, W. F. 1956b. The rotation of liquid helium II, II. The theory of mutual friction in uniformly rotating helium II *Proc. Roy. Soc.* A238, 215–34.

Halley, J. W. and Cheung, A. 1969. Theory of vortex resonance in liquid helium *Phys. Rev.* 168, 209–21.

Halley, J. W. and Ostermeier, R. M. 1977. A new calculation of the ion velocity shift in a vortex resonance experiment in He II *J. Low Temp. Phys.* 26, 877–98.

Hess, G. 1967. Angular momentum of superfluid helium in a rotating cylinder *Phys. Rev.* 161, 189–93.

Hess, G. B. and Fairbank, W. M. 1967. Measurements of angular velocity in superfluid helium *Phys. Rev. Lett.* **19**, 216-8.

Karn, P. W., Starks, D. R. and Zimmermann, W. 1980. Observation of circulation in rotating ^{4}He *Phys. Rev. B* **21**, 1797-805.

Kawatra, M. P. and Pathria, R. K. 1966. Quantized vortices in an imperfect Bose gas and the breakdown of superfluidity in liquid helium II *Phys. Rev.* **151**, 132-7.

Keesom, W. H. 1942. *Helium* Elsevier, Amsterdam.

Keller, W. E. 1969. *Helium-3 and Helium-4* Plenum Press, New York.

Khalatnikov, I. M. 1965. *An Introduction to the Theory of Superfluidity* W. A. Benjamin, Inc., New York.

Kral, S. F. and Zimmermann, W. 1975. Experiments on the superfluid circulation in rotating helium II *Phys. Rev. B* **18**, 3334-47.

Landau, L. D. and Lifshitz, E. M. 1959. *Fluid Mechanics* Pergamon Press, London (second edition, 1987).

Landau, L. D. and Lifshitz, E. M. 1969. *Statistical Physics* Addison-Wesley, Reading.

Lifshitz, E. M. and Pitaevskii, L. P. 1980 *Statistical Physics*, Part 2 (Volume 9 of Landau and Lifshitz, *Course in Theoretical Physics*) Oxford, Pergamon Press.

London, F. 1954. *Superfluids II* John Wiley and Sons, Inc., New York.

Madelung, E. 1927. Quantetheorie in Hydrodynamischer Form *Z. Phys.* **40**, 322-6.

McClintock, P. V. E., Meredith, D. J. and Wigmore, J. K. 1984. *Matter at Low Temperatures* Wiley-Interscience, New York.

Onsager, L. 1949. *Nuovo Cimento Suppl.* **6**, 249-50 (discussion on paper by C. J. Gorter).

Osborne, D. V. 1950. The rotation of liquid helium. *Proc. Phys. Soc.* London **A63**, 909-12.

Packard, R. E. and Sanders, T. M. 1972. Observation on single vortex lines in rotating superfluid helium *Phys. Rev.* **A6**, 799-807.

Pitaevskii, L. P. 1961. Vortex lines in an imperfect Bose gas *Sov. Phys. JETP* **13**, 451-4.

Putterman, S. J. 1974. *Superfluid Hydrodynamics* North-Holland Publishing Co., Amsterdam.

Rayfield, G. W. and Reif, F. 1964. Quantized vortex rings in superfluid helium *Phys. Rev.* **136**, A1194-208.

Roberts, P. H. 1974 in *Proceedings of the International Colloquium on Drops and Bubbles* Vol. 1, California Institute of Technology and Jet Propulsion Laboratory (D. J. Collins, M. S. Plessett and M. M. Saffren, eds), pp. 35-49.

Roberts, P. H. and Donnelly, R. J. 1974. Superfluid mechanics in *Ann. Rev. of Fluid Mechanics* **6** (M. Van Dyke, W. F. Vincenti and J. V. Wehausen, eds.) Annual Reviews Inc., Palo Alto, pp. 179-225.

Roberts, P. H. and Grant, J. 1971. Motions in a Bose condensate I. The structure of the large circular vortex *J. Phys.* **A4**, 55-72.

Sears, V. F., Svensson, E. C., Martel, P. and Woods, A. D. B. 1982. Neutron scattering determination of the momentum distribution and the condensate fraction in liquid ^{4}He *Phys. Rev. Lett.* **49**, 279-82.

Stauffer, D. and Fetter, A. L. 1968. Distribution of vortices in rotating helium II *Phys. Rev.* **168**, 156-9.

Tanner, D. J., Springett, B. E. and Donnelly, R. J. 1965. Ions in rotating He II:

Some exploratory experiments in *Low Temperature Physics LT-9* (J. G. Daunt, D. O. Edwards, F. J. Milford and M. Yaqub, eds.) Plenum Press, New York, pp. 346-8.

Tilley, D. R. and Tilley, J. 1986. *Superfluidity and Superconductivity* Adam Hilger Ltd., Bristol (second edition).

Vinen, W. F. 1957. Mutual friction in a heat current in liquid helium II. I. Experiments on steady heat currents. *Proc. Roy. Soc.* **A240**, 114-27.

Vinen, W. F. 1961. The detection of a single quantum of circulation in liquid helium II *Proc. Roy. Soc.* **A260**, 218-36.

Whitmore, S. C. and Zimmermann, W. 1968. Observation of quantized circulation of superfluid helium. *Phys. Rev.* **166**, 181-96.

Wilks, J. 1967. *The Properties of Liquid & Solid Helium* Clarendon Press, Oxford.

Wilks, J. and Betts, D. S. 1987. *An Introduction to Liquid Helium* Clarendon Press, Oxford (second edition).

Chapter 3

Barenghi, C. F., Donnelly, R. J. and Vinen, W. F. 1983. Friction on quantized vortices in helium II. A review *J. Low Temp. Phys.* **52**, 189-247.

Bendt, P. J. 1967. Attenuation of second sound in helium II between rotating cylinders *Phys. Rev.* **153**, 280-4.

Bendt, P. J. and Donnelly, R. J. 1967. Threshold for superfluid vortex lines in a rotating annulus *Phys. Rev. Lett.* **19**, 214-6.

Donnelly, R. J., Glaberson, W. I. and Parks, P. E. 1967. *Experimental Superfluidity* University of Chicago Press, Chicago.

Feynman, R. P. 1955. Application of quantum mechanics to liquid helium in *Progress in Low Temperature Physics* I (C. J. Gorter, ed.) North-Holland Publishing Co., Amsterdam, Chapter 2.

Hall, H. E. and Vinen, W. F. 1956a. The rotation of liquid helium II, I. Experiments on the propagation of second sound in uniformly rotating helium II *Proc. Roy. Soc.* **A238**, 204-14.

Hall, H. E. and Vinen, W. F. 1956b. The rotation of liquid helium II, II. The theory of mutual friction in uniformly rotating helium II. *Proc. Roy. Soc.* **A238**, 215-34.

Iordanskii, S. V. 1964. On the mutual friction between the normal and superfluid components in a rotating Bose gas *Ann. Phys. (NY)* **29**, 335-49.

Iordanskii, S. V. 1965. On the mutual friction in a superfluid *Phys. Rev. Lett.* **15**, 34-5.

Iordanskii, S. V. 1966, Mutual friction force in a rotating Bose gas *Sov. Phys. JETP* **22**, 160-7.

Lamb, H. 1945. *Hydrodynamics* Dover Publications, New York (sixth edition).

Madelung, E. 1927. Quantetheorie in Hydrodynamischer Form *Z. Phys.* **40**, 322-6.

Martin, K. P. and Tough, J. T. 1983 Evolution of superfluid turbulence in thermal counterflow *Phys. Rev.* **B27**, 2788-99.

Mathieu, P., Marechal, J. C. and Simon, Y. 1980. Spatial distribution of vortices and metastable states in rotating He II *Phys. Rev.* **B22**, 4293-306.

Mathieu, P., Plaçais, B. and Simon, Y. 1984. Spatial distribution of vortices and anisotropy of mutual friction in rotating He II *Phys. Rev.* **B29**, 2489-96.

Milne-Thomson, L. M. 1968. *Theoretical Hydrodynamics* Macmillan Co., New York (fifth edition).

Northby, J. A. and Donnelly, R. J. 1970. Detection of a vortex-free region in rotating helium II *Phys. Rev. Lett.* **25**, 214-17.

Pfotenhauer, J. M. and Donnelly, R. J. 1985. Heat transfer in liquid helium. *Advances in Heat Transfer* Vol. 17 (J. P. Hartnett and T. F. Irvine, Jr., eds.) Academic Press, New York, pp. 65-158.

Rayfield, G. W. and Reif, F. 1964. Quantized vortex rings in superfluid helium *Phys. Rev.* **136**, A1194-208.

Sewell, C. J. T. 1910. The excitation of sound in a viscous atmosphere by small obstacles of cylindrical and spherical form *Phil. Trans. Roy. Soc.* **A210**, 239-70.

Snyder, H. A. and Linekin, D. M. 1966. Measurements of the mutual friction parameters B' in rotating helium II *Phys. Rev.* **147**, 131-9.

Snyder, H. A. and Putney, Z. 1966. Angular dependence of mutual friction in rotating helium II *Phys. Rev.* **150**, 110-17.

Stauffer, D. and Fetter, A. L. 1968. Distribution of vortices in rotating helium II *Phys. Rev.* **168**, 156-9.

Swanson, C. E. 1985. A study of vortex dynamics in counterflowing helium II PhD Thesis, University of Oregon.

Swanson, C. E., Wagner, W. T., Donnelly, R. J. and Barenghi, C. F. 1987. Calculation of frequency and velocity dependent mutual friction parameters in helium II *J. Low Temp. Phys.* **66**, 263-76.

Vinen, W. F. 1957. Mutual friction in a heat current in liquid helium II, III. Theory of mutual friction *Proc. Roy. Soc. London* **A242**, 493-515.

Yarmchuck, E. J. and Glaberson, W. I. 1979. Counterflow in rotating superfluid helium *J. Low Temp. Phys.* **36**, 381-429.

Chapter 4

Cade, A. G. 1965. Binding of positive-ion complexes to vortex rings in liquid-helium II *Phys. Rev. Lett.* **15**, 238-9.

Careri, G., McCormick, W. D. and Scaramuzzi, F. 1962. Ions in rotating liquid helium II *Phys. Lett.* **1**, 61-3.

Chandrasekhar, S. 1943. Stochastic problems in physics and astronomy *Revs. Mod. Phys.*, **15**, 1-89.

Donnelly, R. J. 1965. Theory of the interaction of ions and quantized vortices in rotating helium II *Phys. Rev. Lett.* **14**, 39-41.

Donnelly, R. J. 1974. On 'The Ghost of a Vanished Vortex Ring' in *Quantum Statistical Mechanics in the Natural Sciences* (S. L. Mintz and S. M. Widmayer, eds.) Plenum Press, New York, pp. 359-403.

Donnelly, R. J. and Roberts, P. H. 1969. Stochastic theory of the interaction of ions and vortices in helium II *Proc. Roy. Soc.* **A312**, 519-51.

Donnelly, R. J., Donnelly, J. A. and Hills, R. N. 1981. Specific heat and dispersion curve for helium II *J. Low Temp. Phys.* **44**, 471-89.

Donnelly, R. J., Glaberson, W. I. and Parks, P. E. 1967. *Experimental Superfluidity* University of Chicago Press, Chicago.

Douglass, R. L. 1964. Ion trapping in rotating helium II *Phys. Rev. Lett.* **13**, 791-4.

Fetter, A. L. and Iguchi, I. 1970. Low temperature mobility of trapped ions in rotating He II *Phys. Rev.* **A2**, 2067-74.

Feynman, R. P. 1955. Application of quantum mechanics to liquid helium, in Progress in Low Temperature Physics **I**, Chapter 2 (C. J. Gorter, ed.) North-Holland Publishing Co., Amsterdam.

Ginzburg, V. L. and Pitaevskii, L. P. 1958. On theory of superfluidity *Sov. Phys. JETP* **7**, 858-61.

Glaberson, W. I. 1969. Trapped-ion motion in helium II *J. Low Temp. Phys.* **1**, 289-311.

Glaberson, W. I. and Donnelly, R. J. 1986. Structure, distributions and dynamics of vortices in helium II in *Progress in Low Temperature Physics* **IX** (D. F. Brewer, ed.) North-Holland, Amsterdam, Chapter 1, pp. 1-142.

Glaberson, W. I., Strayer, D. M. and Donnelly, R. J. 1968. Model for the core of a quantized vortex line in helium II *Phys. Rev. Lett.* **21**, 1740-4.

Heiserman, J., Hulin, J. P., Maynard, J. and Rudnick, I. 1976. Precision sound-velocity measurements in He II *Phys. Rev.* **B14** 3862-7.

Hills, R. N. and Roberts, P. H. 1977*a*. Superfluid mechanics for a high density of vortex lines *Arch. Rat. Mech. Anal.* **66**, 43-71.

Hills, R. N. and Roberts, P. H. 1977*b* Healing and relaxation in flows of helium I-II generalization of Landau equations *Int. J. Eng. Sci.* **15**, 305-16.

Hills, R. N. and Roberts, P. H. 1978*a*. Healing and relaxation in flows of helium II. Part II. First, second and fourth sound *J. Low Temp. Phys.* **30**, 709-27.

Hills, R. N. and Roberts, P. H. 1978*b*. Healing and relaxation in flows of helium II, III. Pure superflow *J. Phys.* **C11**, 4485-9.

Huang, W. and Dahm, A. J. 1976. Mobilities of ions trapped on vortex lines in dilute ^{3}He-^{4}He solutions *Phys. Rev. Lett.* **36**, 1466-9.

Johnson, W. W. and Glaberson, W. I. 1974. Size measurement of positive impurity ions in liquid helium. *Phys. Rev.* **10A**, 868-71.

Jones, C. A. and Roberts, P. H. 1982. Motions in a Bose condensate, IV. Axisymmetric solitary waves *J. Phys.* **A15**, 2599-619.

McCauley, J. L. and Onsager, L. 1975*a*. Electrons and vortex lines in He II, I. Brownian motion theory of capture and escape *J. Phys.* **A8**, 203-13.

McCauley, J. L. and Onsager, L. 1975*b*. Electrons and vortex lines in He II, II. Theoretical analysis of capture and release experiments *J. Phys.* **A8**, 882-90.

Muirhead, C. M., Vinen, W. F. and Donnelly, R. J. 1984. The nucleation of vorticity by ions in superfluid ^{4}He, I. Basic theory *Phil. Trans. Roy. Soc.* **A311**, 433-67.

Ohmi, T., Tsuneto, T. and Usui, T. 1969. Phase separation in rotating helium *Prog. Theor. Phys.* **41**, 1395-400.

Ohmi, T. and Usui, T. 1969. Superfluid Vortex, Trapping Neutral Impurities *Prog. Theor. Phys.* **41**, 1400-15.

Ostermeier, R. M. and Glaberson, W. I. 1974. The capture of ions by vortex lines at low temperatures *Phys. Lett.* **49A**, 223-4.

Ostermeier, R. M. and Glaberson, W. I. 1975*a*. The capture of ions by vortex lines in pure ^{4}He and ^{3}He-^{4}He solutions *J. Low Temp. Phys.* **20**, 159-79.

Ostermeier, R. M. and Glaberson, W. I. 1975*b*. The escape of positive ions from quantized vortex lines in superfluid helium *Phys. Lett.* **51A**, 348-50.

Ostermeier, R. M. and Glaberson, W. I. 1975*c*. Instability of vortex lines in the presence of axial normal fluid flow *J. Low Temp. Phys.* **21**, 91-196.

Ostermeier, R. M. and Glaberson, W. I. 1975*d*. Motion of ions trapped on vortices in He II *Phys. Rev. Lett.* **35**, 241–4.

Ostermeier, R. M. and Glaberson, W. I. 1976. The mobility of ions trapped on vortex lines in pure ^{4}He and ^{3}He–^{4}He solutions. *J. Low Temp. Phys.* **25**, 317–51.

Ostermeier, R. M., Yarmchuck, E. J. and Glaberson, W. I. 1975. Condensation of He^3 atoms onto quantized vortex lines in superfluid helium *Phys. Rev. Lett.* **35**, 957–60.

Paintin, J., Vinen, W. F. and Muirhead, C. M. 1990. The motion of a sphere in a superfluid containing quantized vortex lines (preprint).

Parks, P. E. and Donnelly, R. J. 1966. Radii of positive and negative ions in helium II *Phys. Rev. Lett.* **16**, 45–8.

Rayfield, G. W. 1968. Study of the ion-vortex-ring transition *Phys. Rev.* **168**, 222–33.

Rayfield, G. W. and Reif, F. 1964. Quantized vortex rings in superfluid helium *Phys. Rev.* **136**, A1194–208.

Rent, L. S. and Fisher, I. Z. 1969. Adsorption of atomic impurities on quantized vortices in liquid helium II *Sov. Phys. JETP* **28**, 375–9.

Roberts, P. H. 1972. A Hamiltonian theory for weakly interacting vortices *Mathematika* **19**, 169–79.

Roberts, P. H., Hills, R. N. and Donnelly, R. J. 1979. Calculation of the static healing length in helium II *Phys. Lett.* **70A**, 437–40.

Rudnick, I. and Fraser, J. C. 1970. Third sound and the superfluid parameters of thin helium films *J. Low Temp. Phys.* **3**, 225–34.

Scholtz, J. H., McLean, E. D. and Rudnick, I. 1974. Third sound and the healing length of helium II in films as thin as 2.1 atomic layers. *Phys. Rev. Lett.* **32**, 147–51.

Springett, B. E. and Donnelly, R. J. 1966. Pressure dependence of the radius of the negative ion in helium II *Phys. Rev. Lett.* **17**, 364–7.

Springett, B. E., Tanner, D. J. and Donnelly, R. J. 1965. Capture cross sections for negative ions in rotating helium II *Phys. Rev. Lett.* **14**, 585–7.

Steingart, M. and Glaberson, W. I. 1972. Quantized vortex rings and the vortex core radius in He II *J. Low Temp. Phys.* **8**, 61–77.

Tam, W. Y. and Ahlers, G. 1982. Pressure dependence of size effects in superfluid ^{4}He *Phys. Lett.* **92A**, 445–8.

Tanner, D. J. 1966. Negative-ion capture by vortex lines in helium II. *Phys. Rev.* **152**, 121–8.

Williams, G. A. and Packard, R. E. 1978. Ion trapping on vortex lines in rotating ^{3}He–^{4}He mixtures *J. Low Temp. Phys.* **33**, 459–80.

Williams, G. A., Deconde, K. and Packard, R. E. 1975. Positive-ion trapping on vortex lines in rotating He II *Phys. Rev. Lett.* **34**, 924–6.

Chapter 5

Adams, P. W., Cieplak, M. and Glaberson, W. I. 1985. Spin-up problem in superfluid ^{4}He *Phys. Rev.* **32**, 171–7.

Alpar, M. A. 1978. Relaxation of rotating He II following a spin-up of its container: A superfluid Ekman time? *J. Low Temp. Phys.* **31**, 803–15.

Alpar, M. A., Anderson, P. W., Pines, D. and Shaham, J. 1981. Giant glitches and pinned vorticity in the Vela and other pulsars *Ap. J.* **249**, L29–33.

Anderson, P. W., Alpar, M. A., Pines, D. and Shaham, J. 1982. The rheology of neutron stars vortex-line pinning in the crust superfluid *Phil. Mag.* **45A**, 227–38.

Anderson, P. W., Pines, D., Ruderman, M. and Shaham, J. 1978. Questions about rotating superfluid dynamics: Problems of pulsar astrophysics accessible in the laboratory *J. Low Temp. Phys.* **30**, 839–47.

Awschalom, D. D. and Schwarz, K. W. 1984. Observation of a remnant vortex-line density in superfluid helium *Phys. Rev. Lett.* **52**, 49–52.

Barenghi, C. F. and Jones, C. A. 1987. On the stability of superfluid helium between rotating concentric cylinders *Phys. Lett.* **A122**, 425–30.

Barenghi, C. F. and Jones, C. A. 1988. The stability of Couette flow of helium II *J. Fluid Mech.* **197**, 551–9.

Barenghi, C. F., Donnelly, R. J. and Vinen, W. F. 1983. Friction on quantized vortices in helium II. A review *J. Low Temp. Phys.* **52**, 189–247.

Baym, G. and Pines, D. 1969. Spin up in neutron stars: The future of the Vela pulsar *Nature* **224** , 872–4.

Baym, G. and Pines, D. 1971. Neutron starquakes and pulsar speedup *Ann. Phys. (NY)* **66**, 816–35.

Bendt, P. J. 1966. Instability of helium II potential flow between rotating cylinders *Phys. Rev. Lett.* **17**, 680–2.

Bendt, P. J. 1967*a*. Attenuation of second sound in helium II between rotating cylinders *Phys. Rev.* **153**, 280–4.

Bendt, P. J. 1967*b*. Superflow transitions in rotating narrow annuli *Phys. Rev.* **164**, 262–7.

Bendt, P. J. and Donnelly, R. J. 1967. Threshold for superfluid vortex lines in a rotating annulus *Phys. Rev. Lett.* **19**, 214–6.

Campbell, L. J. and Krasnov, Yu. K. 1982. Transient behavior of rotating superfluid helium *J. Low Temp. Phys.* **49**, 377–96.

Campbell, L. J. and Ziff, R. M. 1978. A catalog of two-dimensional vortex patterns, Los Alamos Scientific Laboratory Report No. LA-7384-MS.

Campbell, L. J. and Ziff, R. M. 1979. Vortex patterns and energies in a rotating superfluid *Phys. Rev.* **B20**, 1886–902.

Chandrasekhar, S. 1961. *Hydrodynamic and Hydromagnetic Stability* Clarendon Press, Oxford.

Chandrasekhar, S. and Donnelly, R. J. 1957. The hydrodynamic stability of helium II between rotating cylinders I *Proc. Roy. Soc.* **A241**, 9–28.

Donnelly, R. J. 1959. Experiments on the hydrodynamic stability of helium II between rotating cylinders *Phys. Rev. Lett.* **3**, 507–8.

Donnelly, R. J. and Fetter, A. L. 1966. Stability of superfluid flow in an annulus *Phys. Rev. Lett.* **17**, 747–50.

Donnelly, R. J. and LaMar, M. M. 1988. Flow and stability of helium II between concentric cylinders *J. Fluid Mech.* **186**, 163–98.

Fetter, A. L. 1966. Equilibrium distribution of rectilinear vortices in a rotating container *Phys. Rev.* **152**, 183–9.

Fetter, A. L. 1967. Low-lying superfluid states in a rotating annulus *Phys. Rev.* **153**, 285–96.

Glaberson, W. I. and Donnelly, R. J. 1966. Growth of pinned quantized vortex lines in helium II *Phys. Rev.* **141**, 208–10.

Greenspan, H. P. 1968. *The Theory of Rotating Fluids* Cambridge University Press, Cambridge.

Hegde, S. G. and Glaberson, W. I. 1980. Pinning of superfluid vortices to surfaces *Phys. Rev. Lett.* **45**, 190–3.

Heikkila, W. J. and Hollis Hallett, A. C. 1955. The viscosity of liquid helium II *Can. J. Phys.* **33**, 420–35.

Krasnov, Yu K. 1977. Dynamics of Onsager–Feynman vortices in a rotating superfluid system of the pulsar type *Sov. Phys. JETP* **46**, 181–4.

Mamaladze, Yu G. and Matinyan, S. G. 1963. Stability of rotating of a superfluid liquid *Sov. Phys. JETP* **17**, 1424–25.

Packard, R. E. 1972. Pulsar speed-ups related to metastability of the superfluid neutron-star core *Phys. Rev. Lett.* **28**, 1080–2.

Pines, D. 1971. Inside neutron stars in *Proceedings of the 12th International Conference on Low Temperature Physics-LT12* Academic Press of Japan, Tokyo, pp. 7–17.

Reppy, J. D., Depatie, D. and Lane, C. T. 1960. Helium II in rotation *Phys. Rev. Lett.* **5**, 541–2.

Reppy, J. D. and Lane, C. T. 1961. Vorticity in helium II in *Proceedings of the 7th International Conference on Low Temperature Physics* (G. H. Graham and A. C. H. Hallet, eds.) University of Toronto Press, Toronto, pp. 443–6.

Reppy, J. D. and Lane, C. T. 1965, Angular momentum experiments with liquid helium *Phys. Rev.* **140**, A106–11.

Ruderman, M. 1969. Neutron starquakes and pulsar periods *Nature* **223**, 597–8.

Ruderman, M. 1976. Crust breaking by neutron superfluids and the Vela pulsar *Ap. J.* **203**, 213–22.

Schwarz, K. W. 1981. Vortex pinning in superfluid helium *Phys. Rev. Lett.* **47**, 251–4.

Snyder, H. A. 1974. Rotating Couette Flow of Superfluid Helium in *Proceedings of the 13th International Conference on Low Temp. Physics-LT13* **1** Plenum Press, New York, p. 283.

Stauffer, D. and Fetter, A. L. 1968. Distribution of vortices in rotating helium II *Phys. Rev.* **168**, 156–9.

Swanson, C. E. and Donnelly, R. J. 1987. The appearance of vortices in the flow of helium II between rotating cylinders *J. Low Temp. Phys.* **67**, 185–93.

Tilley, D. R. and Tilley, J. 1986 *Superfluidity and Superconductivity* Adam Hilger Ltd., Bristol (second edition).

Tkachenko, V. K. 1966*a*. On vortex lattices *Sov. Phys. JETP* **22**, 1282–6.

Tsakadze, J. S. and Tsakadze, S. J. 1972. Relaxation phenomena at acceleration of rotation of a spherical vessel with helium II and relaxation in pulsars *Phys. Lett.* **41A**, 197–9.

Tsakadze, J. S. and Tsakadze, S. J. 1973*a*. Measurement of the relaxation time on acceleration of vessels with helium II and superfluidity in pulsars *Sov. Phys. JETP* **37**, 918–21.

Tsakadze, J. S. and Tsakadze, S. J. 1973*b*. Velocity oscillation of a freely rotating vessel with helium II *Sov. Phys. JETP Lett.* **18**. 355–7.

Tsakadze, J. S. and Tsakadze, S. J. 1975. Superfluidity in pulsars *Sov. Phys. Usp.* **18**, 242–50.

Tritton, D. J. 1988. *Physical Fluid Dynamics* Clarendon Press, Oxford (second edition).

Williams, G. A. and Packard, R. E. 1974. Photographs of quantized vortex lines in rotating He II *Phys. Rev. Lett.* **33**, 280-3.

Williams, G. A. and Packard, R. E. 1978. Ion trapping on vortex lines in rotating ^{3}He-^{4}He mixtures *J. Low Temp. Phys.* **33**, 459-80.

Wolf, P. E., Perrin, B., Hulin, J. P. and Elleaume, P. 1981. Rotating Couette flow of helium II *J. Low Temp. Phys.* **44**, 569-93.

Yarmchuck, E. J. and Glaberson, W. I. 1978. Thermorotation effects in superfluid helium *Phys. Rev. Lett.* **41**, 564-8.

Yarmchuck, E. J. and Glaberson, W. I. 1979. Counterflow in rotating superfluid helium *J. Low Temp. Phys.* **36**, 381-430.

Yarmchuck, E. J. and Packard, R. E. 1982. Photographic studies of quantized vortex lines *J. Low Temp. Phys.* **46**, 479-515.

Yarmchuck, E. J., Gordon, M. J. V. and Packard, R. E. 1979. Observation of stationary vortex arrays in rotating superfluid helium *Phys. Rev. Lett.* **43**, 214-7.

Yamauchi, J. and Yamada, K. 1985. Effects of boundary-vortex force in capillary flow of superfluid helium *Physica* **128B**, 45-54.

Chapter 6

Andereck, C. D. and Glaberson, W. I. 1982. Tkachenko waves *J. Low. Temp. Phys.* **48**, 257-96.

Andereck, C. D., Chalupa, J. and Glaberson, W. I. 1980. Tkachenko waves in rotating superfluid helium *Phys. Rev. Lett.* **44**, 33-6.

Andronikashvili, E. L. and Mamaladze, Yu G. 1966. Quantization of macroscopic motions and hydrodynamics of rotating helium II *Rev. Mod. Phys.* **38**, 567-625.

Andronikashvili, E. L. and Tsakadze, D. S. 1960. The propagation of oscillations along vortex lines in rotating helium II *Sov. Phys. JETP* **10**, 227-8.

Ashton, R. A. and Glaberson, W. I. 1979. Vortex waves in superfluid ^{4}He *Phys. Rev. Lett.* **42**, 1062-4.

Barenghi, C. F., Donnelly, R. J. and Vinen, W. F. 1985. Thermal excitation of waves on quantized vortices *Phys. Fluids* **28**, 498-504.

Baym, G. and Chandler, E. 1983. The hydrodynamics of rotating superfluids I. Zero-temperature nondissipative theory *J. Low. Temp. Phys.* **50**, 57-87.

Baym, G. and Chandler, E. 1983. The hydrodynamics of rotating superfluids II. Zero-temperature nondissipative theory *J. Low. Temp. Phys.* **50**, 57-87.

Benjamin, T. B. and Feir, J. E. 1967. The disintegration of wave trains on deep water *J. Fluid Mech.* **27**, 417-30.

Campbell, L. J. 1981. Transverse normal modes of finite vortex arrays *Phys. Rev.* **A24**, 514-34.

Campbell, L. J. and Krasnov, Yu K. 1981. Edge waves of a vortex continuum *Phys. Lett.* **84A**, 75-9.

Cheng, D. K., Cromar, M. W. and Donnelly, R. J. 1973. Influence of an axial heat current on negative ion trapping in rotating helium II *Phys. Rev. Lett.* **31**, 433-6.

Dzyaloshinskii, I. E. and Volovik, G. E. 1980. Poisson brackets in condensed matter physics *Ann. Phys. (NY)* **125**, 67-97.

Fermi, E., Pasta, J. and Ulam, S. 1965. *Collected Papers of Enrico Fermi* (E. Segré, ed.), University of Chicago Press, Chicago.

Glaberson, W. I. and Donnelly, R. J. 1986. Structure, distributions and dynamics of vortices in helium II in *Progress in Low Temperature Physics* IX (D. F. Brewer, ed.) North-Holland, Amsterdam, Chapter 1, pp. 1–142.

Glaberson, W. I., Johnson, W. W. and Ostermeier, R. M. 1974. Instability of a vortex array in He II *Phys. Rev. Lett.* 33, 1197–1200.

Grant, J. 1972. The fluid mechanics of a Bose condensate. Ph.D. thesis. University of Newcastle-upon-Tyne.

Hall, H. E. 1958. An experimental and theoretical study of torsional oscillations in uniformly rotating liquid helium II *Proc. Roy. Soc.* A245, 546–61.

Halley, J. W. and Cheung, A. 1968. Theory of vortex resonance in liquid helium *Phys. Rev.* 168, 209–21.

Halley, J. W. and Ostermeier, R. M. 1977. A new calculation of the ion velocity shift in a vortex resonance experiment in He II *J. Low Temp. Phys.* 26, 877–98.

Havelock, T. H. 1931. The stability of motion of rectilinear vortices in ring formation *Phil. Mag.* 11, 617–33.

Hegde, S. G. and Glaberson, W. I. 1980. Pinning of superfluid vortices to surfaces *Phys. Rev. Lett.* 45, 190–3.

Nadirashvili, Z. S. and Tsakadze, J. S. 1968. Dependence of helium II viscosity properties on oscillation frequency *J. Low Temp. Phys.* 37, 169–77.

Ostermeier, R. M. and Glaberson, W. I. 1975. Instability of vortex lines in the presence of axial normal fluid flow *J. Low Temp. Phys.* 21, 191–6.

Pitaevskii, L. P. 1961. Vortex lines in an imperfect Bose gas *Sov. Phys. JETP* 13, 451–54.

Pocklington, H. C. 1895. The complete system of the periods of a hollow vortex ring *Phil. Trans. Roy. Soc. London* A186, 603–19.

Rajagopal, E. S. 1964. Oscillations of quantized vortices in rotating liquid helium II *Ann. Phys. (NY)* 29, 350–65.

Samuels, D. C. and Donnelly, R. J. 1990. Sideband instability and recurrence at Kelvin waves on vortex cores *Phys. Rev. Lett.* 64, 1385–8.

Samuels, D. C. 1990. The sideband instability and recurrence of vortex waves in superfluid helium II, Ph.D. thesis, University of Oregon.

Sonin, E. B. 1976. Vortex-lattice vibrations in a rotating helium II *Sov. Phys. JETP* 43, 1027–33.

Sonin, E. B. 1983. Low-frequency oscillations of vortices in helium II *JETP Lett.* 37, 100–3.

Sonin, E. B. 1987. Vortex oscillations and hydrodynamics of rotating superfluids *Rev. Mod. Phys.* 59, 87–155.

Swanson, C. E., Barenghi, C. F. and Donnelly, R. J. 1983. Rotation of a tangle of quantized vortex lines in He II *Phys. Rev. Lett.* 50, 190–3.

Tkachenko, V. K. 1966a. On vortex lattices *Sov. Phys. JETP* 22, 1282–6.

Tkachenko, V. K. 1966b. Stability of vortex lattices *Sov. Phys. JETP* 23, 1049–56.

Tkachenko, V. K. 1969. Elasticity of vortex lattices *Sov. Phys. JETP* 29, 945–6.

Tsakadze, D. S. 1976. Oscillations of a vortex system in rotating He II *Sov. Phys. JETP* 44, 398–400.

Tsakadze, D. S. and Tsakadze, S. D. 1973a. Measurement of the relaxation time on acceleration of vessels with helium II and superfluidity in pulsars *Sov. Phys. JETP* **37**, 918-21.

Tsakadze, D. S. and Tsakadze, S. D. 1973b. Velocity oscillations of a freely rotating vessel with helium II *Sov. Phys. JETP Lett.* **18**, 355-7.

Tsakadze, D. S. and Tsakadze, S. D. 1975. Superfluidity in pulsars *Sov. Phys. Usp.* **18**, 242-50.

Volovik, G. E. and Dotsenko, Jr., V. S. 1980. Poisson brackets and continuous dynamics of the vortex lattice in rotating He II *Sov. Phys. JETP Lett.* **29**, 576-9.

Williams, M. R. and Fetter, A. L. 1977. Continuum model of vortex oscillations in rotating superfluids *Phys. Rev.* **B16**, 4846-52.

Yarmchuck, E. J. and Glaberson, W. I. 1979. Counterflow in rotating superfluid helium *J. Low Temp. Phys.* **36**, 381-430.

Yarmchuck, E. J. and Packard, R. E. 1982. Photographic studies of quantized vortex lines *J. Low Temp. Phys.* **46**, 479-515.

Chapter 7

Ahlers, G. 1969. Mutual friction in He II near the lambda transition *Phys. Rev. Lett.* **22**, 54-6.

Awschalom, D. D., Milliken, F. P. and Schwarz, K. W. 1984. Properties of superfluid turbulence in a large channel *Phys. Rev. Lett.* **53**, 1372-5.

Baehr, M. L. and Tough, J. T. 1985. Dissipation in combined normal and superfluid flows of He II: A unified description *Phys. Rev. B* **32**, 5632-8.

Barenghi, C. F. 1982. Experiments on quantum turbulence Ph.D. thesis, University of Oregon.

Barenghi, C. F., Swanson, C. E. and Donnelly, R. J. 1982. Induced vorticity fluctuations in counterflowing He II *Phys. Rev. Lett.* **48**, 1187-9.

Bon Mardion, C., Claudet, G. and Seyfert, P. 1978. Steady state heat transport in superfluid helium at 1 bar in *Proceedings 7th International Cryogenic Conference*, IPC Science and Technology Press, Guildford, pp. 214-21.

Carey, R. F., Rooney, J. A. and Smith, C. W. 1978. Ultrasonically generated quantized vorticity in He II *Phys. Lett.* **65A**, 311-13.

Courts, S. C. and Tough, J. T. 1988. Transition to superfluid turbulence in two-fluid flow of He II *Phys. Rev.* **B38**, 74-80.

Courts, S. C. and Tough, J. T. 1989. Dissipation in two fluid flow of He II *Phys. Rev. B* **39**, 8924-33.

Cummings, J. C. 1974. Development of a high performance cryogenic shock tube *J. Fluid Mech.* **66**, 177-87.

d'Humierès, D. D. and Libchaber, A. 1978. Self diffusion of superfluid turbulence through porous media *J. Physique* C6 **39**, 156-7. See also Hulin, J. P., d'Humierès, D. D. , Perrin, B. and Libchaber, A. 1974. Critical velocities for superfluid helium flow through a small hole *Phys. Rev.* **A9**, 885-92.

Dimotakis, P. E. and Broadwell, J. E. 1973. Local temperature measurements in supercritical counterflow in liquid helium II *Phys. Fluids* **16**, 1787-95.

Donnelly, R. J. and Swanson, C. E. 1986. Quantum turbulence *J. Fluid Mech.* **173**, 387-429.

Feynman, R. P. 1955. Applications of quantum mechanics to liquid helium in *Progress in Low Temperature Physics*, I (C. J. Gorter, ed.) North Holland, Amsterdam, pp. 17-53.

Foreman, L. R. and Snyder, H. A. 1979. The penetration of superfluid turbulence through porous filters *J. Low Temp. Phys.*, **34**, 529-38.

Giordano, N. 1984*a*. Vibrating superleak second-sound transduceers. Theory and experiment *J. Low. Temp. Phys.* **55**, 495-526.

Giordano, N. 1984*b*. Observations of critical velocity effects in vibrating superleak second-sound transducers in *Proc. 17th Intl. Conf. on Low Temperature Physics* LT-17, Vol. 1, North Holland, Amsterdam, pp. 307-8.

Giordano, N. and Musikar, P. 1984. Theory of critical velocity effects in vibrating superleak second-sound transducers in *Proc. 17th Intl. Conf. on Low Temperature Physics* LT-17, Vol. 1, North Holland, Amsterdam, pp. 309-10.

Gorter, C. J. and Mellink, J. H. 1949. On the irreversible processes in liquid helium II *Physica* **15**, 285-304.

Griswold, D., Lorenson, C. P. and Tough, J. T. 1987. Intrinsic fluctuations of the vortex line density in superfluid turbulence *Phys. Rev.* **B35** 3149-61.

Hoch, H., Busse, L. and Moss, F. 1975. Noise from vortex line turbulence in He II *Phys. Rev. Lett.* **34**, 384-7.

Jones, C. A. and Roberts, P. H. 1982. Motions in a Bose condensate IV. Axisymmetric solitary waves *J. Phys. A* **15**, 2599-619.

Laguna, G. 1975. Second-sound attenuation in a supercritical counterflow jet *Phys. Rev. B* **12**, 4874-81.

Liepmann, H. W. 1952. Deflection and diffusion of a light ray passing though a boundary layer Douglas Rep. S.M. 14397.

Liepmann, H. W. and Laguna. G. A. 1984. Nonlinear interactions in the fluid mechanics of helium II in *Annual Review of Fluid Mechanics*, Vol. 16 (M. Van Dyke, W. G. Vincenti and J. V. Wehausen, eds.) Annual Reviews, Inc. Palo Alto, pp. 139-77.

Liepmann, H. W., Cummings, J. C. and Rupert, V. C. 1973. Cryogenic shock tube *Phys. Fluids* **16**, 332-3.

Mantese, J., Bischoff, G. and Moss, F. 1977. Vortex line density fluctuations in turbulent superfluid helium *Phys. Rev. Lett.* **39**, 565-8.

Martin, K. P. and Tough, J. T. 1983. Evolution of superfluid turbulence in thermal counterflow *Phys. Rev. B* **27**, 2788-99.

Milliken, F. P., Schwarz, K. W. and Smith, C. W. 1982. Free decay of superfluid turbulence *Phys. Rev. Lett.* **48**, 1204-7.

Opatowsky, L. B. and Tough, J. T. 1981. Homogeneity of turbulence in pure superflow *Phys. Rev. B* **24**, 5420-1.

Ostermeier, R. M., Cromar, M. W., Kittel, P. and Donnelly, R. J. 1980. Fluctuations in turbulent He II counterflow *Phys. Lett.* **77A**, 321-4.

Schwarz, K. W. 1978. Turbulence in superfluid helium: steady homogeneous counterflow *Phys. Rev. B* **18**, 245-62.

Schwarz, K. W. 1982. Generating superfluid turbulence from simple dynamical rules *Phys. Rev. Lett.* **49**, 283-5.

Schwarz, K. W. 1983. Critical velocity for a self-sustaining vortex tangle in superfluid helium *Phys. Rev. Lett.* **50**, 364-7.

338   *References*

Schwarz, K. W. 1985. Three-dimensional vortex dynamics in superfluid ^{4}He, I. Line-line and line-boundary interactions *Phys. Rev. B* **31**, 5782–804.

Schwarz, K. W. 1988. Three-dimensional vortex dynamics in superfluid ^{4}He *Phys. Rev. B* **38**, 2398–417.

Schwarz, K. W. and Smith, C. W. 1981. Pulsed-ion study of ultrasonically generated turbulence in superfluid ^{4}He *Phys. Lett.* **82A**, 251–4.

Slegtenhorst, R. P., Marees, G. and van Beelen, H. 1982a. Steady flow of helium III in the presence of a heat current *Physica* **113B**, 341–66.

Slegtenhorst, R. P., Marees, G. and van Beelen, H. 1982b Transient effects in superfluid turbulence *Physica* **113B**, 367–79.

Smith, C. W. and Tejwani, M. J. 1984. Fluctuations of a negative ion current in turbulent He II *Phys. Lett.* **104A**, 281–4.

Swanson, C. E. 1985. A study of vortex dynamics in counterflowing helium II PhD thesis, University of Oregon.

Swanson, C. E. and Donnelly, R. J. 1985. Vortex dynamics and scaling in turbulent counterflowing helium II *J. Low Temp. Phys.* **61**, 363–99.

Tough, J. T. 1982. Superfluid turbulence in *Progress in Low Temperature Physics*, **VIII** (D. F. Brewer, ed.) North-Holland, Amsterdam, pp. 133–219.

Vinen, W. F. 1957a. Mutual friction in a heat current in liquid helium II, I. Experiments on steady heat currents *Proc. Roy. Soc. Lond.* **A240**, 114–27.

Vinen, W. F. 1957b. Mutual friction in a heat current in liquid helium II, II. Experiments on transient effects *Proc. Roy. Soc. Lond.* **A240**, 128–43.

Vinen, W. F. 1957c. Mutual friction in a heat current in liquid helium II, III. Theory of the mutual friction *Proc. Roy. Soc. Lond.* **A242**, 493–515.

Vinen, W. F. 1958. Mutul friction in a heat current in liquid helium II, IV. Critical heat currents in wide channels *Proc. Roy. Soc. Lond.* **A243**, 400–13.

Wang, R. T., Swanson, C. E. and Donnelly, R. J. 1987. Anisotropy and drift of a quantum vortex tangle *Phys. Rev. B* **36**, 5240–4.

Chapter 8

Adams, P. W. 1986. A direct measurement of vortex diffusivity in thin films of ^{4}He Ph.D Thesis, Rutgers University.

Adams, P. W. and Glaberson, W. I. 1987. Vortex dynamics in superfluid helium films *Phys. Rev. B* **35**, 4633–52.

Agnolet, G., McQueeney, D. F. and Reppy, J. D. 1989. Studies of the Kosterlitz-Thouless transition in helium films *Phys. Rev. B* **39**, 8934–58.

Agnolet, G., Teitel, S. L. and Reppy, J. D. 1981. Thermal transport in a ^{4}He film at the Kosterlitz–Thouless transition *Phys. Rev. Lett.* **47**, 1537–9.

Allum, D. R., McClintock, P. V. E., Phillips, A. and Bowley, R. M. 1977. The breakdown of superfluidity in liquid ^{4}He: An experimental test of Landau's theory *Phil. Trans. Roy. Soc. Lond.* **A284**, 179–224.

Ambegaokar, V. and Teitel, S. 1979. Dynamics of vortex pairs in superfluid films *Phys. Rev. B* **19**, 1667–70.

Ambegaokar, V., Halperin, B. I., Nelson, D. R. and Siggia, E. D. 1978. Dissipation in two-dimensional superfluids *Phys. Rev. Lett.* **40**, 783–9.

Ambegaokar, V., Halperin, B. I., Nelson, D. R. and Siggia, E. D. 1980. Dynamics of superfluid films *Phys. Rev. B* **21**, 1806–26.

Anderson, P. W. 1966. Considerations on the flow of superfluid helium *Rev. Mod. Phys.* **38**, 298-310.

Andreev, A. F. 1966. Surface tension of weak helium isotope solutions *Sov. Phys. JETP* **23**, 939-41.

Barenghi, C. F., Donnelly, R. J. and Vinen, W. F. 1985. The nucleation of vorticity by ions in superfluid ^{4}He II. Theory of the effect of dissolved ^{3}He *Proc. Roy. Soc. Lond.* **A402**, 225-43.

Bishop, D. J. and Reppy, J. D. 1978 Study of the superfluid transition in two-dimensional ^{4}He films *Phys. Rev. Lett.* **40**, 1727-30.

Bishop, D. J. and Reppy, J. D. 1980. Study of the superfluid transition in two-dimensional ^{4}He films *Phys. Rev. B* **22**, 5171-85.

Bowley, R. M. 1976. Vortex nucleation by negative ions in liquid ^{4}He *J. Phys. C: Solid State Phys.* **9**, L367-70.

Bowley, R. M. and Sheard, F. W. 1977. Motion of negative ions at supercritical drift velocities in liquid ^{4}He at low temperatures *Phys. Rev. B* **16**, 244-54.

Bowley, R. M., McClintock, P. V. E., Moss, F. E. and Stamp, P. C. E. 1980. Vortex nucleation in isotopically pure superfluid ^{4}He *Phys. Rev. Lett.* **44**, 161-4.

Bowley, R. M., McClintock, P. V. E., Moss, F. E., Nancolas, G. G. and Stamp, P. C. E. 1982. The breakdown of superfluidity in liquid ^{4}He III. Nucleation of quantized vortex rings *Phil. Trans. Roy. Soc. Lond.* **A307**, 201-60.

Bowley, R. M., Nancolas, G. G. and McClintock, P. V. E. 1984. Vortex nucleation in ultradilute superfluid ^{3}He-^{4}He solutions *Phys. Rev. Lett.* **52** 659-62.

Brooks, J. S. and Donnelly, R. J. 1977. The calculated thermodynamic properties of superfluid helium-4 *J. Phys. and Chem. Ref. Data* **6**, 51-104.

Browne, D. A. and Doniach, S. 1982. Vortex pinning and the delay of persistent currents in unsaturated superfluid helium films *Phys. Rev. B* **25**, 136-50.

Caldeira, A. O. and Leggett, A. J. 1981. Influence of dissipation on quantum tunneling in macroscopic systems *Phys. Rev. Lett.* **46**, 211-4.

Campbell, L. J. 1979 Rotational speedups accompanying angular deceleration of a superfluid *Phys. Rev. Lett.* **43**, 1336-9.

Careri, G. 1961. Helium ions in liquid helium II in *Progress in Low Temperature Physics* **III** (C. J. Gorter, ed.) North-Holland, Amsterdam, pp. 58-79.

Careri, G., Cunsolo, S., Mazzoldi, P. and Santini, M. 1965. Experiments on the creation of charged quantized vortex rings in liquid helium at 1 K *Phys. Rev. Lett.* **15**, 392-6.

Dahm, A. J. 1969. Evidence for condensation of He^3 atoms on the surface of bubbles in liquid He^4 *Phys. Rev.* **180**, 259-62.

Donnelly, R. J. and Roberts, P. H. 1971. Stochastic theory for the nucleation of quantized vortices in superfluid helium *Phil. Trans. Roy. Soc. Lond.* **A271**, 41-100.

Donnelly, R. J., Hills, R. N. and Roberts, P. H. 1979. Superflow in restricted geometries *Phys. Rev. Lett.* **42**, 725-8.

Ekholm, D. T. and Hallock, R. B. 1979. Behavior of persistent currents under conditions of strong decay *Phys. Rev. Lett.* **42**, 449-52.

Ekholm, D. T. and Hallock, R. B. 1980. Studies of the decay of persistent currents in unsaturated films of superfluid ^{4}He *Phys. Rev. Lett. B* **21**, 3902-12.

Edwards, D. O. and Saam, W. F. 1978. The free surface of liquid helium" in *Progress in Low Temperature Physics* **VIIA** (D. F. Brewer, ed.) North-Holland, Amsterdam, pp. 283-369.

Ellis, T., McClintock, P. V. E., Bowley, R. M. and Allum, D. R. 1980. The breakdown of superfluidity in liquid ^{4}He—II: An investigation of excitation emission from negative ions travelling at extreme supercritical velocities *Phil. Trans. Roy. Soc. Lond.* **A296**, 581-95.

Feynman, R. P. 1955. Application of quantum mechanics to liquid helium in *Progress in Low Temperature Physics* **I** (C. J. Gorter, ed.) North-Holland, Amsterdam, Chapter 2.

Finotello, D. and Gasparini, F. M. 1985. Universality of the Kosterlitz-Thouless transition in ^{4}He film as a function of thickness *Phys. Rev. Lett.* **55**, 2156-9.

Gillis, K. A., Volz, S. and Mochel, J. M. 1985. Superfluid flow dissipation in two dimensions *J. Low Temp. Phys.* **61**, 173-84.

Glaberson, W. I. and Donnelly, R. J. 1986. Structure, distributions and dynamics of vortices in helium II, in *Progress in Low Temperature Physics* **IX** (D. F. Brewer, ed.) North-Holland, Amsterdam, Chapter 1, pp. 1-142.

Gopal, E. S. R. 1963. Motion and stability of vortices in a finite chemical: Application to liquid helium II *Ann. Phys.* **25**, 196-220.

Hendry, P. C., Lawson, N. S. McClintock, P. V. E., Williams, C. D. H. and Bowley, R. M. 1988. Macroscopic quantum tunneling of vortices in He II *Phys. Rev. Lett.* **60**, 604-7.

Hohenberg, P. C. 1967. Existence of long-range order in one and two dimensions *Phys. Rev.* **158**, 383-6.

Huberman, B. A., Myerson, R. J. and Doniach, S. 1978 Dissipation near the critical point of a two-dimensional superfluid *Phys. Rev. Lett.* **40**, 780-2.

Iordanskii, S. V. 1965. Vortex ring formation in a superfluid *Zh. Eksp. Teor. Fiz.* **48**, 708.

Jones, C. A. and Roberts, P. H. 1982. Motions in a Bose condensate IV. Axisymmetric solitary waves *J. Phys.* **A15**, 2599-619.

Keller, W. E. and Hammel, E. F. 1968. Isothermal flow of liquid He II through narrow channels *Physica* **2**, 221-45.

Kim, M. and Glaberson, W. I. 1984*a*. Vortex diffusity in two-dimensional helium films *Phys. Rev. Lett.* **52**, 53-6.

Kim, M. and Glaberson, W. I. 1984*b*. Vortex motion in helium films in *Proceedings of the 17th International Conference on Low Temperature Physics-LT17* Part I (U. Eckern, A. Schmid, W. Weber and H. Wühl, eds.) North Holland, Amsterdam, pp. 299-300.

Kojima, H., Veith, W., Guyon, E. and Rudnick, I. 1974. Decay of saturated and unsaturated persistent currents in superfluid helium in *Low Temperature Physics-LT-13* Plenum Press, New York, Chapter 1, pp. 279-282.

Kosterlitz, J. M. and Thouless, D. J. 1973. Ordering metastability and phase transitions in two-dimensional systems *J. Phys.* **C6**, 1181-203.

Kosterlitz, J. M. and Thouless, D. J. 1978. Two dimensional physics in *Progress in Low Temperature Physics* **VIIB** (D. F. Brewer, ed.) North-Holland. Amsterdam, Chapter 5, pp. 371-433.

Kuchnir, M., Ketterson, J. B. and Roach, P. R. 1972. Ion motion in dilute ^{3}He-^{4}He solutions at ultralow temperatures *Phys. Rev. A* **6**, 341–55.

Kukich, G. 1970. The decay of persistent currents of superfluid helium PhD Thesis, Cornell University.

Langer, J. S. and Fisher, M. E. 1967. Intrinsic critical velocity of a superfluid *Phys. Rev. Lett.* **19**, 560–3.

Langer, J. S. and Reppy, J. D. 1970. Intrinsic critical velocities in superfluid helium in *Progress in Low Temperature Physics* **6** (C. J. Gorter, ed.) North-Holland, Amsterdam, pp. 1–35.

Long, E. and Meyer, L. 1950. Superfluidity and thermomechanical effect in the absorbed helium II film *Phys. Rev.* **79**, 1031–2.

McCauley, J. P. 1977, Dissociation of a two-dimensinal coulomb gas at low temperatures *J. Phys. C* **10**, 689–92.

McClintock, P. V. E., Moss, F. E., Nancolas, G. G. and Stamp, P. C. E. 1981. Profound influence of isotopic impurities on vortex nucleation in He II *Physica* **107B**, 573–4.

McQueeney, D. F. 1988. Superfluidity in the pure ^{3}He and ^{3}He-^{4}He mixture films Ph.D thesis, Cornell University.

Meyer, L. and Reif, F. 1961. Ion motion in superfluid liquid helium under pressure *Phys. Rev.* **123**, 727–31.

Muirhead, C. M., Vinen, W. F. and Donnelly, R. J. 1984. The nucleation of vorticity by ions in superfluid ^{4}He. I. Basic theory *Phil. Trans. Roy. Soc.* **A311**, 433–67.

Muirhead, C. M., Vinen, W. F. and Donnelly, R. J. 1985. The nucleation of vorticity by ions in superfluid ^{4}He, II. Theory of the effect of dissolved ^{3}He *Proc. Roy. Soc. Lond.* **A402**, 225–43.

Nancolas, G. G., Bowley, R. M. and McClintock, P. V. E. 1985. The breakdown of superfluidity in liquid ^{4}He. IV Influence of ^{3}He isotopic impurities on the nucleation of quantized vortex rings *Phil. Trans. Roy. Soc. Lond.* **A313**, 537–610.

Neeper, D. A. 1968. Vortex-ring formation by negative ions in He II under pressure *Phys. Rev. Lett.* **21**, 274–5.

Neeper, D. A., and Meyer, L. 1969. Ion motion and vortex-ring formation in pure liquid He^4 and He^3-He^4 solutions between 0.05 and 0.5 K *Phys. Rev.* **182**, 223–4.

Nelson, D. R. 1983. *Phase Transitions* Academic Press, London.

Nelson, D. R. and Kosterlitz, J. M. 1977. Universal jump in superfluid density of two-dimensional superfluids *Phys. Rev. Lett.* **39**, 1201–5.

Petschek, R. G. and Zippelius, A. 1981. Renormalization of the vortex diffusion constant in superfluid films *Phys. Rev. B* **23**, 3483–93.

Rayfield, G. W. 1966. Roton emission from negative ions in helium II *Phys. Rev. Lett.* **16**, 934–6.

Rayfield, G. W. 1967. Evidence for a peeling model of vortex ring formation by ions in liquid helium *Phys. Rev. Lett.* **19**, 1371–3.

Rayfield, G. W. 1968a. Vortex-ring creation by negative ions in dilute mixtures of He^3-He^4 *Phys. Rev. Lett.* **20**, 1467–8.

Rayfield, G. W. 1968*b*. Study of the ion-vortex-ring transition *Phys. Rev.* **168**, 222-3.

Rayfield, G. W. and Reif, F. 1964. Quantized vortex rings in superfluid helium *Phys. Rev.* **136**, A1194-208.

Rudnick, I. 1978. Critical surface density of the superfluid component in ^{4}He films *Phys. Rev. Lett.* **40**, 1454-5.

Rutledge, J. E., McMillan, W. L., Mochel, J. M. and Washburn, T. E. 1978. Third sound, two-dimensional hydrodynamics and elementary excitations in very thin helium films *Phys. Rev. B* **18**, 2155-68.

Schwarz, K. W. and Jang, P. S. 1973. Creation of quantized vortex rings by charge carriers in superfluid helium *Phys. Rev. A* **8**, 3199-210.

Senbetu, L. 1978. Effect of the ^{3}He impurities on the structure of quantized vortex lines and the lifetime of negative ions trapped on vortices *J. Low Temp. Phys.* **32**, 571-88.

Shikin, V. B. 1973. Interaction between impurity excitation and negative ions in liquid helium *Sov. Phys. JETP* **37**, 718-22.

Springett, B. E. and Donnelly, R. J. 1966. Pressure dependence of the radius of the negative ion in helium II *Phys. Rev. Lett.* **17**, 364-7.

Strayer, D. M. 1971. Structure and nucleation of quantized vortices in helium II PhD Thesis, University of Oregon (unpublished).

Strayer, D. M., and Donnelly, R. J. 1971. Measurement of the probability of nucleation of a vortex ring in helium II *Phys. Rev. Lett.* **26**, 1420-3.

Titus, J. A. and Rosenshein, J. S. 1973. The dependence of vortex ring creation in He II. *Phys. Rev. Lett.* **31**, 146-9.

Tkachenko, V. K. 1966*b*. Stability of vortex lattices *Sov. Phys. JETP* **23**, 1049-56.

Wilks, J. 1967. *Liquid and Solid Helium* Clarendon Press, Oxford.

Vinen, W. F. 1963. *Liquid Helium: Proceedings of the International School of Physics Enrico Fermi*, Course XXI, Academic Press, New York, pp. 336-55.

Zoll, R. 1976. Study of the vortex-ring transition in superfluid ^{4}He *Phys. Rev. B* **14**, 2913-26.

Zoll, R. and Schwarz, K. W. 1973. New features of the vortex-ring transition *Phys. Rev. Lett.* **31**, 1440-3.

Index